Designing Multi-Agent Systems

*Principles, Patterns, and Implementation
for AI Agents*

Victor Dibia

California   multiagentbook.com

Designing Multi-Agent Systems
Principles, Patterns, and Implementation for AI Agents

by Victor Dibia

Published by Victor Dibia
Website: https://multiagentbook.com

First Edition: November 2025

ISBN: 979-8-9931012-0-0

For questions, corrections, or feedback, please visit:
https://github.com/victordibia/designing-multiagent-systems

Printed and bound in the United States of America.

Table of contents

Preface

Writing this book has been an exploration of one of the fastest-evolving fields in technology, and I'd like to start by sharing my personal experience with AI agents.

In early 2022, the core capabilities of generative AI models began showing incredible promise in solving various text generation tasks out of the box (question answering, summarization, and code generation, for example). It became clear that the potential for creating new types of applications addressing new tasks was immense. At the time, I was working with the Human/AI Experiences (HAX) group at Microsoft Research, partnering with GitHub to improve offline evaluation for an early version of GitHub Copilot — arguably the first example of a modern LLM working in real time at scale to provide large code completions to developers in an Integrated Development Environment (IDE). Our work identified the need for granular metrics (Dibia et al. 2022) to better capture code correctness and proposed improvements to the UX for Copilot. Early results showed that Copilot effectively doubled productivity for developers (Peng et al. 2023)!

Soon afterwards, I created LIDA (Dibia 2023)— one of the first systems for automatic data visualization using language models—in July 2022, months before ChatGPT's public release. The system demonstrated how users with no visualization skills could produce high-quality visualizations, and how authoring time could be drastically reduced even for experts. LIDA was implemented as a 4-stage pipeline: a data summarizer, visualization goal generator, visualization code generator (with code execution, error correction and retries), and an optional infographics generator. Ideas from LIDA are now integrated across Microsoft products (Excel, Fabric, Purview, and internal tools). Since then, exploring agents—AI models that not only generate artifacts but can also *act* (call APIs, execute code, etc.)—has become standard practice.

While creating pipelines encourages reliability (as each step can be independently tested), it also requires effort and assumes that the correct task decomposition is known, the expertise to implement each step is available, and the task is static and predictable. These assumptions often don't hold true for many complex, real-world tasks that require planning, diverse expertise, or adaptation in dynamic environments. These observations raised an important question: How can we create systems that *independently adapt* to dynamic environments and generalize to solve multiple disparate task types?

At the time, a few colleagues at Microsoft Research were exploring conversation-based multi-agent systems, which felt like a natural way to extend or generalize the rigid pipeline nature of LIDA. The overall promise was compelling: define discrete agents, give them access to models and tools, and let them *self-organize* through conversation to solve *general* problems. Even better, these systems could improve across multiple task categories simply through *improvements to the underlying models*.

As part of this work, we built a framework around this concept — AutoGen (Wu et al. 2023)— which was released in May 2023. We identified use cases where conversational and iterative reasoning capabilities improved system performance, but we also learned hard reliability lessons. These included agents failing to follow instructions (early models were so polite that they would thank each other endlessly across multiple turns!), the importance of planning and careful tool selection for performance (see the Magentic One project (Fourney et al. 2024)), how to build effective developer tooling for these systems (Dibia et al. 2024; Mozannar et al. 2025; Epperson et al. 2025), and the security challenges associated with autonomous agents.

Through this work, I've had the privilege of advising dozens of internal Microsoft teams, as well as customers—including Fortune 500 companies—on implementing multi-agent systems in production environments. Additionally, I've answered hundreds of questions from startups in the open-source AutoGen community. This experience has given me a unique perspective on what the building blocks of multi-agent systems are, where teams typically struggle, and what practical guidance they need most.

Fifty thousand GitHub stars later, a major API redesign, thousands of open source issues resolved, and multiple research papers published (Fourney et al. 2024; Mozannar et al. 2025; Dibia et al. 2024), AutoGen has helped shape the AI agent landscape and refine our collective understanding of multi-agent system design patterns. As the project matured, it evolved into the Microsoft Agent Framework—a production-ready successor that builds on AutoGen's learnings with enhanced reliability, better tooling, and enterprise-grade features.

The excitement around autonomous multi-agent systems as *one way* (certainly not the only way) to build applications has exploded across the entire industry. Today, there are over a dozen multi-agent frameworks (CrewAI, OpenAI Agents SDK, Google Agent Development Kit, Pydantic AI, and many others), accompanied by a dizzying array of buzzwords and concepts that we'll explore throughout this book—from foundational building blocks like agents, multi-agent systems, agent middleware, memory, and tools, to advanced capabilities like computer use, OpenTelemetry Gen-AI semantic conventions, and distributed agent protocols. Everyone wants to use this technology, but there's limited clarity on how to do it well.

As the dust settles, clear patterns are emerging that can guide the development of effective multi-agent systems—whether you're building simple agent interactions or complex orchestrated workflows. This book focuses on identifying these patterns and providing practical guidance for applying them effectively.

> 💡 A Note on AI Hype and Reality
>
> In my experience, there are two camps of well-intentioned AI practitioners. The first fixate on *what AI cannot do today*, dismissing it as hype. The second explores what AI can do today, acknowledging limitations while focusing on solving real problems.
>
> When I first showed early LIDA prototypes to researchers, one pointed to a visualization error and declared AI unsuitable for data visualization based on the 20% error rate at the time. A month later, with improved prompts and a new OpenAI model release, errors dropped to 3%—suddenly, the reception was entirely different. Meanwhile, product teams who saw the same demos immediately began experimenting and released features that improved alongside model updates.
>
> The lesson: take a pragmatic engineering approach. The best AI systems today combine traditional software engineering practices—proper task decomposition, systematic evaluation, and optimization—with AI capabilities. This book explores both the practical applications available today and the longer-term vision of autonomous AI, giving you the foundation to build effectively in this rapidly evolving landscape.

About This Book

This book teaches you to build effective multi-agent systems from first principles, regardless of your chosen tools or frameworks. You'll learn not just *how* these systems work, but *why* they work—including how to translate business problems into multi-agent architectures.

A note on scope: While this book focuses on multi-agent systems, they're not the right solution for every problem. Throughout these pages, you'll develop not just the skills to build sophisticated agent systems, but the judgment to recognize when simpler approaches will serve you better.

This book covers:

- **Multi-Agent Fundamentals**: Core concepts, design patterns, and user experience principles
- **Implementation from Scratch**: Building agents, workflows, and orchestration by creating a complete Python library called `picoagents`
- **Evaluation and Optimization**: Testing, measuring performance, and optimizing for reliability and scale
- **Production Deployment**: Security, ethics, and responsible AI practices for real-world applications
- **Domain Applications**: Complete implementations for information processing, data analysis, and software engineering use cases

By the end, you'll know how to choose the right multi-agent architecture for any task and how to build your own systems from scratch (should you choose to do so).

Book Content at a Glance

	Count
Code Snippets	186
Callout Boxes	77
References	73
Figures & Diagrams	52
Tables	26

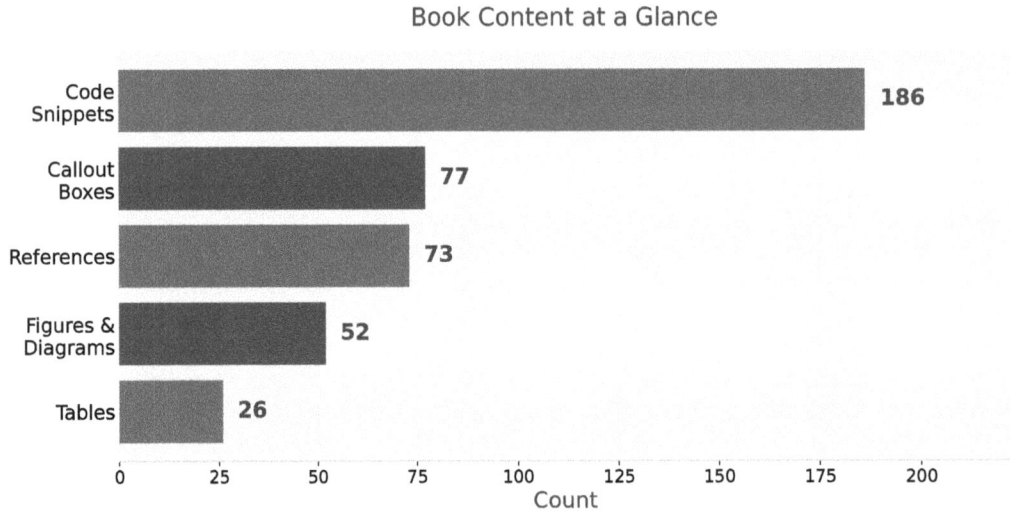

Figure 1. Book content breakdown by type.

The book contains 15 chapters with 56,232 words, 186 code snippets, 52 figures, 26 tables, 77 callout boxes, and 73 references (Figure 1).

Who This Book Is For

This book is designed for **technical practitioners** building AI-powered systems—whether you're just getting started with agents or implementing them in production.

Primary audience:

- **System architects and software engineers** designing AI applications with multiple intelligent components
- **Technical leaders** making architectural decisions about multi-agent implementations

- **AI engineers** transitioning from individual LLM models to orchestrated agent systems
- **Product managers** needing to understand multi-agent capabilities and trade-offs

You'll get the most value if you: - Want to understand **fundamental concepts** rather than just framework-specific tutorials - Are interested in **implementation details from scratch**—critical for evaluating tools and making architectural decisions - Need **practical, worked examples** you can adapt to real-world problems

Building effective multi-agent systems is roughly **40% theory** (understanding models, evaluation, translating business problems) and **60% systems engineering** (architecture, code, deployment, scaling). This book reflects that balance, providing both conceptual foundations and hands-on implementation guidance.

What Makes This Book Different

This book takes a unique approach compared to other multi-agent system resources:

- **Thorough treatment of concepts and theory**—Part I provides comprehensive foundations that executives, designers, and technical teams all need to understand multi-agent systems
- **Building production-ready systems from scratch**—You'll build agents and multi-agent orchestration patterns from first principles, so you understand every component and design decision
- **Focus on evaluation and optimization**—Moving beyond simply applying LLMs to building reliable systems requires evaluation skills often missing from traditional software engineering curricula
- **Framework and technology agnostic**—All examples use Python and are built from scratch, but can be reimplemented in any programming language or framework of your choice
- **Complete end-to-end approach**—This book covers theory (Part I), implementation (Part II), optimization (Part III), and real-world examples (Part IV), with reusable sample code throughout

Prerequisites

To get the most out of this book, you should have:

- Basic Python programming experience
- Understanding of machine learning fundamentals (helpful, but not required)
- Experience with command-line tools

Implementation Philosophy

This book takes a **fundamentals-first approach**, guiding you through implementing core multi-agent concepts from scratch. As part of this, we will build a multi-agent library— `picoagents` —from scratch and use it to implement a set of complex use cases. You'll learn to build:

- **Individual agents** by configuring them with AI models, tools, and memory components to solve tasks
- **Multi-Agent coordination** through both deterministic workflows and autonomous orchestration patterns
- **Evaluation and optimization approaches** for measuring and improving system per-

formance

- **End-user integrations** that seamlessly embed multi-agent capabilities into applications complete with the *right* user interface and user experience

By implementing these fundamental building blocks yourself, you'll gain deep understanding of the design decisions and trade-offs involved in multi-agent systems. You'll learn how and when to use existing frameworks, and how to architect systems that meet your exact requirements. PicoAgents reflects the same foundational principles and architectural patterns found in production frameworks like AutoGen, Microsoft Agent Framework, Google ADK, and Pydantic AI. Understanding these core concepts prepares you to work effectively with any multi-agent framework.

How This Book Is Organized

This book is divided into four parts that systematically build your expertise in multi-agent systems, following a **theory -> build -> optimize -> apply** progression:

Part I: Foundations of Multi-Agent Systems

Part I establishes the theoretical foundation you need to understand multi-agent systems. This part is purely conceptual, focusing on core principles without getting into implementation details:

- **Chapter 1**: Understanding multi-agent systems—defining agents, multi-agent systems, and how complex tasks benefit from multi-agent architectures;

- **Chapter 2**: Multi-agent patterns—organizing orchestration patterns along a spectrum from explicit control (workflows: sequential, conditional, parallel, supervisor) to emergent control (autonomous: plan-based, handoff, conversation-driven)
- **Chapter 3**: User experience of multi-agent systems—framing the shift from direct manipulation to delegation design; tracing evolution from Software 1.0 to Software 3.0+; establishing four essential UX principles (capability discovery, cost-aware delegation, observability and provenance, interruptibility)

Part II: Building Multi-Agent Systems from Scratch

Part II takes you from theory to practice, teaching you to build multi-agent systems from first principles. This hands-on section is framework-agnostic, focusing on core implementation concepts:

- **Chapter 4**: Building your first agent—implementing the agent execution loop with async-first architecture, event-based streaming, and graceful cancellation; establishing the agent abstraction pattern (agent = Agent(model, tools, memory)); adding middleware for control and observability
- **Chapter 5**: Computer use agents—extending agents to interact with user interfaces

through three core components (action sequence generation, interface representation, action execution); demonstrating implicit vs. explicit planning for action sequences; enabling hierarchical agent composition

- **Chapter 6**: Building multi-agent workflows—implementing workflow patterns as deterministic computational graphs with type-safe steps, conditional edges, parallel execution, and streaming observability; adding automatic checkpointing with structure hash validation for safe resume after failures
- **Chapter 7**: Autonomous multi-agent orchestration—establishing the orchestrator loop (select agent, execute turn, check termination, repeat); building composable termination conditions; implementing round-robin, AI-driven selection, and plan-based orchestration patterns
- **Chapter 8**: Building modern agent UX applications—establishing the two-component architecture (backend with FastAPI + Server-Sent Events, frontend with vanilla JavaScript/React); implementing UX principles (observability through event streaming, capability discovery, interruptibility); comparing SSE versus WebSockets for stateless, horizontally-scalable deployments
- **Chapter 9**: Multi-agent frameworks—identifying ten core capabilities that distinguish effective frameworks from basic implementations (intuitive developer experience, async-first architecture, observability, state management, declarative configuration, guardrails/middleware, pattern support, etc.); providing evaluation criteria and decision frameworks for assessing any framework against specific needs

Part III: Evaluation, Optimization, and Responsible AI

Part III addresses the critical challenges of making multi-agent systems work reliably and responsibly through systematic evaluation, optimization, security, and ethical frameworks:

- **Chapter 10**: Evaluating multi-agent systems—testing, benchmarks, and measuring success through trajectory-based evaluation
- **Chapter 11**: Optimizing multi-agent systems—identifying common failure modes and optimization strategies for production reliability
- **Chapter 12**: Protocols for distributed agents—exploring Model Context Protocol (MCP) for standardized tool integration with agentic capabilities (progress notifications, elicitation, sampling) and Agent-to-Agent Protocol (A2A) for cross-organizational collaboration; demonstrating practical MCP integration
- **Chapter 13**: Ethics and responsible AI—examining societal-scale challenges including agentic noise, platform imbalance, occupational disruption, and distributed responsibility; addressing security when agents can act through jailbreaks and the Rule of Two framework

Part IV: Real-World Applications

The final part brings everything together through comprehensive real-world case studies that demonstrate complete implementations:

- **Chapter 14**: Answering business questions from unstructured data—creating systems that extract insights from large volumes of text using cost-optimized workflows
- **Chapter 15**: Software engineering agent—developing agents that assist with code generation, testing, and development workflows through tools, prompts, and memory

Code Examples and Resources

All code examples in this book are available in the companion GitHub repository:

https://github.com/victordibia/designing-multiagent-systems

The repository includes:

- Complete source code for all examples
- Additional exercises and challenges
- Community discussions and updates

> 💡 💻 Working Code References
>
> Throughout this book, you'll see special **Working Code** callouts—like this one—that point to complete, runnable examples in the companion repository. These examples demonstrate the concepts in action and provide a foundation for your own implementations. Look for the 💻 icon to find these practical code references.

References and Academic Sources

Throughout the book, you'll find citations to relevant research papers, as well as many concepts that have been explored in blog posts, conference presentations, and open-source projects. Where possible, original sources are cited—consider exploring these materials for deeper theoretical understanding and alternative perspectives.

Stay Connected

The AI agent space is evolving rapidly, and this book is written with that in mind. See the Epilogue for information about the companion digital platform (with quarterly updates, complete application samples, and deployment guides—available separately) and free open-source resources. All readers can follow the book GitHub page for code updates.

- **Book Website**: multiagentbook.com. The book website also contains a **labs** section with useful resources such as an implementation of multi-agent usecases using multiple frameworks.
- **GitHub**: Book Code Repository

- **Author:** Victor Dibia

Note: If you find any issues or have suggestions, please open an issue in the book's GitHub repository.

Acknowledgments

Writing this book has been shaped by countless conversations, collaborations, and insights from colleagues, friends, and the broader AI community. This work stands on the shoulders of many researchers and practitioners whose contributions are cited and acknowledged throughout these pages. I am deeply grateful for their insights and generosity in knowledge-sharing.

My understanding of human-AI interaction, AI agents, and multi-agent orchestration patterns has also been profoundly shaped by conversations and collaborations with members of the Human/AI Experiences (HAX) group at Microsoft Research: Saleema Amershi, Adam Fourney, Gagan Bansal, Jingya Chen, Suff Syed, and Hussein Mozannar—thank you for the many collaborations! I am equally grateful to the leaders of the AI Frontiers lab at MSR—Ece Kamar and Ahmed Awadallah—who have been incredible mentors and encouraged me to write this book. My thanks also go to Mike Kistler, Caitie McCaffrey, Maria Naggaga Nakanwagi - the many conversations we had on MCP, A2A greatly informed how I have written about protocols for distributed agents in chapter Chapter 12.

Special thanks to Gonzalo Ramos, Steven Druker, Dan Marshall, Dave Brown, and Nathalie Riche of the Visualization Group at Microsoft Research, with whom I shared early excitement and collaborated on generative AI for data visualization ideas. Thank you to the many reviewers who took the time to provide feedback on early drafts of the book, including Tamer Abuelsaad, Becky Whitney, Valliappa Lakshmanan, David Adamo, Piali Choudhury, Sasa Junosovic, Andrew Reed, Shipi Dhanorkar, Ade Famoti, Adewale Akinfaderin, Justin Norman, Ryan Sweet, Elisa Piccin, Tara Walker, Edidiong-Abasi Anwanane, Innocent Dibia, Raimondas Lencevicius and many others. Thank you to Happy Thorp who helped with editing and proofreading.

I am deeply grateful to the AutoGen community—from Chi Wang and Qingyun Wu (who started the AutoGen project) to Jack Gerrits and Eric Zhu (who stepped up to help lead it, brought engineering rigor, and helped make it a success). Many of the implementation ideas conveyed in this book (e.g., task cancellation, component serialization, memory interface etc) emerged from many design discussions with Jack and Eric. Thank you to all the contributors, early adopters, and community members who have shared ideas, reported issues, provided feedback, and contributed code. The broader open-source AI community has been instrumental in shaping the ideas in this book—through countless discussions, thousands of issues on GitHub, and numerous community calls. Thank you!

Finally, I am grateful to my family. This book is as much yours as it is mine—to my parents Juliana, Paulinus, Karine, and Norayr for laying the foundation that made this journey possible, to my wife Hermine for all the support and brilliant feedback along the way, and to my little boy Liam whose hugs provided the energy I needed when things got tough. Thank you!

Victor Dibia

Part I

Part I: Foundations of Multi-Agent Systems

Chapter 1

Understanding Multi-Agent Systems

This chapter covers:

- The complexity challenge: understanding when individual AI models become insufficient and why agents are needed
- What defines an agent: core capabilities (reason, act, communicate, adapt) and essential components (model, tools, memory)
- Complex tasks that motivate multi-agent systems: planning, diverse expertise, extensive context, and adaptive solutions
- When to use (and when not to use) multi-agent approaches: Multi-agent systems are not always the *right* solution; know when to use them

1.1   Introduction

Generative AI (GenAI) models today have demonstrated remarkable abilities in modeling complex relationships within the vast amount of data on which they are trained. For example, models such as GPT-4, Claude, Gemini, etc.—also known as large language models (LLMs) (Vaswani et al. 2017), excel at text processing tasks—for example: summarizing passages, extracting entities, generating code, etc. Given the capabilities of these models, they have now been applied directly within applications to solve tasks in various industries. For instance, LLMs have been integrated into products for generating marketing copy, reviewing legal documents for compliance, providing advanced natural language understanding to chatbots and virtual assistants, and assisting researchers in the scientific discovery process. And adoption is growing rapidly across both individual users and enterprises. A recent report states that **75%** of knowledge workers are already using AI in the workplace today (Microsoft

AI Blog 2024), and a rising number of startups are building applications that leverage these models to solve real-world problems.

---

i Quantifying the Shift to AI Agents

It's widely acknowledged that interest in AI agents is growing rapidly, but by how much? To answer this question with concrete data, Chapter 14 demonstrates how to build a multi-agent workflow that analyzes startup data from Y Combinator—one of North America's largest startup incubators.

The workflow we will design analyzes descriptions from 5,622 YCombinator companies. We found that startups building AI agents grew from 6.1% in 2020 to 47.7% in 2025—a 7.8x increase in just five years. This rapid growth signals where the industry sees opportunity: moving beyond AI-assisted tools to fully autonomous systems that can handle complex, multi-step tasks.

This analysis itself exemplifies the multi-agent workflow patterns we'll explore in Chapter 2, with the complete implementation detailed in Chapter 14. Feel free to jump ahead if you're curious about the methodology!

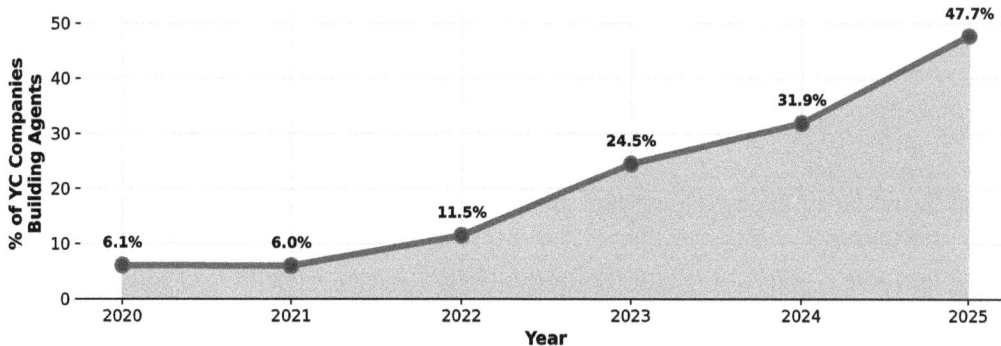

Figure 1.1. Growth in Y Combinator startups building AI agents (2020-2025). This dramatic trend demonstrates the industry shift toward autonomous agentic AI systems.

However, as tasks become more *complex* —those that are long-running, involve multiple steps, require consideration of various perspectives, and necessitate exploration of a dynamic solution space—these models alone are insufficient. To illustrate this progression and highlight the need for different types of AI systems, consider the following task examples shown in Table 1.1, representing three distinct levels of complexity:

Table 1.1. Task complexity levels: Model-level tasks require only information retrieval, agent-level tasks require planning and tool use, and multi-agent tasks require diverse expertise and iterative development.

Complexity Level	Task Example	Description
Model-Level	"What is the height of the Eiffel Tower?"	Direct information retrieval from training data
Agent-Level	"Tell me the stock price of NVIDIA today."	Requires current data, planning, and tool use
Multi-Agent	"Build a mobile application that helps users view stock prices, buy stock and file taxes."	Involves multiple domains of expertise and iterative development

> ℹ Definitions: Agents, Multi-Agent Systems, and Tools
>
> - **Agent:** An AI system that can reason, act with tools, communicate, and adapt
> - **Multi-Agent System:** Multiple agents working together, each with specialized capabilities
> - **Tools:** External capabilities like APIs, code execution, web search that extend agent abilities

Task 1 from Table 1.1 can be reliably addressed by retrieving general knowledge facts from a model's training data or existing systems such as web search engines. This represents what a **model** can accomplish - straightforward information retrieval without the need for planning, tools, or adaptation.

Task 2, however, reveals where models alone become insufficient. The stock price of NVIDIA *today* represents information that an LLM is unlikely to have seen during training; thus, the results it generates are likely to be incorrect (hallucination). Task 2 requires understanding the query, *planning* (e.g., decomposing the task into steps such as fetching current stock data and analyzing the data to answer the question), *action* (executing each step using tools like web search or APIs), and presentation of final results. This is where we need an **agent** - a system that combines the reasoning capabilities of these models with tools that enable them to act on our behalf. Current GenAI applications explore this approach by building prescribed solution pipelines that include Large Language Models (LLMs) which drive tools for action.

Task 3 represents a level of complexity that goes beyond what even a capable single agent can handle effectively. It involves multiple types of expertise (interface design, Android development, API usage, integration testing, etc.) and requires an iterative solution where actions are taken and resulting outcomes determine the next steps toward a solution (e.g., retrieving appropriate SDK documentation, writing multiple versions of each module in the app, testing them to identify errors, repairing and integrating until a working application is crafted). This is where **multiple agents** become essential - each bringing specialized capabilities while collaborating to solve tasks that exceed any single system's scope.

In this book, we will focus on exploring how to address these complex tasks through the development of applications that implement elements of planning, action and communication across *multiple agents*. In turn, these applications hold potential to enable radically new, and general-purpose digital interfaces, solve new classes of tasks or address existing tasks in a manner that reduces human toil and effort.

The potential extends beyond generic task automation. When agents understand user context (current activity, environment, recent interactions) and preferences (explicitly defined settings and behavioral patterns), they can provide highly personalized assistance with tasks such as scheduling meetings, drafting email responses, booking flights, making e-commerce purchases, and filing taxes.

To realize this potential, we need to understand the fundamental building blocks: What exactly makes a system an "agent"? How do we build systems where multiple agents work together effectively? Let's start by examining these core concepts.

The progression from models to agents to multi-agent systems is visualized in Figure 1.2.

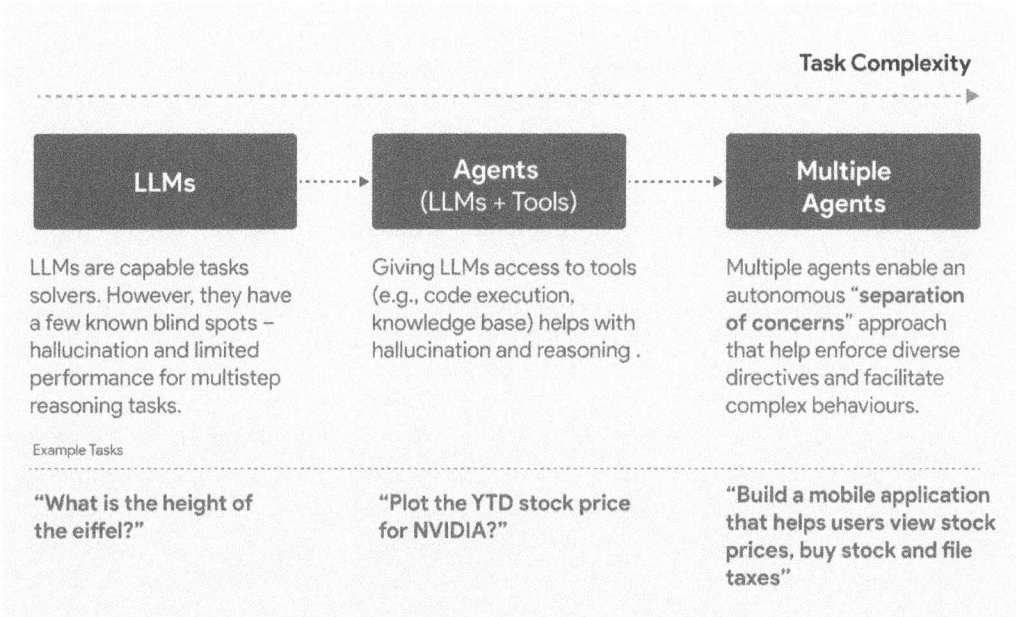

Task Complexity

LLMs	Agents (LLMs + Tools)	Multiple Agents
LLMs are capable tasks solvers. However, they have a few known blind spots – hallucination and limited performance for multistep reasoning tasks.	Giving LLMs access to tools (e.g., code execution, knowledge base) helps with hallucination and reasoning .	Multiple agents enable an autonomous **"separation of concerns"** approach that help enforce diverse directives and facilitate complex behaviours.

Example Tasks

"What is the height of the eiffel?"	"Plot the YTD stock price for NVIDIA?"	"Build a mobile application that helps users view stock prices, buy stock and file taxes"

Figure 1.2. Generative AI models (e.g., Large Language Models) can address a wide range of tasks including answering general knowledge questions, translating text, summarizing passages, generating code, etc. However, as tasks become more complex, requiring reasoning, planning, and action, the limitations of these models become apparent. We can address these increasingly complex tasks by providing LLMs with access to tools that enable them to act as agents on our behalf and enabling collaboration across multiple agents.

> 💡 Single vs Multi-Agent: A Blurry Boundary
>
> In practice, the boundary between "a single agent" and "a multi-agent system" is often blurry. An agent might use tools that are themselves agents (as we'll explore in Section 4.11 in Chapter Chapter 4), or what appears as one agent to a user might internally coordinate multiple specialized sub-agents working together.
>
> Throughout this book, we illustrate explicit multi-agent architectures—where coordination between agents is visible and intentional—because these patterns make design principles most clear. However, the orchestration patterns, evaluation approaches, and ethical considerations we explore apply broadly to any autonomous agentic system, regardless of whether its architecture presents as "one" agent or "many."

1.2 A Primer on Generative AI

Generative models are a subset of deep neural networks skilled at *identifying complex patterns within datasets, enabling them to generate new, similar data points*. These models are distinct from discriminative models, which focus on differentiating between data types or classifying them. Generative models can be trained on data from various modalities. For instance, models trained on written text in human languages, such as the GPT series, are known as Large Language Models (LLMs); those trained on images, like the DALL-E series, are termed Image Generation Models (IGMs); and models trained on both, such as GPT4-V, are often referred to as Large Multimodal Models (LMMs) . This book will concentrate on Generative Models that generate text output, including both large language models and large multimodal models.

The fundamental concept for training generative models is relatively simple. In the case of LLMs, these models are trained to predict the next word in a sequence or to fill in blanks for masked or hidden sections of text - aka *sequence prediction* . By applying this training objective to a sufficiently large dataset, the models learn representations of the world as depicted through text. In turn, the models can leverage these representations in generating text that is coherent, relevant, and contextually appropriate.

To *apply* these models to solve tasks, early efforts focused on framing multiple tasks as sequence prediction problems. For example, to classify text, a description of the classification problem is provided, followed by a list of labels and a prompt asking, " `<problem>` `<labels>` Which of the provided labels is the correct class?". Similarly, to summarize a given passage, given a prompt structured as " `<passage>` The summary of the passage above is ..", the model is induced to generate a summary of the passage. It turns out that framing tasks as sequence prediction problems allows the models to leverage knowledge gained during training to perform a wide range of tasks (e.g., language translation, sentiment analysis, question answering, dialogue generation, named entity recognition, syntax parsing, code synthesis, paraphrasing, grammar correction etc.) beyond their initial training objectives.

This background is important to understand the capabilities and limitations of these models when applied to complex tasks. For example, the *reasoning* capabilities observed in these models (e.g. 2 apples + 10 apples = 12 apples) are often limited to scenarios well represented in training data, with failures to inherently generalize to rare or unseen scenarios (e.g. solving linear equations in base 3).

The practice of creating sequences that increase the likelihood of successfully completing a task is known as *prompt engineering* . Techniques such as few-shot prompting (Brown et al. 2020), where examples of the task and solution are included in the sequence, chain-of-thought prompting (J. Wei et al. 2022), where examples of tasks broken down into solution steps are included, and ReAct prompting (Yao et al. 2022), which combines reasoning and acting in language models, have become leading prompt engineering approaches. Beyond prompt engineering, fine-tuning models on task-specific instruction data and explicitly collected human feedback data (Ouyang et al. 2022) have proven to enhance model alignment - efforts aimed at ensuring that generated outputs align with human intentions, preferences, and values.

Finally, while LLMs can be instructed to generate text sequences that solve tasks, the amount of text they can process at a time is fixed, constrained by what is known as the *context window* . The context window size defines the maximum number of tokens (words or word pieces) the model can handle in a single input and output sequence. This limit is dictated by the model's architecture, particularly its computational and memory constraints. Understanding the context window is crucial for developers as it affects how input data should be structured and how tasks should be framed. Strategies like truncation, sliding window techniques, summarization, chunking, and optimized prompt engineering are essential to operate effectively within these constraints, maximizing the model's potential in various applications while balancing computational resources.

1.2.1   Getting Access to Generative AI Models

State-of-the-art models like GPT-4 or Claude are primarily accessible through cloud-based APIs due to their computational requirements and proprietary nature. Model providers including OpenAI, Anthropic, Google, and Microsoft Azure offer API-level access through software development kits (SDKs). The typical workflow involves creating a developer account, configuring billing (most providers offer usage-based pricing with free tiers), and generating API keys for authentication.

Throughout this book, we standardize on the OpenAI API specification for a pragmatic reason: it has emerged as a de facto standard adopted across multiple providers. The OpenAI SDK natively supports both OpenAI's services and Azure OpenAI—Microsoft's managed enterprise deployment of OpenAI models that offers additional security, compliance, and regional availability guarantees. This standardization means code written for OpenAI's API can work with minimal changes across providers like Together AI, Anyscale, and even local model deployments.

For local development and privacy-sensitive applications, tools like vLLM (vLLM Team 2023) enable self-hosting of open-source models while maintaining OpenAI API compatibility. This architectural pattern—where different providers conform to a common API specification— allows the multi-agent systems we build to remain agnostic to the underlying model provider, whether cloud-based or locally deployed.

1.3   Why Multiple Agents? Complex Task Characteristics

While a single agent can effectively handle tasks like retrieving current stock prices, even capable agents encounter fundamental limitations when facing highly complex challenges. Consider our Task 3 from Table 1.1: "Build a mobile application that helps users view stock prices, buy stock and file taxes." A single agent attempting this task would need to understand application requirements across multiple domains, design user interfaces and user experiences, write Android development code, integrate stock market APIs, implement tax filing functionality, and handle testing, debugging, and deployment.

Figure 1.3 illustrates the complexity of this multi-domain task.

Even with access to powerful models and comprehensive tools, a single agent faces several challenges. The extensive context required quickly exceeds practical limits - existing studies (Liu et al. 2024) have shown that while LLMs can process increasingly longer text, they are likely to attend to instructions at the beginning and end of the text, while neglecting data in the middle. More fundamentally, this task demands diverse types of expertise that are difficult to encode effectively in a single agent's instructions.

Complex tasks that motivate multi-agent approaches typically exhibit four key characteristics, as shown in Figure 1.4. While the presence of any one of these characteristics can increase the complexity of a task, it is often their combination that presents the greatest opportunity for multi-agent systems.

1.3.1   Planning

Complex tasks often require a high level plan, which involves decomposing the task into steps that must be completed successfully. Given some context, a plan prescribes a set of actions that should be executed to achieve some target success state. This is a well-known and studied property in the robotics planning literature and we will borrow concepts from this domain across this book.

1.3.2   Diverse Expertise

Decomposing complex tasks into multiple steps often results in steps that can benefit from specific expertise. For instance, consider the app development task where the objective is to build a mobile app for viewing and purchasing stocks and filing taxes. This task may be broken down into steps such as analyzing app requirements, designing the user interface, implementing required functionalities, integrating necessary APIs for stock data and tax filing, and testing and deploying the app. In turn, these tasks map into specific roles that can be

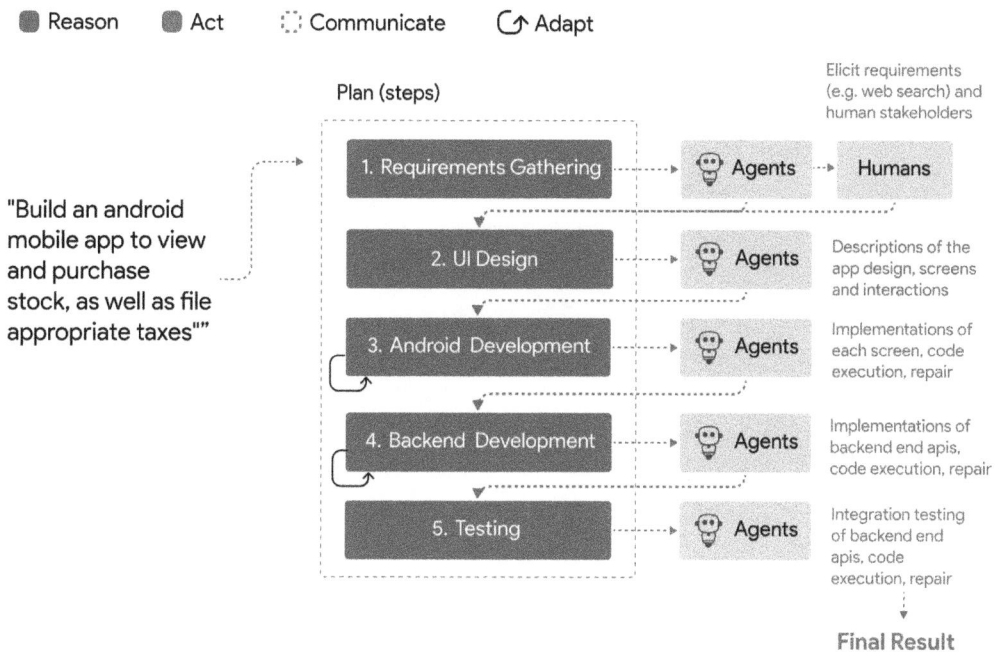

Figure 1.3. *Consider the following complex task - "build an app to view and purchase stock, as well as file appropriate taxes". To address such a task, a multi-agent application must derive a plan (e.g., gather requirements, design an interface, write android code etc), optionally delegate steps in the plan to various entities with specialized relevant capabilities (agents and humans), communicate effectively and adapt (recover from errors, explore new approaches) to solve the task.*

addressed by dedicated agents (e.g. a UX agent, Android App agent etc), each possessing specialized knowledge and skills.

The separation of expertise allows for a separation-of-concerns approach that single-agent systems may struggle to achieve following a generic instruction set. From the application development perspective, creating agents with specific directives provides a useful abstraction for cleanly mapping responsibilities to specific entities, enabling domain-driven design of applications.

1.3.3 Extensive Context

Complex tasks often require extensive context that need to be processed to solve the task. For example, in our app development task, the agents may conduct web search queries as well as ask for human feedback to assemble requirements, across multiple turns to assemble the right initial context. Such lengthy instructions or context present a significant challenge for single-agent systems. Existing studies (Liu et al. 2024) have shown that while LLMs can

What makes a task complex?

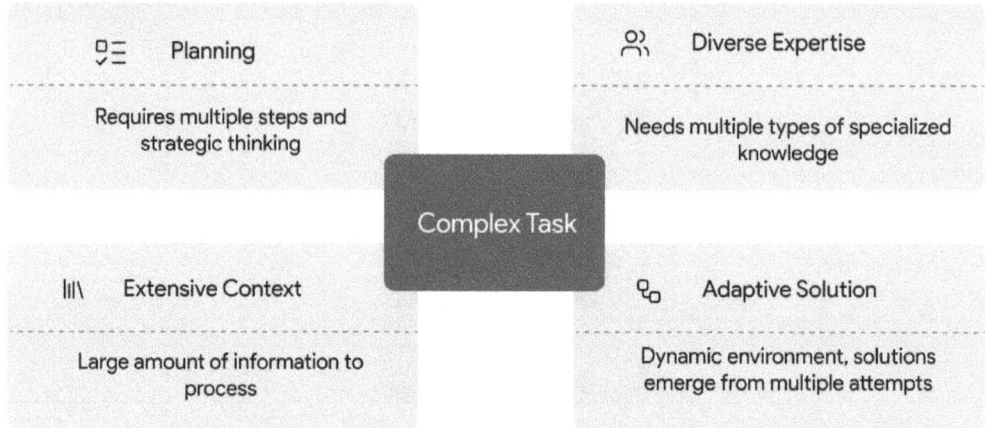

Planning	Diverse Expertise
Requires multiple steps and strategic thinking	Needs multiple types of specialized knowledge

Complex Task

Extensive Context	Adaptive Solution
Large amount of information to process	Dynamic environment, solutions emerge from multiple attempts

Figure 1.4. Complex tasks exhibit four key characteristics that make them suitable for multi-agent systems: Planning (requiring multiple steps and strategic thinking), Diverse Expertise (needing specialized knowledge across different domains), Extensive Context (processing large amounts of information), and Adaptive Solutions (where solutions emerge from multiple attempts). Tasks become increasingly complex when they combine multiple characteristics, making them ideal candidates for multi-agent approaches.

process increasingly longer text, they are likely to attend to instructions at the beginning and end of the text, while neglecting data in the middle. This perspective is also informed by theories from cognitive science, such as cognitive load theory (Sweller 1988), which suggests that the human brain has a limited working memory capacity. When presented with lengthy or underspecified instructions, this capacity becomes overwhelmed, leading to decreased comprehension and performance.

A multi-agent perspective can help address this limitation by selective context provisioning - i.e., only sections of available context (for example, a history of actions performed so far) relevant to a task are provided to an agent focused on that task. By structuring instructions and managing context in a way that minimizes unnecessary cognitive load, we can significantly improve task performance in AI systems.

1.3.4 Adaptive Solutions

Complex tasks are often situated in dynamic environments where the exact solution is unknown or uncertain until some actions are taken. For example, specific data sources may be unavailable at a given time requiring the exploration of other data sources or each action may have unexpected side effects etc. For example, in the app development task, the initial request to the API may fail (e.g., due to a network issue, incorrect argument format or general changes to the API), requiring the agent to adapt by retrying the request with different parameters based on the *feedback* (e.g., error message received from the API request or seeking explicit human assistance). In these cases, the solution emerges from *adaptation* - iteratively reasoning across some initial plan, actions and outcomes of agents as they solve the task.

This property is also inspired by insights from metacognition studies, which suggest that an agent's awareness and regulation of their own cognitive processes can significantly influence the problem-solving process. Metacognitive skills, which are challenging for single-agent systems today, enable agents to reflect on, evaluate, and adjust their strategies in real-time, leading to more adaptive and potentially innovative solutions within the emergent framework of multi-agent collaboration.

Given these adaptive requirements, it becomes more challenging to write deterministic pipelines (e.g., specific prompts, tools), hence the need for agents that can self-orchestrate (with some autonomy) to address the task.

Table 1.2 shows some examples of complex tasks and how they map to each of complex task properties.

Table 1.2. Examples of complex tasks and their defining characteristics: Planning, Diverse Expertise, Extensive Context, and Adaptive Solutions.

Task	Planning	Diverse Expertise	Extensive Context	Adaptive Solutions
Web App Development	Requirements, design, compliance, APIs	Developers, designers, security, data experts	Legacy systems, docs, workflows, standards	Security updates, feedback, changes
Financial Reporting	Data collection, analysis, reporting	Analysts, statisticians, writers, domain experts	Market data, tools, standards, trends	New sources, feedback, market shifts
Tax Filing	Data gathering, analysis, filing, compliance	Tax advisors, analysts, legal, jurisdiction experts	Records, tax codes, regulations, structures	Law changes, queries, corrections
Presentation Design	Content review, design, interactivity	Designers, editors, presentation experts	Content, audience, goals, guidelines	Feedback, content updates, format changes

Furthermore, our analysis of real world data from YCombinator reveals that 47.7% of Y Combinator startups building AI agents in 2025 are focused on the following domains:

productivity (25.5%), software (18.4%), finance (15.6%), health (11.6%), and e commerce (6.8%) where planning and diverse expertise are essential. See the in-depth analysis in Chapter 14.

Reflecting on these patterns provides context for understanding *when* and *why* a multi-agent approach is beneficial. When tasks are complex, dynamic, and require diverse expertise, a multi-agent approach can offer the adaptability and collaborative problem-solving capabilities necessary to manage these tasks effectively. Recent research studies have also highlighted benefits of multi-agent approaches. For instance, (Du et al. 2023) find that a *society of mind* approach (Minsky 1986) - a theory proposing that human intelligence arises from the interaction of multiple simple agents working together - improves reasoning and factuality when multiple agents debate outcomes over multiple rounds of conversation turns. Similarly, (Liang et al. 2023) demonstrate that using multiple agents with separate roles (e.g., agents that perform a task and agents that adjudicate the quality of the task) can improve the diversity and quality of generated outcomes.

1.4   What is an Agent?

Definitions of agents vary across the AI literature, from early concepts in robotics to more recent ideas around entities driven by AI models. A good starting point is the *classic* work of Russell and Norvig (Russell and Norvig 2020) that define an agent as anything that perceives its environment through sensors and acts upon that environment through actuators, operating through an agent function that maps perceptions to actions. This classical definition emphasizes perception-action *loops* driven by logical reasoning.

To illustrate this classical approach, consider a robotic vacuum cleaner: it perceives its environment through sensors (detecting obstacles, dirt levels, battery status), applies logical rules (if obstacle detected, turn; if dirt detected, increase suction; if battery low, return to dock), and acts through actuators (motors for movement, suction mechanisms, brush rotation). The agent function follows predetermined logic: obstacle -> avoid, dirt -> clean, low battery -> recharge. This creates a continuous perception-action loop where each sensor reading triggers specific programmed responses.

We can build on this foundation but extend it for the generative AI era. While classical agents operate through predefined functions, generative AI agents maintain the same perception-action cycle but add sophisticated capabilities: complex reasoning powered by large language models, dynamic tool use, natural language communication, and adaptive behavior based on outcomes. These enhanced capabilities make them particularly suitable for the collaborative problem-solving we explore in this book.

For this book, we define agents as entities that can reason, act, communicate, and adapt to solve problems. Consider our NVIDIA stock price example (Task 2 from Table 1.1): an effective agent must understand the information request, reason about obtaining current stock data, take action using appropriate tools (web search or financial APIs), and adapt if the initial approach fails. This adaptation might involve trying alternative data sources or adjusting the query strategy.

The core components of an agent are shown in Figure 1.5.

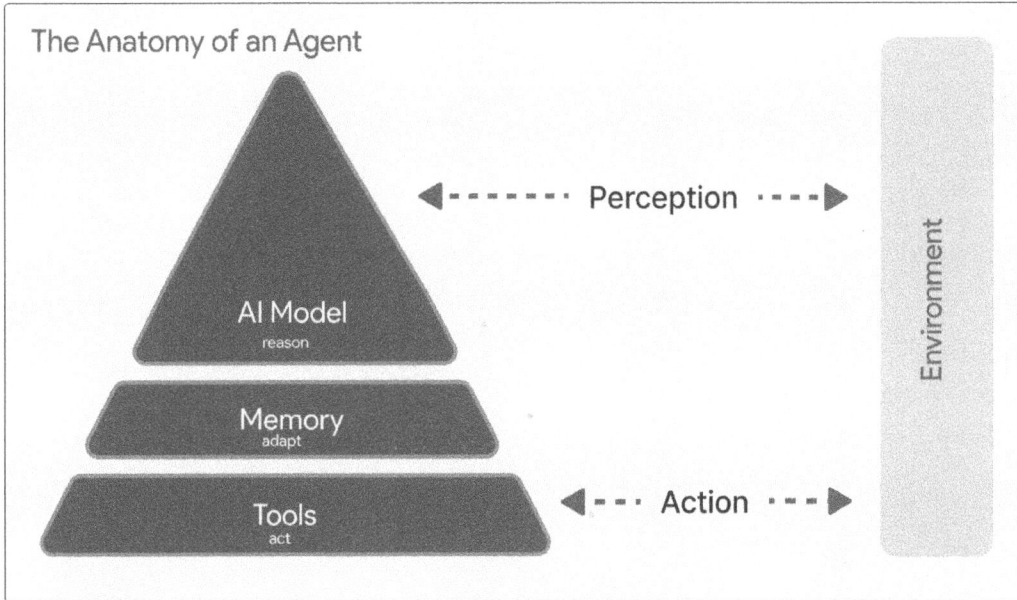

Figure 1.5. An agent is a software entity that possesses core components - a generative AI model that enables reasoning, tools that enable the agent to act, and memory that enables the agent to recall and reuse information.

1.4.1 Core Capabilities

An agent possesses four fundamental capabilities that distinguish it from basic models or traditional programs:

Reason: Agents can synthesize new information by applying rules or logic to available context. This reasoning may be deductive, inductive, or abductive and can be driven by a Generative AI model, custom processing functions, or a combination of both. In our stock price example, reasoning involves understanding that "NVIDIA today" requires current market data, not historical information from training data.

Act: Agents can take concrete actions to affect their environment or gather information. This goes beyond generating text responses - agents can execute code, call APIs, search the web, or interact with external systems. For the stock price task, acting means actually calling a financial data API or performing a web search to retrieve current pricing information.

Communicate: Agents can effectively exchange information with users, other agents, and external systems. This includes understanding natural language inputs, formatting appropriate responses, and knowing when and how to request additional information or clarification.

Adapt: Agents can modify their approach based on feedback, changing conditions, or new information. If an initial API call fails due to rate limiting, an effective agent might wait and retry, switch to an alternative data source, or adjust its approach based on the error message received.

1.4.2 How Agents Work

Agents operate through a fundamental **action-perception loop** shown in Figure 1.6. Unlike models that generate single responses, agents work through iterative cycles—taking action, perceiving results, and adapting based on outcomes until the task is resolved. Multi-agent systems build upon this principle, with multiple agents running coordinated action-perception cycles to solve complex tasks that exceed any single agent's capabilities.

Agent Action Perception Loop

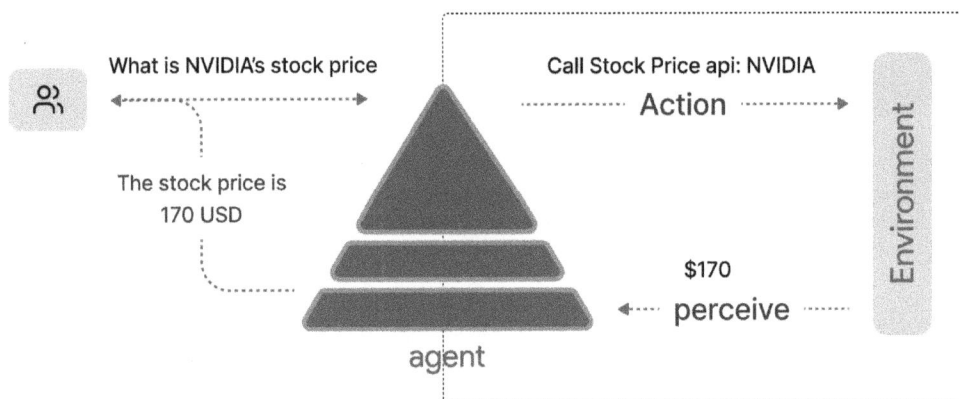

Agents take **action**, and then **perceive** the results over an iterative loop. The response may be an error that the agent has to handle with a **retry** or error response to the user.

Figure 1.6. The agent action-perception loop: Agents operate through iterative cycles where they take action (such as calling an API), perceive the results (processing the response), and adapt their approach based on outcomes. When successful, agents provide natural language responses; when errors occur, they may retry with different parameters or inform users of limitations.

These capabilities are enabled by the three core components shown in Figure 1.5 working together:

1.4.2.1 Model

The reasoning engine that enables decision-making and planning. This is typically a large language model (LLM) or large multimodal model (LMM) that can understand context, generate plans, and determine appropriate actions. The model serves as the "brain" of the agent, processing inputs and deciding what to do next.

For agents to successfully accomplish tasks, they must reason over the task state and determine the appropriate actions to take. This core decision-making capability can be driven by generative AI models, predefined logic, or explicit human input. Besides generative AI models, agents can also make use of custom logic (e.g., logic to call an API whenever a message is received based on some heuristics), or explicit human input to drive their actions and reasoning. Implementations that intelligently combine multiple drivers—using generative AI models to reason over the task state and requesting just-in-time human input to provide feedback—can significantly enhance the quality and adaptability of agent responses.

1.4.2.2 Tools

Tools, also known as skills or plugins, are specific implementations of logic designed to carry out particular tasks. They serve as the primary method for agents to *act* on tasks. Tools can be grouped into two categories: *general-purpose* tools and *domain-specific* tools.

General-purpose tools enable a broad range of capabilities, such as a code executor that allows agents to complete any task expressible as code, or a UI interface driver that allows agents to carry out tasks formulated as a sequence of UI interactions. In contrast, domain-specific tools are designed to address a specific task or a set of related tasks (e.g., calling a weather API with particular parameters). Providing agents with access to tools can significantly impact the range (diversity) and complexity of tasks that agents can address.

1.4.2.3 Memory

For agents to effectively perform tasks and improve over time, they need memory—the ability to recall and reuse information from past interactions. This enables agents to learn from experience and apply lessons to future tasks.

Short-Term Memory acts as working memory for the current task, tracking recent actions, conversation history, and temporary information needed to complete the immediate objective. In multi-agent systems, short-term memory often includes shared context so agents can coordinate effectively.

Long-Term Memory stores accumulated knowledge, successful strategies, and learned patterns that persist across different tasks and sessions. This allows agents to build expertise over time and apply past experience to new situations.

Beyond these temporal distinctions, memory systems also differ in **who controls them**. Application-managed memory is controlled by the developer—information is stored and retrieved automatically by the application code based on predefined strategies. Agent-managed memory gives agents direct control—they explicitly decide what to store, when to retrieve information, and how to organize their knowledge base through tools. This distinction between application-controlled and agent-controlled memory proves important when building agents that need to actively curate their own knowledge. We explore these memory architectures and the tradeoffs between them in Section 4.8.

These components work together in a continuous cycle: the model reasons about the current state and determines what action to take, selects and uses appropriate tools to execute that action, observes the results, updates its memory with new information, and adapts its approach for the next step.

---

i Conversational Programming

The action-perception loop described above raises a practical question: how do we implement such iterative reasoning and action-taking in code?
A powerful approach is *conversational programming* , where each step in the agent cycle—reasoning, acting, and processing results—is represented as messages in a conversation. This aligns naturally with chat-fine-tuned models, which are designed to process conversation histories and generate contextually appropriate responses.
This paradigm enables agents to seamlessly blend natural language interaction with concrete actions like code execution or API calls, and forms the foundation for multi-agent frameworks like AutoGen (Wu et al. (2023)), where agents coordinate through message passing—a pattern we'll explore in detail in later chapters.

1.4.3 Agent vs. Model: A Clear Distinction

While both agents and models can process natural language and generate responses, agents differ fundamentally in their ability to interact with the world beyond text generation:

- **Models** excel at text processing, analysis, and generation based on their training data
- **Agents** can take actions, use tools, maintain context across interactions, and adapt their behavior based on results

Our NVIDIA stock price example (Task 2 from Table 1.1) perfectly illustrates this distinction: a model might generate a plausible-sounding but potentially incorrect stock price based on its training data, while an agent would recognize the need for current information, use appropriate tools to fetch real-time data, and provide an accurate, up-to-date response.

This ability to move beyond text generation into action-taking and adaptation is what makes agents powerful tools for addressing complex, real-world tasks that require more than just information synthesis.

1.5 What is a Multi-Agent System?

Building on our agent definition, we define multi-agent systems as applications that involve a *group of agents*, each with diverse capabilities and specialized objectives, *collaborating to solve tasks*. While these agents could be embodied in physical robots, this book focuses on agents that exist primarily in the digital world.

As shown in Figure 1.7, multi-agent systems can be organized through two distinct approaches.

Multi-Agent Orchestration **Workflows**

Workflow (Graph) Ochrestration

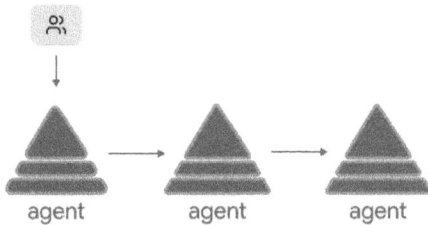

Autonomous (AI Driven) Orchestration

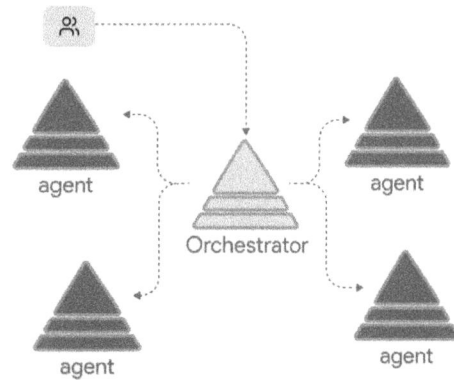

Control flow is predefined (typically modelled as a graph)

Control flow is Driven by an AI model at runtime across several patterns

Plan-based   AI Turn Taking   Handoff

Figure 1.7. Two approaches to multi-agent orchestration: predefined workflows (left) versus AI-driven autonomous orchestration (right). The key difference lies in whether control flow is predetermined or emerges dynamically from AI-driven decisions at runtime.

---

i Multi-Agent System

A multi-agent system (or multi-agent application) is a collection of agents that collaborate to solve tasks. Each agent maintains specific capabilities—reasoning, acting, and communicating—and can adapt to changes in the task or environment. The key distinguishing feature of multi-agent systems is their *orchestration mechanisms*: the patterns that determine how agents communicate, when they act, and how they share data and control flow during task execution.

> **i** Orchestration
>
> Across this book, we will repeatedly use the term orchestration and it is rather helpful to define what we mean by the term. **Orchestration** refers to the mechanisms and patterns that enable multiple agents to work together effectively toward shared goals. This encompasses two main aspects - how they share information, and who controls the flow of execution (the order in which they act).
>
> While academic literature and emerging practices sometimes uses "coordination" and "orchestration" interchangeably, we use "orchestration" as the primary term throughout this book to align with industry standards and common usage in AI/ML frameworks.

As shown in Figure 1.7, multi-agent systems can be organized through two distinct approaches. Multi-agent workflows (left) use predefined control flow where agents follow established sequences and handoffs, creating predictable and reliable processes. Autonomous multi-agent systems (right) use AI-driven orchestration where the control flow is determined dynamically at runtime, enabling adaptive responses to complex or unpredictable tasks. While both approaches use agents with the same core anatomy—model, memory, and tools—they differ fundamentally in how orchestration decisions are made.

1.5.1   Two Approaches to Multi-Agent Orchestration

Multi-agent systems can be organized through two fundamentally different approaches, each suited to different types of problems:

Multi-Agent Workflows (Defined Orchestration) These systems follow pre-defined collaboration patterns where each agent has clearly specified roles, responsibilities, and handoff points. The orchestration logic is explicitly programmed, creating predictable and repeatable processes. For example, a document processing workflow might have agents that specialize in text extraction, analysis, and formatting, working in a predetermined sequence with defined inputs and outputs for each stage.

Autonomous Multi-Agent Orchestration (AI-Driven Orchestration) These systems use AI models to drive orchestration decisions, allowing agents to dynamically negotiate responsibilities and adapt their collaboration based on task requirements and intermediate results. The orchestration emerges from agent interactions rather than being pre-programmed. This approach is particularly valuable for complex tasks where the optimal solution strategy cannot be predetermined and must evolve through exploration and adaptation.

The choice between these approaches directly impacts system behavior: workflows provide predictability and reliability, while autonomous systems offer adaptability and innovation. We explore the specific patterns within each approach in detail in Chapter 2, including sequential and supervisor patterns for workflows, and group chat and handoff patterns for autonomous systems.

> **i** An Adaptive Multi-Agent Application Example
>
> Consider this scenario: You need to create an application that books flights to specific destinations at the best price. However, your target airline, specialairlines.com, doesn't offer an API—only a web interface and mobile app designed for human users.
> This creates several challenges beyond simply finding and booking flights:
> **Interface Navigation**: The system must understand interface content (HTML elements, visual layouts), determine appropriate actions (clicking buttons, filling forms), and verify success (checking for confirmation messages). Each action changes the interface state, affecting subsequent possible actions.
> **Dynamic Adaptation**: The system must handle interface changes (button relocations, updated form fields) and recover from errors (failed bookings, network issues) through iterative problem-solving.
> **Multi-Agent Orchestration**: Success requires collaboration between specialized agents—perhaps a navigation agent for interface interaction, a payment agent for transactions, and a monitoring agent for verification—along with communication with human users when assistance is needed.
> This scenario illustrates a task where the solution cannot be predetermined but must emerge through a series of adaptive actions. It exemplifies the complex problems that require reasoning, acting, adapting, and communicating across multiple agents— tasks that exceed the capabilities of current single-agent systems and represent ideal candidates for multi-agent approaches.

1.6 Why Now?

Multiple factors highlight the importance of multi-agent Generative AI applications today, including advances in AI reasoning capabilities, economic opportunities through time arbitrage, self-improving systems, opportunities to tackle complex tasks, the increasing demand for reliable automation, platform economics, and the growing need for ethical AI deployment. These elements emphasize the significance of multi-agent systems in addressing contemporary challenges.

1.6.1 Advances in Generative AI Reasoning Capabilities

The general premise of multi-agent systems is not new. This topic has been extensively researched both from the perspective of understanding how humans collaborate to solve tasks (collective intelligence, crowdsourcing) as well as how artificial agents collaborate with themselves and humans (robot planning, robot navigation, human-robot collaboration, swarm intelligence, etc.). However, the development of artificial agents has been limited due to the lack of a *reasoning engine* (or *artificial brain*) that can adapt to context, synthesize plans that address tasks, and drive actions required as part of such plans. Recent advances in the demonstrated reasoning capabilities of Generative AI models, such as GPT-3.5 and GPT-4, change that status quo and now provide a critical component that enables new experiments and progress in creating truly autonomous multi-agent systems.

1.6.2   Economic Value Through Time Arbitrage

Multi-agent systems offer compelling economic advantages through time arbitrage . While an agent might take longer to complete a task than a human in absolute terms, the economics favor delegation when considering the value of human attention. For example, a task that costs you $30-100 of time (1 hour at your hourly value) might cost $1-2 for an agent to complete over 2-3 hours. As GenAI inference costs continue to drop while human time becomes more valuable, this equation only improves.

1.6.3   The Promise of Self-Improving Systems

Unlike traditional software that requires engineering effort to improve, multi-agent systems *can* automatically benefit from advances in underlying AI models. This creates a unique opportunity to build systems that compound in value over time with minimal additional investment.

> **i** Note
>
> Across my career, I have seen multiple instances where the performance of systems where capabilities are delegated to AI agents improve as the underlying model improves. A good example is my work with LIDA (Dibia (2023)) - a workflow based multi-iagent system for automatically generating visualizatons from data. The initial version of this system built with the davinci model family from openai had a 20% error rate on the initial evaluation harness. Simply switching to GPT 3.5 turbo which became available a few months after led to an error rate reduction form 20% to 3% across the board. I still see these sorts of system level improvements as new advanced models become available.

1.6.4   Opportunity to Address Tacit Knowledge Tasks

Many repetitive problems today have been difficult to automate or support with reliable software tools. Existing research suggests that some of the automation challenges arise because these tasks require *tacit knowledge* — a form of knowledge that is challenging to articulate or codify but more readily transferred through experience or practice. However, groups of agents enabled by LLMs that encode vast amounts of knowledge and can collaborate with humans and other agents provide a new opportunity to address these complex tasks. This approach can help automate tasks that were previously considered infeasible to automate, thus reducing human effort and potentially errors.

1.6.5 Increasing Demand for Reliable Automation

Businesses and individuals seek AI solutions that can autonomously execute sophisticated tasks while maintaining high levels of reliability and safety. Humans should still remain in the oversight loop, ensuring that the automated solutions reliably and safely deliver cost and effort savings. Multi-agent systems can provide the reliability and adaptability needed for complex, real-world automation challenges.

1.6.6 Platform Economics and Integration Value

Multi-agent systems enable "everything app" economics - creating value through integration rather than specialized point solutions. This allows for capturing user preferences, learning across domains, and reducing switching costs. The same system can handle diverse tasks without requiring specialized configuration or domain knowledge from users. Platform economics benefits include creating value through integration, saving time by being the platform and avoiding switching costs, and the ability to capture user preferences, learn, and become even more valuable over time.

1.6.7 Ethical AI Deployment

As AI deployment expands and generative AI models are applied across various problem domains, the need for responsible, secure, and human-centric multi-agent frameworks becomes increasingly critical. Ensuring ethical considerations are met in AI development and deployment helps to build trust in these systems and reduces potential negative consequences.

> 💡 Distributed Agents: Agents that live across multiple organizational boundaries
>
> Depending on the background of an engineer, the term "multi-agent systems" can evoke different mental models. For some, the focus is on software entities that collaborate within a single application/thread/process with similar permission structures. For others, the focus is on a *distributed internet of agents* where agents are primarily unaware of each other (requiring some discovery protocol), permissions (security and access) are markedly different, and communication must support both synchronous and asynchronous modes at scale.
>
> In this book, we will focus primarily on the first scenario. While the second is important, it is frequently the case that an early focus on distributed agents is overengineering—an increase in complexity—and there are standard methods that can help us transition to a distributed setup as the need arises. In the version 0.4 rewrite/redesign of AutoGen, we specifically provided a runtime concept that could be single process or distributed, allowing developers to create agents that could run in either mode without changing the agent logic itself. We found that the vast majority of use cases were well served by single process/thread applications.
>
> Specifically, in a later chapter (see chapter Chapter 12), we will discuss emerging protocols for distributed agent-to-agent communication, the design choices, and how

existing agents can be adapted to work in a distributed setting.

1.7 Choosing the Right AI Agent Architecture

Building AI applications requires choosing the right level of complexity from a spectrum of architectures: simple **models** for text generation, **single agents** for action-taking, **multi-agent workflows** for structured deterministic collaboration, and **autonomous multi-agent systems** for exploratory problem-solving. While autonomous multi-agent systems offer the most sophisticated capabilities, they introduce significant challenges including unpredictable outcomes, reproducibility issues across model versions, increased error potential, and additional cost and latency.

The decision framework shown in Figure 1.8 helps you select the appropriate architecture by following a logical progression of questions about your task requirements.

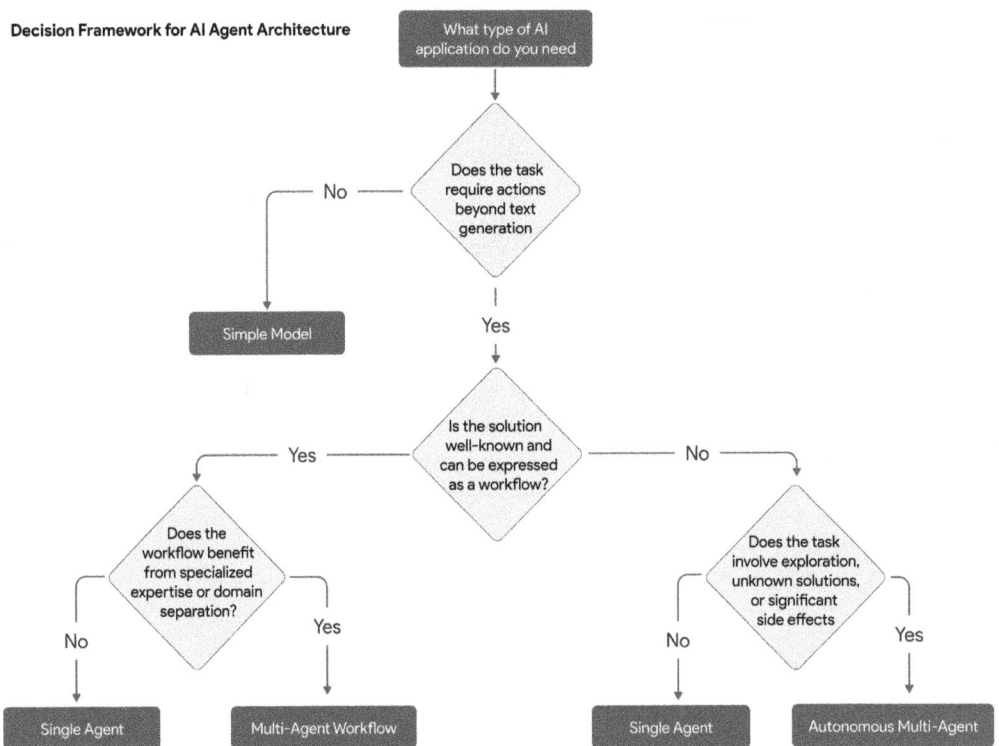

Figure 1.8. Decision framework for choosing the right AI agent architecture based on task requirements.

1.7.1 Decision Framework

The choice of architecture follows a logical progression based on your task characteristics.

Start with the fundamental question: Does your task require actions beyond text generation? If your task involves only text processing, analysis, or generation—such as answering questions from training data, document summarization, or code generation from specifications—then directly calling an AI model to generate a response is sufficient.

For tasks requiring action-taking, the next consideration is whether your solution approach is well-known and can be expressed as a workflow. When you have a clear understanding of the steps needed and how they should be executed, you're choosing between single agents and multi-agent workflows.

If your workflow benefits from specialized expertise—where different parts of the task require distinct domain knowledge—a multi-agent workflow is appropriate. This approach works best when the collaboration pattern between specialized agents is well-understood and each agent's role and handoffs follow established processes. For example, financial analysis pipelines require separate market data, risk assessment, and portfolio optimization agents working in a defined sequence. However, if the task can be handled effectively by one type of expertise, a single agent with the necessary tools is more efficient.

For tasks where the solution approach is not well-defined, you're choosing between single agents and autonomous multi-agent systems based on the level of exploration required. If the task involves known action sequences but requires relatively simple action-perception loops—such as API calls with result validation—a single agent is suitable. However, autonomous multi-agent systems become necessary when solutions emerge through exploration and dynamic agent interaction, when agents must be defined and improved independently, when agents must adapt their collaboration based on intermediate results, or when performance improves through iterative learning from successes and failures. For example, application development where requirements and implementation approaches evolve through experimentation.

A guiding principle: Choose the simplest architecture that effectively addresses your requirements. Each level of sophistication introduces additional complexity, development time, and potential failure points while offering greater capability and adaptability. When in doubt, start simple and evolve your architecture as you better understand your task's true requirements.

These principles are validated by comprehensive evaluation results in Section 10.6. When comparing direct model calls, single agents with tools, and multi-agent systems across diverse task types, the data confirms that direct model calls excel on simple reasoning tasks (9.7/10 with **24x** better token efficiency than multi-agent), while multi-agent systems justify their overhead only when tasks require tool coordination or specialized expertise. For example, research tasks requiring web search show direct models scoring **3.2/10** compared to tool-enabled agents scoring 9.0/10—demonstrating when architectural complexity becomes necessary.

1.8   Building Your First Multi-Agent System

Oh boy! If you have read this far, you are probably excited to get your hands dirty and start building your first agent. Starting in Chapter 4, we will collaboratively build a multi-agent library called `picoagents` from scratch.

As a quick sneak peek, let us take a look at some of the functionality we will build and also use this as an opportunity to run your first multiagent application by creating a simple poet-critic collaboration system step by step. We will define two agents - a poet that writes haikus and a critic that provides feedback on the quality of poems. The idea is that feedback from the critic can iteratively improve the poet's output.

1.8.1   Environment Setup

The first thing we will need to do before we write and run any code is to ensure we have our python environment setup correctly. Luckily, we only need to do this once and can reuse the same environment for all the examples in this book.

I always recommend using a virtual environment to manage dependencies - this means we want to ensure that we install the right version of python that we need, and that all the libraries we install are isolated from other projects on the same machine. You can create one using `venv` (built-in with Python) or `conda` if you prefer. Here, we'll use `venv` for simplicity.

```
# Create and activate a virtual environment
python -m venv multiagent_env
source multiagent_env/bin/activate
# On Windows: multiagent_env\Scripts\activate

pip install picoagents # install from PyPI
```

1.8.2   Step 1: Import and Model Setup

Let's start with the essential imports (`Agent` and `OpenAIChatCompletionClient` classes).

```
import asyncio
from picoagents import Agent
from picoagents.llm import OpenAIChatCompletionClient
import os
```

Now, we need to setup our language model client. Here, we will use OpenAI's GPT-4.1-mini model via the `OpenAIChatCompletionClient` class. We can also use other model clients such as `AzureOpenAIChatCompletionClient` if we want to use Azure OpenAI service - or really any other model client that conforms to the expected interface.

```
# Setup the language model client
client = OpenAIChatCompletionClient(
    model="gpt-4.1-mini",
    api_key=os.getenv("OPENAI_API_KEY")
)
```

Note that you will need to either pass in your API key directly or set it as an environment variable named `OPENAI_API_KEY` (recommended).

1.8.3   Step 2: Create Your First Agent

Next, let's create a single agent with a rather simple task - generate a haiku (a poem with a 5-7-5 syllable structure):

```
# Create a haiku poet
poet = Agent(
    name="poet",
    description="Haiku poet.",
    instructions="You are a haiku poet.",
    model_client=client
)
```

While we are building up towards a multi-agent system, it is always a good idea to test each component as you build it. Let's test our poet agent:

```
# Test the poet
async def test_poet():
    response = await poet.run("Write a haiku about cherry blossoms in spring")
    print(f"Poet says: {response}")
asyncio.run(test_poet())
```

1.8.4   Step 3: Add a Critic Agent

Let's create a poetry critic to provide feedback:

```
# Create a poetry critic
critic = Agent(
    name="critic",
    description="Poetry critic who provides constructive feedback on haikus.",
    instructions="You are a haiku critic. \
        When you see a haiku, provide 2-3 specific, actionable \
        suggestions for improvement. . Be constructive and brief. \
        If satisfied with the haiku, respond with 'APPROVED'",
    model_client=client
)
```

Similarly, we can also test our critic agent:

```
# Test the critic
async def test_critic():
    haiku = ("Cherry blossoms fall\n"
             "Petals dancing in spring breeze\n"
             "Nature's gentle song")
    response = await critic.run(f"Please critique this haiku: {haiku}")
    print(f"Critic says: {response}")
asyncio.run(test_critic())
```

1.8.5   Step 4: Round-Robin Orchestration

Now for the exciting part - let's make them collaborate using round-robin orchestration (a simple pattern where agents address tasks by taking turns in a shared conversation until a termination condition is met):

```
from picoagents.orchestration import RoundRobinOrchestrator
from picoagents.termination import MaxMessageTermination, TextMentionTermination

# Create termination conditions
termination = (MaxMessageTermination(max_messages=8) |
               TextMentionTermination(text="APPROVED"))

# Create the orchestrator
orchestrator = RoundRobinOrchestrator(
    agents=[poet, critic],
    termination=termination,
    max_iterations=4
)
```

Now, lets run the orchestrator with a task:

```
# Run orchestration
async def run_orchestration():
    task = "Write a haiku about cherry blossoms in spring"
    stream = orchestrator.run_stream(task)
    async for message in stream:
        print(f"{message}")
asyncio.run(run_orchestration())
```

The `RoundRobinOrchestrator` manages a turn-taking loop, while termination conditions ensure the conversation concludes appropriately - here either reaching the maximum number of messages or the critic approving the haiku with a specific text mentioned.

The output of the code above looks like the following:

```
⬚ Task: Write a haiku about cherry blossoms in spring
⬚ Poet and Critic collaboration:

[user]: 22:58:25
================================================
Write a haiku about cherry blossoms in spring

[poet]: 22:58:27
================================================
Petals drift like dreams,
Whispers of a fleeting past,
Spring's soft blush unfolds.

[critic]: 22:58:28
================================================
1. Consider incorporating a more vivid description of the cherry
   blossoms to strengthen the imagery—what do they look or smell like?...
APPROVED

⬚ RoundRobinOrchestrator: duration: 3.6s, tokens: in:232, out:121,
calls: 2 . Stop reason: Text mention found: 'APPROVED'
```

Congratulations - you just built and ran your first multi-agent app!

> 💡 💻 Working Code: First Multi-Agent System
>
> The complete working example for this haiku system is available in
> `examples/orchestration` .
>
> Run it with `python round-robin.py` or add the `--web` flag to launch the Web UI.

1.8.6   Visualizing Your Multi-Agent System

Watching agents collaborate through console output is helpful, but seeing the interaction visually makes debugging and understanding much easier. PicoAgents includes a web interface that lets you interact with your multi-agent system through a browser.

To launch the haiku system with a web interface, wrap the orchestrator with the `serve()` function:

```
from picoagents.webui import serve

# Use the orchestrator we created above
serve(entities=[orchestrator], port=8070, auto_open=True)
```

This single line starts a web server with a complete interface where you can see the conversation unfold between the poet and critic agents, inspect individual messages, and explore the orchestration pattern in real time (Figure 1.9).

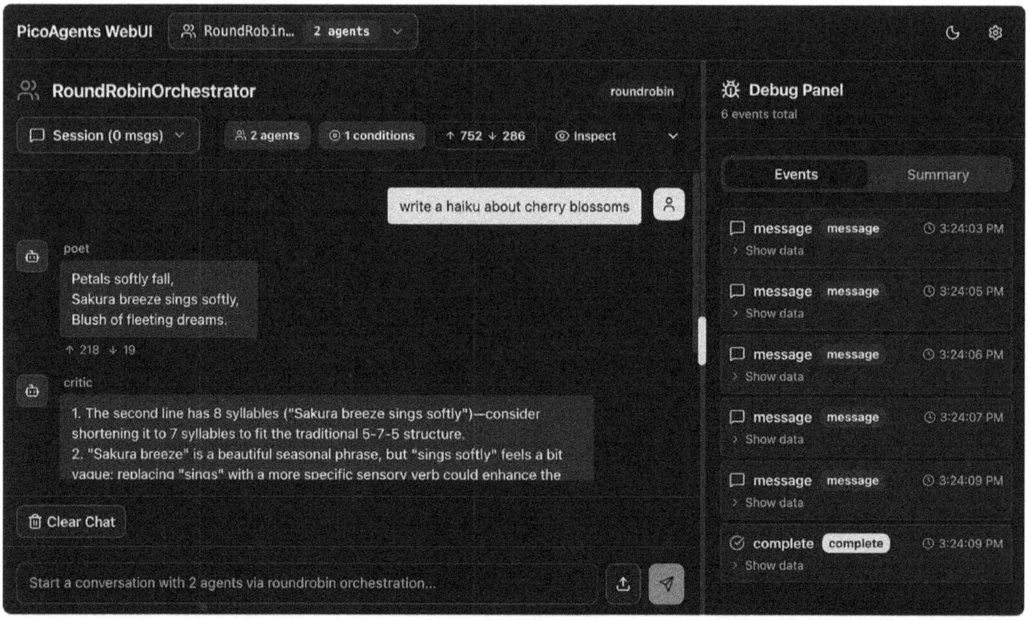

Figure 1.9. PicoAgents Web UI showing the round-robin orchestrator with poet and critic agents. The interface displays participating agents, termination conditions, the conversation flow, and a debug panel for inspecting agent interactions.

The Web UI provides several useful features: view message history between agents, inspect tool calls and responses, monitor termination conditions, and debug multi-agent interactions. We will explore how to build similar web interfaces that integrate agents and orchestrators from scratch in Chapter 8.

Depending on your level of expertise with AI models, you might have questions - how is the `Agent` class implemented? How does the `RoundRobinOrchestrator` work? What is a `model_client`? How do termination conditions work? Don't worry - we will explore all these components - we will build them up from scratch in chapter Chapter 4 and subsequent chapters.

1.9   Summary

In summary, here are the key takeaways from this chapter:

- **The Complexity Challenge**: Generative AI models excel at many tasks but face limitations as complexity increases. Tasks requiring reasoning, planning, and action beyond text generation need agents - systems that combine models with tools and can take actions in the world.

- **Agent Definition and Components**: Agents are entities that can reason (synthesize information using logic), act (take concrete actions using tools), communicate (exchange information effectively), and adapt (modify approaches based on feedback). These capabilities are enabled by three core components working together: a model (reasoning engine, typically an LLM), tools (enable action-taking), and memory (short-term for immediate tasks, long-term for accumulated knowledge and experiences).

- **Complex Task Characteristics**: Multi-agent systems are most beneficial for tasks exhibiting four key characteristics: requiring planning (multi-step decomposition), needing diverse expertise (specialized knowledge domains), involving extensive context (large amounts of information), and requiring adaptive solutions (dynamic environments where solutions emerge iteratively).

- **Decision Framework**: Use multi-agent approaches when tasks are complex, require diverse expertise, involve dynamic environments, or benefit from adaptation. Avoid them for simple, deterministic tasks with minimal expertise diversity or static environments.

- **Current Opportunity**: Multiple factors make this an opportune time for multi-agent system development: advances in generative AI reasoning capabilities providing the foundation for autonomous agents, economic value through time arbitrage (agents cost $1-2 vs $30-100 human time), self-improving systems that benefit from AI model advances without re-engineering, opportunities to address previously unautomatable tacit knowledge tasks, increasing demand for reliable automation, platform economics enabling "everything app" integration value, and the need for ethical AI deployment frameworks.

Chapter 2

Multi-Agent Patterns

This chapter covers:

- A taxonomy of multi-agent orchestration patterns: understanding the spectrum from explicit workflow control to autonomous emergent orchestration
- Workflow patterns: sequential, conditional, parallel, and supervisor patterns with developer-defined execution paths and predictable behavior
- Autonomous patterns: group chat, handoff, and hierarchical patterns where agents determine orchestration through runtime reasoning and communication
- Pattern selection guidance: comparing trade-offs between reliability, autonomy, and complexity to choose the right approach for your task

As a software engineer, you've likely encountered *patterns* throughout your career. Whether learning design patterns like singleton and observer in computer science courses, or discovering algorithmic patterns like two-pointer and sliding window techniques while preparing for interviews, the underlying principle remains consistent: through experimentation and practice, our field identifies what works best and codifies these insights into reusable solutions.

Patterns provide structure for managing complexity and prevent us from reinventing solutions to well-understood problems. Multi-agent systems are no different. In Chapter 1, we defined agents as entities that can reason, act, communicate, and adapt to solve tasks, and explored the benefits of using multiple agents for complex tasks with planning requirements, diverse expertise needs, and adaptive solutions. This chapter explores the orchestration patterns that have emerged for coordinating multiple agents *effectively*.

The architecture of multi-agent orchestration profoundly impacts system behavior. A poorly chosen pattern can lead to infinite loops, excessive costs, or brittle solutions that break under real-world conditions. Conversely, some patterns result in inherently more reliable (but constrained) systems, while others enable scalable systems that gracefully adapt to handle complex tasks. Importantly, agents work on tasks, and selecting the right task management pattern—i.e., when to step, delegate to a human, or retry—is crucial for preventing runaway execution and ensuring that agents remain productive.

This chapter presents a taxonomy of multi-agent design patterns organized along a spectrum from explicit control to emergent behavior. Figure 2.1 provides a visual overview of this spectrum.

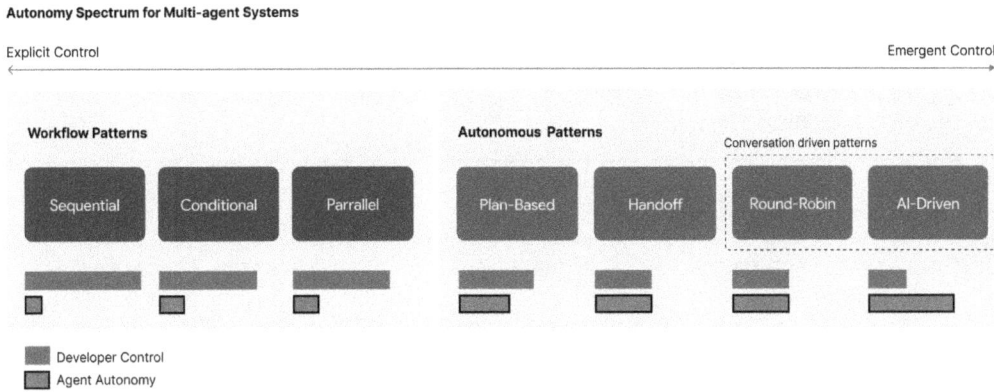

Autonomy Spectrum for Multi-agent Systems

Explicit Control   Emergent Control

Workflow Patterns   **Autonomous Patterns**
                                                                        Conversation driven patterns

| Sequential | Conditional | Parrallel | | Plan-Based | Handoff | | Round-Robin | AI-Driven |

Developer Control
Agent Autonomy

Figure 2.1. Autonomy spectrum for multi-agent orchestration patterns, showing the progression from explicit workflow control to emergent autonomous coordination.

2.1 A Taxonomy of Multi-Agent Orchestration Patterns

If you read multiple articles on how agents should be orchestrated, you'll encounter various terms like "sequential patterns," "supervisor patterns," "hierarchical patterns," and "swarm patterns." While these descriptions vary, the underlying patterns can be categorized along a fundamental spectrum: the level of autonomy granted to agents in determining their orchestration.

At one end of this spectrum are **workflow patterns** - where developers explicitly define execution paths and control flow. At the other end are **autonomous patterns** - where agents determine orchestration through runtime reasoning and communication. Understanding this spectrum is crucial because each approach offers distinct trade-offs between predictability and flexibility.

Multi-agent coordination builds on decades of distributed systems research. Traditional workflow engines like Apache Airflow, Temporal, and Prefect remain excellent foundations for implementing workflow patterns through DAG-based dependency management and robust state synchronization. However, AI agents introduce fundamental differences: **semantic reasoning** using natural language rather than rules/heuristics, **dynamic planning** that adapts based on intermediate results, **probabilistic outputs** requiring new error detection approaches, and **token-based economics** where communication costs scale with conversation length. These differences become most pronounced in autonomous patterns, where AI-driven coordination enables capabilities beyond traditional workflow systems.

2.1.1 Implementation Roadmap

Before diving into each pattern, here's how this chapter's concepts map to concrete implementations in subsequent chapters (see Table 2.1). This roadmap helps you connect theoretical patterns to working code as you progress through the book.

Table 2.1. Pattern-to-implementation mapping showing how theoretical concepts in this chapter connect to concrete code in subsequent chapters.

Pattern Category	Specific Pattern	Implementation Chapter	Key Classes/Concepts
Workflow	Sequential	Chapter 6	`Workflow`, `FunctionStep`, `.chain()`
Workflow	Conditional	Chapter 6	`Edge` with conditions, branching logic
Workflow	Parallel	Chapter 6	DAG execution, fan-out/fan-in
Autonomous	Round-Robin	Chapter 7	`RoundRobinOrchestrator`
Autonomous	AI-Driven	Chapter 7	`AIOrchestrator` with LLM selector
Autonomous	Plan-Based	Chapter 7	`PlanBasedOrchestrator` with explicit plans
Autonomous	Handoff	Chapter 7	Peer-to-peer delegation patterns

Navigating This Chapter:

- **Theory-first readers**: Read all pattern sections sequentially to build a complete mental model, then proceed to implementation chapters with full context.
- **Learn-by-building readers**: Read each pattern section, then immediately explore the corresponding implementation chapter to see the concept in action.
- **Reference readers**: Use Table 2.1 to jump between theoretical patterns and their code implementations as needed.

The remainder of this chapter explores each pattern's characteristics, trade-offs, and use cases. As you read, remember that **workflow patterns** become the `picoagents.workflow` module (explicit control via computational graphs), while **autonomous patterns** become the `picoagents.orchestration` module (emergent control via agent reasoning).

2.2   Workflow Patterns (Explicit Control)

Workflow patterns provide developer-defined execution paths with predictable behavior. A common, emergent practice is to borrow concepts from graph theory and model multi-agent orchestration as computational graphs where **nodes** represent computational units (a single function, agent, or even an agent team) and **edges** define control flow between nodes.

The key concepts in this approach include:

- Node: A computational unit within the graph (e.g., an agent or function). Nodes typically have defined inputs which they can process and generate an output data structure. In addition, nodes may maintain their own internal state, read and modify some shared state. They interact with the rest of the system through their output and shared state.
- Edge: A connection between nodes that defines control flow or transitions between two nodes. For an edge E that defines a transition from node A to node B, the implied behavior is that the output of A becomes the input of B. Edges may be conditional—e.g., the transition from node A to node B only occurs if a certain condition is met. The condition may be based on the output of the previous node, some shared state, some external signal, arbitrary logic, human input, or a combination of these.

In turn, a graph approach enables several key benefits: validation of execution paths before runtime, visualization of agent interactions (ease of reasoning about system behavior), and deterministic execution behavior.

2.2.1   Sequential Workflows

Sequential workflows are orchestration patterns that implement linear execution where each node's output feeds into the next node (A -> B -> C), ensuring ordered processing with predictable execution timing. For example, a news summarization pipeline shown in Figure 2.2:

Sequential Workflow

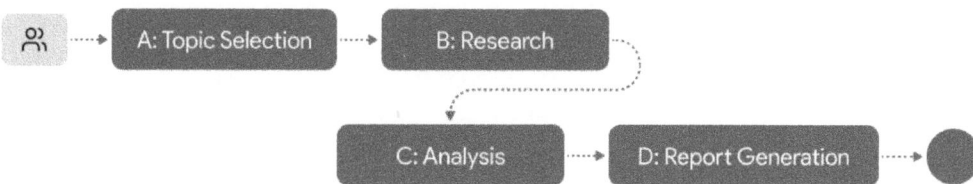

Figure 2.2. Sequential Workflow Pattern showing linear execution flow from user request through topic selection, research, analysis, to report generation.

Each step must complete before the next begins, ensuring ordered processing of information. Sequential workflows are ideal for tasks with natural sequential dependencies where the output of one stage becomes the essential input for the next. They provide predictable execution timing and clear error isolation - if one stage fails, you know exactly where the breakdown occurred. In Chapter 6, you'll implement this pattern using `workflow.chain(step_a, step_b, step_c)` to connect steps sequentially.

2.2.2   Conditional Workflows

Conditional workflows are orchestration patterns that use logic-based edges to determine the next node based on conditions, enabling branching execution paths and dynamic routing. A classic example is code development shown in Figure 2.3:

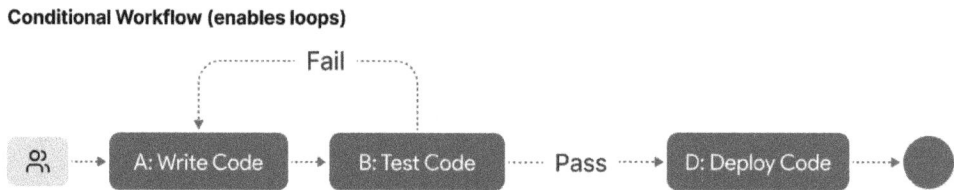

Figure 2.3. Conditional Workflow Pattern showing branching logic where code testing can either pass (leading to deployment) or fail (leading back to code writing).

This pattern extends naturally to **supervisor-style** workflows , a conditional workflow variant where a central control node evaluates requests and routes tasks to specialized agents based on task characteristics. Consider a customer service system that routes inquiries to appropriate departments:

The supervisor evaluates each incoming request and delegates to the appropriate specialist agent, with results flowing back for final processing. In Chapter 6, conditional routing uses `Edge` objects with conditions to enable branching based on validation results:

```
workflow.add_edge(
    "validate",
    "process",
    condition={
        "type": "output_based",
        "field": "is_valid",
        "value": True
    }
)
```

When each node itself contains a workflow, we get hierarchical patterns that enable complex systems like multi-tier support desks. The main supervisor delegates to department supervi-

Hierarchical / Supervisor Workflows

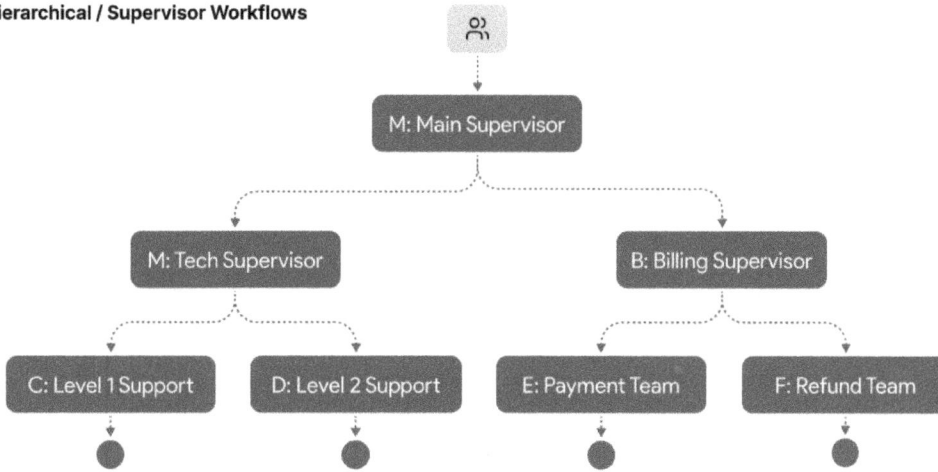

Figure 2.4. Supervisor Workflow Pattern showing a control node routing tasks to specialized agents based on inquiry type.

sors, who in turn manage their specialized teams—creating multiple layers of orchestration for handling complex organizational workflows.

2.2.3   Parallel Workflows

Parallel workflows are orchestration patterns that enable concurrent execution of independent tasks using Directed Acyclic Graphs (DAGs) , with fan-out phases that split work and fan-in phases that combine results. The pattern has a fan-out phase (splitting work) and fan-in phase (combining results):

Parallel execution is ideal for tasks where work can be parallelized without side effects, like processing multiple data sources or running independent analyses. This pattern provides significant performance benefits when individual tasks are time-consuming and can run simultaneously. However, it requires careful orchestration at the fan-in stage to ensure all parallel branches complete before proceeding. The implementation in Chapter 6 uses DAG execution with automatic fan-in detection—when multiple edges point to the same step with "always" conditions, the runner waits for all dependencies to complete.

The primary advantage of workflow patterns is reliability - you know exactly what will happen and when. However, this predictability comes at the cost of flexibility. Workflow patterns struggle with tasks that require dynamic adaptation or where the optimal sequence of actions cannot be predetermined.

From an implementation perspective, there is the concept of a build phase where the graph is constructed, and a run phase where the graph is executed.

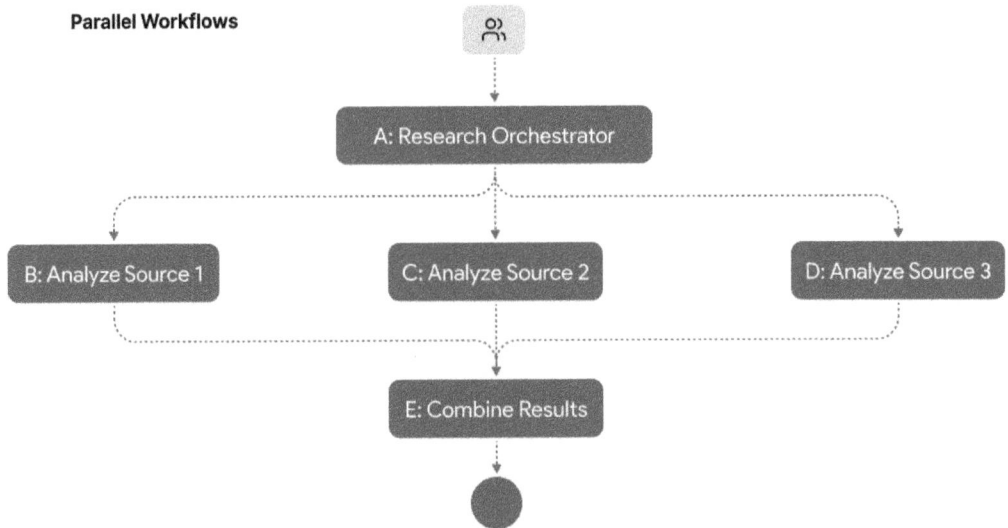

Figure 2.5. *Parallel Workflow Pattern showing concurrent execution of independent tasks with fan-out and fan-in phases.*

The experienced reader will likely observe that these workflow patterns are fundamentally similar to well-known pipeline patterns used extensively in data processing and software development. Beyond incorporating intelligent computations (e.g., LLMs in nodes or nested workflows), the core concepts remain largely unchanged. This similarity is actually advantageous—it allows us to leverage existing ideas from software engineering, scaling, and testing to build reliable multi-agent systems.

However, this approach relies on several assumptions: that the solution to the problem is known, meaning the correct task decomposition is identified; that the necessary resources, such as time and developer expertise, are available to create the appropriate sequence of actions; and that the task is static and predictable, without the need for real-time adjustments or changes. These assumptions often do not hold in many dynamic, real-world situations where task parameters can change and solutions require flexibility and adaptability that a fixed, predefined pipeline cannot provide. In addition, many teams may lack the resources needed to craft specific solutions for a wide array of tasks.

When these assumptions don't hold—when tasks are dynamic, solutions are unknown, or resources are limited—autonomous patterns offer an alternative approach that shifts control from developers to the agents themselves.

2.3   Autonomous Patterns (Emergent Control)

Autonomous patterns enable runtime-determined execution based on task state and agent reasoning. A critical idea here is that the **flow of control is driven by an AI model** and hence dynamically determined at runtime. Rather than following prescribed paths, agents orchestrate through communication and shared understanding of the current task context. These patterns exist on a spectrum of control, from structured orchestration to fully emergent behavior.

> **i** Agents All the Way Down
>
> An important design principle applies to both workflow and autonomous patterns: any "agent" may itself be a multi-agent system internally. These composite agents present as a single entity to the broader system while their internal conversations and orchestration remain private.
>
> For example, what appears as a single "Coder agent" in a plan-based orchestration might internally use a sequential workflow (Research▯Code▯Test steps). Similarly, a "Research Team" agent in a handoff pattern might internally coordinate specialists through AI-driven conversation. A workflow node labeled "Legal Review" could internally be a parallel workflow distributing documents across multiple reviewers simultaneously.
>
> This hierarchical approach provides clear boundaries and reduces communication noise in the overall system. It also allows you to use different orchestration patterns at different abstraction layers—choosing the right pattern for each level's specific coordination needs rather than forcing a single pattern everywhere.

2.3.1   Plan-Based Orchestration Pattern

The plan-based orchestration pattern is an autonomous pattern that employs a single orchestrator agent to manage entire task execution through explicit plan creation, dynamic task assignment, and centralized progress monitoring. This agent acts as a "project manager", creating plans, assigning work, reviewing progress, and orchestrating between specialized agents.

Control flow characteristics:

- **Plan Management**: Orchestrator maintains explicit task plans with assignments and dependencies
- **Visibility**: Orchestrator sees all context; other agents receive only relevant information
- **Task Assignment**: Explicit work distribution by the orchestrator based on the current plan
- **State Management**: Centralized in the orchestrator (plan, progress monitoring, result evaluation).

Consider a software development task: the orchestrator creates a development plan, assigns

Plan Based Orchestration Pattern

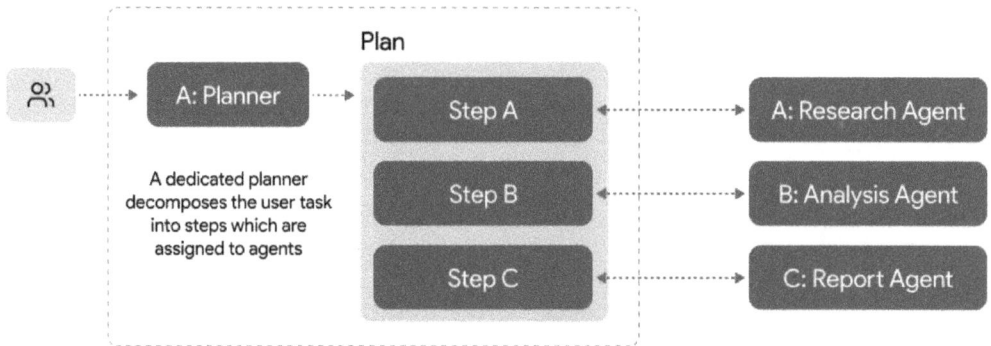

Figure 2.6. Plan-Based Orchestration Pattern showing a single orchestrator managing task execution across multiple agents.

research to Agent A, monitors the output, then assigns coding to Agent B with only the relevant context from Agent A's research. This selective information sharing can help prevent context overload and maintains focus.

Orchestration patterns excel at complex task decomposition and resource management. They can efficiently orchestrate diverse specialists while maintaining oversight. However, they create a single point of failure and may bottleneck on the orchestrator's reasoning capabilities.

In Chapter 7, the `PlanBasedOrchestrator` implements this pattern with LLM-generated plans and automatic step evaluation:

```
from picoagents.orchestration import PlanBasedOrchestrator

orchestrator = PlanBasedOrchestrator(
    agents=[researcher, coder, reviewer],
    model_client=model,
    max_retries=2
)

result = await orchestrator.run("Build a data analysis tool")
```

The orchestrator autonomously creates an execution plan (Step 1: research, Step 2: code, Step 3: review), evaluates each step's success before proceeding, and retries failed steps with enhanced instructions based on the failure analysis. This automated planning and recovery makes the pattern resilient to transient failures while maintaining oversight.

> **ℹ Addressing Complex Tasks with Plan-Based Orchestration**
>
> In November 2023, a group of researchers from HuggingFace and Meta wrote a paper introducing the GAIA benchmark (Mialon et al. 2023). The core purpose was to introduce tasks that require capabilities such as reasoning, multi-modality handling, web browsing, and generally tool-use proficiency - tasks that are conceptually simple for humans yet challenging for most advanced AIs. At the time, I got to collaborate with my colleague Adam Fourney who led a project where our goal was to use AutoGen to design a multi-agent system that did well on GAIA. Over many iterations and after winning the #1 spot on the leaderboard, we arrived at a design that was published in a paper titled "Magentic-One: Building a Multi-Agent System with LLMs for Complex Task Completion" (Fourney et al. 2024). Magentic-One uses a multi-agent architecture where a lead agent, the Orchestrator, directs four other agents to solve tasks through explicit Task and Progress Ledgers, achieving statistically competitive performance to the state-of-the-art on multiple challenging agentic benchmarks. Similarly, Anthropic's Research system (Hadfield et al. 2025) employs an orchestrator-worker pattern, where a lead agent coordinates the process while delegating to specialized subagents that operate in parallel, with the lead agent dynamically spawning 1-10+ subagents based on query complexity, resulting in a 90.2% performance improvement over single-agent approaches.

While orchestration patterns maintain centralized control, handoff patterns distribute decision-making across agents while preserving structured orchestration.

2.3.2 Handoff Pattern

Handoff patterns are autonomous patterns that enable agents to operate with limited, local knowledge while still achieving coordinated behavior through peer-to-peer delegation. Agents make local decisions about when and to whom they should transfer control based on their understanding of the task and knowledge of other available agents.

Control flow characteristics:

- **Visibility:** Each agent knows only a subset of other agents and their capabilities
- **Turn-taking:** Direct handoff via explicit transfer mechanisms

- **Decision making:** Local decisions based on task needs and known agents
- **State management:** Explicitly passed between agents during handoff

A customer service scenario illustrates this pattern: a Customer Service agent can handoff to either a Technical Specialist or Billing agent based on the issue type. The Technical Specialist, in turn, can handoff to a Senior Engineer or return control to Customer Service. Each agent makes handoff decisions based on their local assessment of the situation and passes relevant context along.

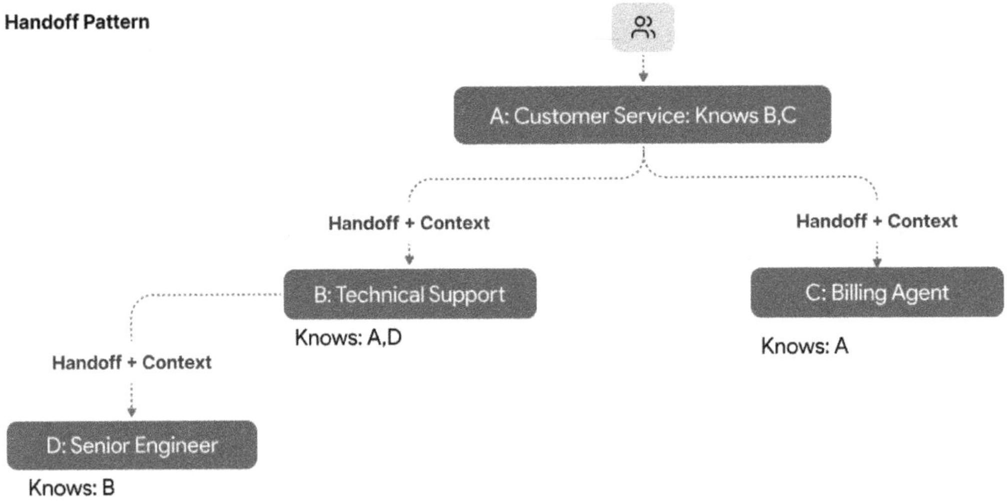

Handoff Pattern

Figure 2.7. Handoff Pattern showing agents transferring control and context based on task needs.

Handoff patterns provide scalability and specialization while avoiding central bottlenecks. They're particularly effective for well-defined domains where handoff criteria can be clearly established. From an implementation perspective, this pattern is relatively simple - agents can be represented as tools; for example, Agent A sees Agents B and D as available tools, and handoffs become tool calls. However, they require careful design to prevent agents from getting stuck or cycling ineffectively.

Moving further along the autonomy spectrum, conversation-driven patterns remove even the structured handoff constraints, allowing orchestration to emerge naturally through shared dialogue.

2.3.3 Conversation-Driven Pattern (Group Chat)

In conversation-driven patterns (also known as group chat patterns), all agents participate in a shared conversation where orchestration emerges through turn taking as part of a dialogue rather than explicit plans or structured handoffs. Turn taking is the mechanism by which agents take sequential turns contributing to the shared conversation. This approach mirrors human team collaboration where everyone can see the discussion and contribute when appropriate.

> ℹ Why Conversation-Driven Patterns Work Naturally
>
> Most modern AI models are trained for "chat completions" - taking turns within conversations by reading message history and generating appropriate responses. This makes

LLMs inherently well-suited for conversation-driven coordination, as they can naturally participate in multi-turn dialogues and determine appropriate contributions based on conversational context.

Implementation simplicity: The orchestration mechanism is essentially "all agents append to the same shared conversation list" - requiring minimal orchestration logic compared to complex workflow graphs.

Automated improvements: System performance improves automatically as upstream AI models become more capable, providing a "free upgrade" path without code changes.

Conversation Based Patterns (Group Chat)

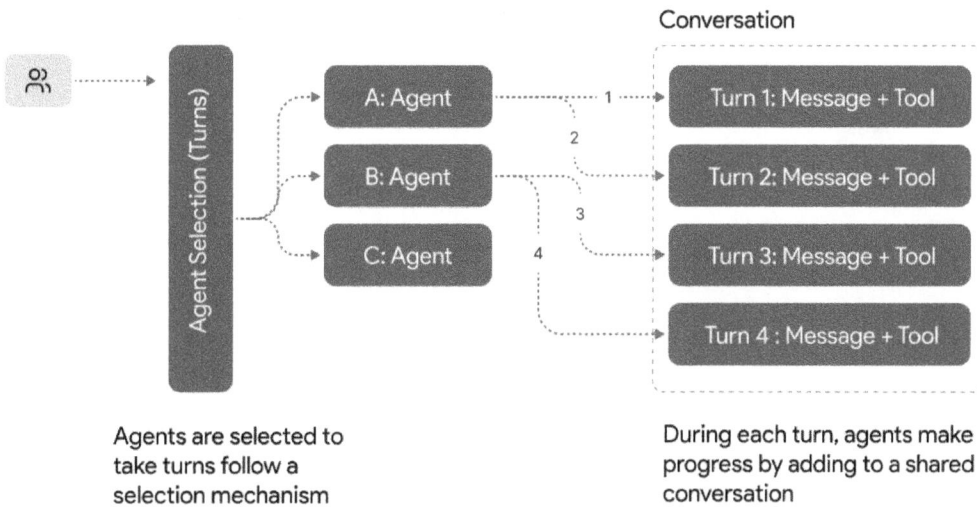

Figure 2.8. In conversation-Driven patterns, agents primarily collaborate by appending to a shared conversation.

Control flow characteristics:

- **Visibility**: Broadcast - all agents observe all messages in the shared conversation
- **Turn-taking**: Determined by selection mechanisms (round-robin, random, or AI-driven)
- **Decision making**: Next speaker selected based on conversation context, not predetermined plans
- **State management**: Implicit in the conversation history

2.3.3.1   Round-Robin Conversation Pattern

In the round-robin conversation pattern, agents take turns in a fixed, repeating order, until some termination condition is met. The conversation flows in a circular manner, with each agent contributing their input before passing the turn to the next agent.

Round Robin Conversation Pattern (Group Chat)

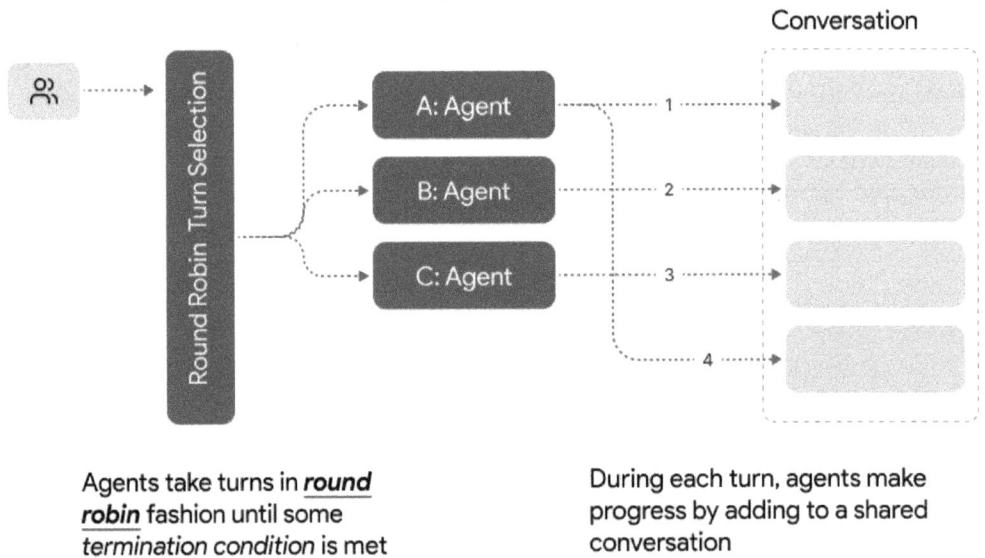

Agents take turns in *round robin* fashion until some *termination condition* is met

During each turn, agents make progress by adding to a shared conversation

Figure 2.9. In the round-robin conversation pattern, agents take turns in a fixed order.

In Chapter 7, the `RoundRobinOrchestrator` implements this pattern by cycling through agents in order:

```
from picoagents.orchestration import RoundRobinOrchestrator
from picoagents.termination import MaxMessageTermination

orchestrator = RoundRobinOrchestrator(
    agents=[researcher, writer, critic],
    termination=MaxMessageTermination(12)  # Stop after 12 messages
)

async for message in orchestrator.run_stream("Research solar energy trends"):
    print(message)  # researcher -> writer -> critic -> researcher -> ...
```

2.3.3.2 AI-Driven Conversation Pattern

In an AI-driven conversation pattern, an LLM or other AI model selects the agent to take the next turn based on the current conversation context. This selection can be based on the task state, agent expertise, or other dynamic factors. The turn selection mechanism itself can be another LLM prompt that evaluates the conversation history and decides which agent should speak next.

For example, consider a research task where a Research agent gathers information, a Writer

AI Driven Conversation Pattern (Group Chat)

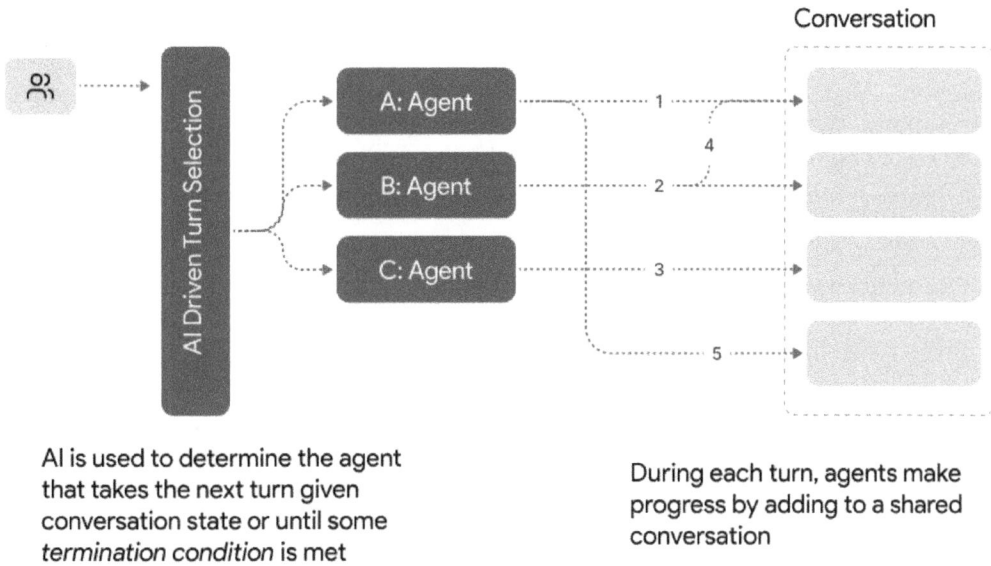

Figure 2.10. In the AI-driven conversation pattern, an LLM dynamically selects the next agent based on context.

crafts content, and a Critic provides feedback. In round-robin selection, each agent takes turns in a predetermined order. In AI-driven selection, an LLM selector determines who should take the next turn based on the current conversation state: "The research is complete, so the Writer should draft content" or "The draft needs improvement, so the Critic should provide feedback."

In Chapter 7, the `AIOrchestrator` uses an LLM to intelligently select the next speaker:

```
from picoagents.orchestration import AIOrchestrator

orchestrator = AIOrchestrator(
    agents=[researcher, writer, critic],
    model_client=model,  # Used for agent selection
    termination=MaxMessageTermination(12)
)

async for message in orchestrator.run_stream("Write article on AI trends"):
    print(message)
```

The LLM analyzes the conversation context and agent capabilities to select the next speaker dynamically. For example, after research completes, the LLM selects the writer to draft content; after the draft, it selects the critic to provide feedback. This adaptive selection enables more natural collaboration compared to fixed turn-taking.

The strength of conversation-driven patterns lies in their transparency and emergent collaboration. All agents maintain shared context and can build naturally on each other's contributions. With AI-driven selection, sophisticated behaviors emerge organically - for example, a Coder agent might be selected again after code execution results appear in the conversation history, enabling natural retry loops, branching logic, and in-situ exploration without explicit programming.

> ℹ Connection to Single-Agent Patterns
>
> Conversation-driven patterns naturally enable well-known single-agent research patterns like ReAct (Yao et al. 2022) and Reflexion (Shinn et al. 2023) through turn-taking and shared conversation history.
> **ReAct** creates a Thought-Action-Observation loop where reasoning traces guide action plans. When answering complex questions, the agent alternates between thinking ("I need to search for X"), acting (searching Wikipedia), and observing results to plan next steps.
> **Reflexion** converts failures into verbal feedback stored as episodic memory. When tasks fail, agents generate reflections explaining mistakes and corrections—like "I incorrectly assumed the pan was in the stove burner. Next time I should verify its location first"—then apply these lessons in subsequent attempts.
> In conversation-driven patterns, these behaviors emerge naturally: specialized agents (Thinker, Actor, Observer) implement the Think-Act-Observe cycle through turn-taking, while past mistakes in conversation history enable Reflexion-like learning. The conversation itself becomes the episodic memory, enabling both reasoning loops and learning from failure.

However, conversation-driven patterns face significant limitations. Token usage grows linearly with conversation length, making them expensive for extended tasks. Context windows can overflow, causing agents to lose important early context. As highlighted by research from Cognition (Yan 2025), agents working from the same conversation may still make "conflicting decisions" due to incomplete understanding of each other's reasoning, potentially leading to fragmented outcomes despite shared visibility.

> 💡 🖥 Working Code: Orchestration Patterns
>
> Explore complete implementations of the patterns discussed:
> - **Round-Robin & AI-Driven**: `orchestration/` - Round-robin turns and intelligent agent selection
> - **Plan-Based**: `orchestration/plan-based.py` - Centralized orchestrator with dynamic planning

- **Workflow Patterns**: `workflows/` - Sequential, conditional, and parallel execution

Having explored the full spectrum of coordination patterns—from explicit workflow control to emergent autonomous behavior—we can now compare their characteristics and understand how to choose the right approach for specific use cases.

2.4 Pattern Selection and Comparison

Having explored the full spectrum of multi-agent orchestration patterns, we can now compare each approach across four key dimensions: **Control Flow** (how execution moves between agents), **Autonomy** (how flexibly the system can explore solution paths not explicitly programmed by developers), **Control** (how much predictable control developers have over system behavior), and **Complexity** (the implementation and operational complexity required).

2.4.1 Comparative Analysis

Table 2.2 summarizes the trade-offs between patterns. Notice how control and autonomy move inversely: workflows give you predictability at the cost of flexibility, while conversation-driven patterns enable exploration but with less deterministic behavior.

*Table **2.2**. Comparative analysis of multi-agent orchestration patterns showing trade-offs between control flow mechanisms, agent autonomy, developer control, and implementation complexity.*

Pattern	Control Flow	Autonomy	Control	Complexity
Sequential Workflow	Linear sequence (A->B->C)	Low	High	Low
Conditional Workflow	Conditional routing	Low	High	Low-Medium
Parallel Workflow	Concurrent execution	Low	High	Medium
Plan-Based Orchestration	Dynamic orchestration	Medium	High	Medium
Handoff	Peer-to-peer delegation	Medium	Medium	Low
Round-Robin Conversation	Fixed rotation	Low	High	Low
AI-Driven Conversation	Context-driven selection	High	Low	Medium

2.4.2 Selection Criteria

Rather than asking "which pattern is best?", ask "what does my task need?" Different tasks have fundamentally different requirements. A data pipeline with known steps needs predictability; an open-ended research task needs exploration. Match the pattern to the task's inherent characteristics.

Task Characteristics:

- **Well-defined, repeatable processes** -> Workflow patterns (Sequential, Conditional, Parallel)
- **Dynamic, exploratory tasks** -> Autonomous patterns (Conversation-driven)
- **Complex planning required** -> Plan-Based Orchestration
- **Domain expertise needed** -> Handoff patterns

System Requirements:

- **High predictability needed** -> Workflow patterns
- **Maximum autonomy required** -> AI-Driven Conversation
- **Resource constraints** -> Handoff patterns (minimal coordination overhead)
- **Scalability concerns** -> Parallel Workflows or Handoff patterns

Implementation Considerations:

- **Developer resources available** -> Workflow patterns
- **Rapid prototyping needed** -> Conversation-driven patterns
- **Production reliability critical** -> Workflow patterns with explicit task management
- **Human oversight required** -> Any pattern + Human Delegation

2.4.3 Hybrid Approaches

Most production systems benefit from **hybrid approaches** that combine multiple patterns based on task decomposition:

- Use **workflow patterns** for predictable, well-understood components
- Apply **autonomous patterns** where flexibility and adaptation are essential

- Implement **task management patterns** consistently across all orchestration types
- Design **hierarchical compositions** where complex agents internally use different patterns

The key principle is **pattern-to-task alignment**: match the orchestration pattern to each task's inherent characteristics and your system's reliability requirements, rather than forcing all orchestration through a single pattern type.

2.5 Task Management Patterns

Beyond orchestration patterns, successful multi-agent systems require explicit task management strategies to ensure productive outcomes and prevent runaway execution. These cross-cutting concerns apply to both workflow and autonomous patterns.

Task management patterns address critical operational concerns that transcend the orchestration taxonomy. While workflow patterns often embed some task management logic within their graph structure, autonomous patterns particularly benefit from explicit task management strategies.

2.5.1 Termination Patterns

Effective termination strategies prevent infinite loops and excessive resource consumption:

- **Budget-based termination** : Set limits on time, cost, or iteration count
- **Semantic termination** : Use LLM reasoning to detect task completion
- **External signals**: Enable human intervention or API-triggered stops

For instance, a research task might terminate after 10 iterations OR when an LLM determines the quality threshold has been met, whichever comes first.

2.5.2 Human Delegation Patterns

Complex or sensitive tasks often require human oversight . Two primary approaches enable effective human delegation:

LLM-Based Delegation relies on agent reasoning to determine escalation needs. Agents assess factors like risk, complexity, and confidence levels to decide when human input is required. For example: "This customer complaint mentions potential legal action and could impact our reputation - escalating to human review."

Rule-Based Delegation uses explicit triggers defined in code. These provide deterministic, auditable escalation based on hard thresholds: financial transactions exceeding $10,000, requests involving medical or legal domains, or three consecutive failed attempts at task completion.

2.6 Summary

- **Orchestration Spectrum**: Multi-agent patterns exist on a spectrum from explicit control (workflow patterns) to emergent behavior (autonomous patterns). The level of autonomy granted to agents fundamentally determines system predictability versus flexibility trade-offs.

- **Workflow Patterns (Explicit Control)**: Sequential workflows provide linear execution (A->B->C), conditional workflows enable branching logic and supervisor delegation, and parallel workflows support concurrent execution via DAGs. These patterns offer high reliability and predictability but limited adaptability to dynamic situations.

- **Autonomous Patterns (Emergent Control)**: Plan-based orchestration uses a central orchestrator for dynamic planning, handoff patterns enable peer-to-peer delegation with local decision-making, and conversation-driven patterns allow orchestration through shared dialogue. These patterns provide greater flexibility but require more sophisticated orchestration mechanisms.

- **Pattern Selection Criteria**: Choose workflow patterns for well-defined, repeatable processes requiring high predictability. Select autonomous patterns for dynamic, exploratory tasks needing flexibility. Consider hybrid approaches that combine multiple patterns based on task decomposition and system requirements.

- **Task Management Essentials**: Successful multi-agent systems require explicit task management strategies including termination patterns (budget-based, semantic, or external signals), human delegation patterns (LLM-based or rule-based), and error handling mechanisms that transcend specific orchestration approaches.

- **Implementation Guidance**: Match orchestration patterns to task characteristics rather than forcing all orchestration through a single pattern type. Use hierarchical composition where complex agents internally implement different patterns while presenting unified interfaces to the broader system.

Chapter 3

UX Principles for Multi-Agent Systems

This chapter covers

- Examining how the user interface landscape has evolved from the command line to multimodal, multi-agent systems
- Tracing the shift in software engineering practices from traditional development to autonomous multi-agent systems
- Discerning between end-user and developer personas in understanding the user
- Spelling out UX design principles for multi-agent applications, including capability discovery, cost-aware action delegation, interruptibility, observability, and provenance

The ultimate goal of building a multi-agent system (or any software tool) is to solve a specific problem for a user. This system may be utilized as components by developers (such as software developers via SDKs or APIs etc.) to create larger systems or directly used by end-users to accomplish particular tasks. Regardless of the target user, it is crucial that we develop systems with an optimal user experience—systems that are *effective*, meaning they help users achieve their tasks with minimal friction or effort, and *enjoyable*, making them engaging and delightful to use. To achieve this, we must understand the user and their behaviors, motivations, attitudes, and goals, and then design systems that cater to these needs.

Multi-agent systems represent a fundamental shift from **interface design** to **delegation design** —a UX paradigm where users provide open-ended task instructions to autonomous systems rather than directly manipulating specific interface controls. Traditional applications offer direct manipulation—clicking "Send Email" executes a predictable, programmer-defined routine. Multi-agent systems operate differently through three key properties: **autonomy** (AI models decide actions from vast action spaces), **consequences** (real-world impact based on AI reasoning), and **duration** (complex, long-running coordination across systems).

Instead of controlling specific actions, users delegate open-ended tasks to autonomous systems that can execute any combination of available capabilities. A travel booking system with access to email, calendar, and booking platforms could take hundreds of different action sequences based on how AI interprets a single natural language instruction.

While UX principles apply to all software, multi-agent systems present unique challenges that amplify their importance. Unlike traditional applications that respond to user input with predefined behaviors, multi-agent systems contain autonomous agents that can take actions with real-world consequences based on AI decision-making. When multiple agents coordinate, individual AI decisions can cascade across the system, creating compound effects that users must understand and control. This shift from predefined routines to AI-determined action sequences requires new UX approaches for communicating agent capabilities, action costs, coordination patterns, and intervention points.

In this chapter, we will explore the user experience (UX) of multi-agent systems, focusing on how this delegation paradigm creates new needs and expectations for both end users and developers. We will discuss how this fundamental shift from direct manipulation to task delegation creates unique challenges and opportunities. By understanding how users must now trust AI reasoning rather than programmer logic, we can derive design principles that help users effectively delegate tasks while maintaining appropriate control and understanding.

> i Do End Users See Multiple Agents or a Single System?
>
> End users typically perceive multi-agent systems as a single entity, issuing high-level commands that underlying agents execute collaboratively. This abstraction provides a seamless experience, but building trust requires optional visibility into the system's internal workings—especially for non-trivial decisions with significant consequences. The observability principle (Section 3.4) addresses this transparency need.

3.1 Background: The Evolution of Interfaces and Software Development

To understand the unique UX challenges of multi-agent systems, it's helpful to examine how both user interfaces and software development practices have evolved. This evolution directly shapes the design principles we'll explore later in the chapter.

3.1.1 User Interfaces - From Command-Line to Multimodal, Multi-Agent Interfaces

The evolution of user interfaces has progressed from text-based command lines to today's intelligent multimodal systems, with each shift reducing cognitive load while introducing new interaction possibilities. This progression culminates in multi-agent interfaces that can interpret natural language, images, and voice input while performing complex actions autonomously.

Traditional interfaces required users to learn specific commands (CLI) or visual metaphors (GUI), limiting accessibility to those with specialized knowledge. Immersive interfaces (AR/VR) introduced spatial interactions but remained constrained by predefined capabilities.

Consider email as an illustrative example of this evolution. Command-line email required memorizing syntax (`mail -s "Subject" recipient@example.com`), GUI clients introduced visual composition with drag-and-drop organization, and today's intelligent interfaces enable natural language requests: "Send an email to the team about Monday's meeting and attach the agenda." The system can interpret intent, locate relevant files, and generate appropriate content—but users must understand when to trust these capabilities.

Intelligent multimodal, multi-agent interfaces represent the latest advancement, offering several key capabilities:

- **Natural input processing**: Free-form text, voice, images, and video across multiple languages
- **Dynamic output generation**: Context-appropriate responses as text, audio, video, or code
- **Autonomous action**: Agents with tools can perform multi-step tasks on users' behalf
- **Adaptive interaction**: Systems adjust based on user context and preferences

However, this flexibility introduces new challenges. Unlike a GUI with clearly defined buttons and capabilities, intelligent interfaces can potentially address vast ranges of tasks with varying degrees of reliability. This creates a **capability discovery problem**—users struggle to understand what the system can do effectively versus where it might fail.

This evolution emphasizes the continuous drive toward more natural and adaptive interactions, while highlighting the need for transparent communication about system capabilities and limitations.

ℹ What Makes Intelligent Interfaces Different

Generative AI introduces three transformative capabilities that distinguish modern interfaces from their predecessors:
- **More Accessible Interfaces**: Intelligent multimodal interfaces enable natural language, voice, image, and video interaction across hundreds of languages, making systems accessible to broader audiences. This inclusivity allows users to engage with systems in ways that align with their preferences and abilities, reducing barriers to entry.
- **Support for Fuzzy Input**: Generative models excel at interpreting ambiguous or error-prone inputs, processing imprecise commands and natural language nuances while still providing coherent responses. However, this flexibility introduces risks like potential misunderstandings or plausible but incorrect information, requiring

---

robust validation mechanisms.

- **Dynamic Content Generation**: Generative AI creates and adapts content dynamically based on user input and task requirements, enabling more personalized and contextually relevant interactions. This reduces extensive predesign needs, accelerating innovation while providing uniquely adaptive user experiences.

3.1.2 From Software 1.0 (Rules) to Software 3.0+ (Autonomous Multi-Agent Systems)

The software engineering discipline is undergoing a transformation from explicit rule-based programming to intelligent, adaptive systems (see Figure 3.1). This evolution directly impacts how we build user-facing applications.

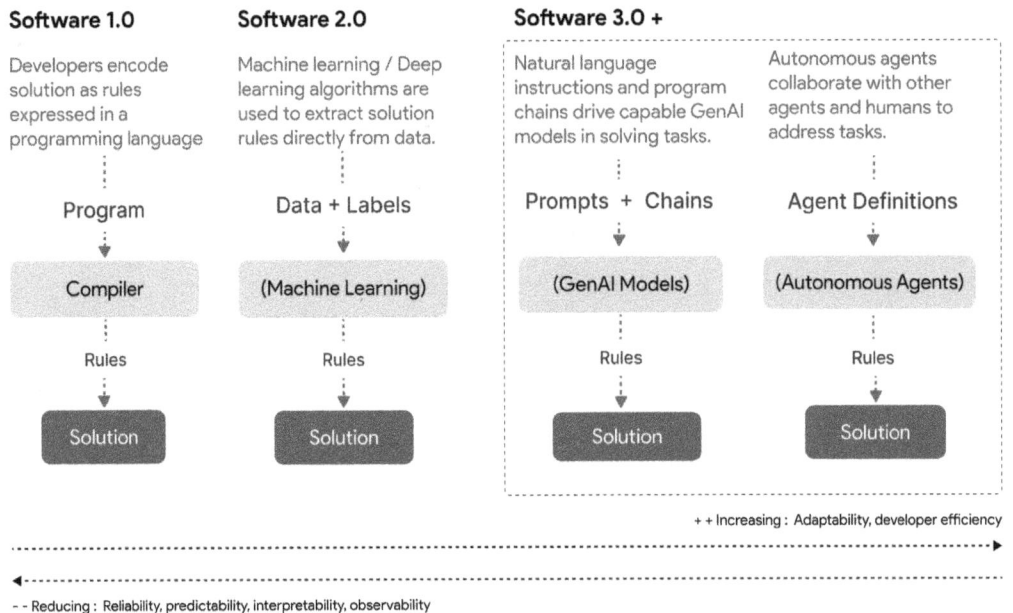

Software 1.0

Developers encode solution as rules expressed in a programming language

Program
↓
Compiler
↓
Rules
↓
Solution

Software 2.0

Machine learning / Deep learning algorithms are used to extract solution rules directly from data.

Data + Labels
↓
(Machine Learning)
↓
Rules
↓
Solution

Software 3.0 +

Natural language instructions and program chains drive capable GenAI models in solving tasks.

Prompts + Chains
↓
(GenAI Models)
↓
Rules
↓
Solution

Autonomous agents collaborate with other agents and humans to address tasks.

Agent Definitions
↓
(Autonomous Agents)
↓
Rules
↓
Solution

+ + Increasing : Adaptability, developer efficiency

‑ ‑ Reducing : Reliability, predictability, interpretability, observability

Figure 3.1. The software engineering discipline is evolving from traditional software development (software 1.0) paradigms where developers write explicit code to more dynamic, adaptive, and intelligent systems (software 3.0) where autonomous agents can reason, communicate, and act to solve complex tasks.

Software 1.0 involves writing explicit code with fixed rules and logic. A spam filter, for example, uses predefined keyword dictionaries to classify emails.

Software 2.0 introduces machine learning systems that learn patterns from data. The same spam filter now trains on large datasets to recognize spam patterns, improving accuracy through data-driven classification rather than hand-coded rules.

Software 3.0+ employs generative models and autonomous agents that can reason, communicate, and act adaptively. An email system might now include agents that automatically draft responses, check calendars for availability, propose meeting times, and even schedule appointments—all while adapting to user preferences and writing styles over time.

The key insight is that each paradigm doesn't completely replace the previous one. Email sending protocols remain efficiently handled by Software 1.0 logic, while features benefiting from pattern recognition (spam filtering) leverage Software 2.0 approaches, and complex, adaptive interactions utilize Software 3.0+ capabilities. The challenge for developers shifts from writing explicit code to defining agent capabilities and orchestrating intelligent systems that can address vast arrays of tasks.

3.2   The Multi-Agent Reliability Challenge

Multi-agent systems inherit the reliability paradox of individual AI models while introducing new coordination challenges that can amplify unpredictability. A model like GPT-4 can extract entities from complex JSON schemas and write sophisticated React components, yet fail at simple comparisons like "which number is bigger, 9.11 or 9.9?" When multiple such models coordinate through agents, these individual inconsistencies compound in unexpected ways.

When agents work together on complex tasks, coordination breakdowns can occur even when individual agents perform well in isolation. The dynamic nature of agent interactions makes it difficult to predict which tasks will succeed and which will fail, creating unreliable user experiences that vary based on factors invisible to the user.

This creates a **jagged frontier** (Dell'Acqua et al. 2023)—a phenomenon where AI capabilities are highly uneven, with some tasks that appear difficult being performed well while seemingly easy tasks fail unexpectedly. The jagged nature means that within the same workflow, some tasks may fall on the side of the frontier where AI excels, while others of similar apparent difficulty may fall on the side where AI struggles. In multi-agent systems, this unreliability is amplified not just by individual model limitations, but by coordination breakdowns. Tasks of seemingly comparable complexity may succeed or fail based on how well agent coordination handles implicit assumptions and context sharing.

UX Implications: For users, this means multi-agent systems can exhibit sophisticated behavior in some scenarios while failing catastrophically in others that appear similar. Unlike traditional software with predictable failure modes, the dynamic nature of agent coordination makes it difficult for users to develop accurate mental models of system capabilities and limitations.

3.3   Multi-Agent User Personas and Challenges

Multi-agent systems create unique user experience challenges that don't exist with single-agent or traditional software. Understanding these challenges from both end-user and developer perspectives is essential for designing effective interfaces. These personas represent two

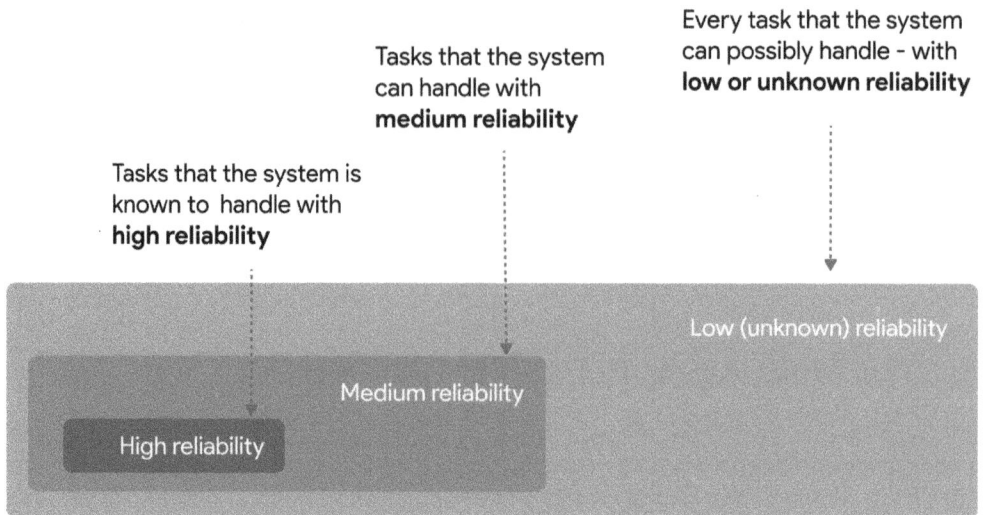

Figure 3.2. The Reliability Paradox: Multi-agent systems can address a broad range of tasks, but coordination challenges create additional reliability variations beyond individual model limitations. UX design must communicate this uncertainty while building trust through transparency.

distinct user types with fundamentally different needs for transparency and control.

3.3.1 End User Persona: Jane - Functional Transparency Needs

Consider Jane booking a complex business trip involving flights, hotels, and car rentals. In a multi-agent system, her request might be handled by specialized agents:

- **Travel Agent:** Searches flights and hotels
- **Calendar Agent:** Checks schedule conflicts and preferences

- **Expense Agent:** Ensures compliance with company policies
- **Booking Agent:** Completes transactions

Jane's Core Needs:

- **Understanding What's Happening:** Jane needs to know what the system is doing without technical coordination details ("Travel Agent found 3 flights, checking your calendar for conflicts")
- **Understanding Why Decisions Are Made:** When agents make choices, Jane needs simple explanations ("Flight selected based on your preference for morning departures")
- **Simple Control Mechanisms:** Jane wants easy ways to provide input, correct assumptions, or change direction without restarting workflows

- **Progressive Disclosure**: Jane prefers starting with basic information and accessing more details only when needed
- **Trust Through Clarity**: Clear communication about what worked, what didn't, and why builds Jane's confidence in delegating tasks

Jane represents end users who need **functional transparency** —understanding outcomes and having simple controls without being overwhelmed by technical complexity.

3.3.2   Developer Challenges: Coordination Design

Mike, building the travel system, faces multi-agent specific technical challenges:

- **Coordination Pattern Selection**: Should he use conversation-driven patterns (transparent but potentially chaotic) or handoff patterns (clean but opaque to users)?
- **Error Attribution**: When the system fails, Mike needs debugging tools that distinguish coordination failures from individual agent errors.
- **Mental Model Design**: Mike must surface agent coordination in ways that help users understand the system without overwhelming them with complexity.
- **Profiling and Optimization:** Mike needs mechanisms to profile agent actions to understand inefficiencies (e.g., redundant expensive steps), bottlenecks, and optimization opportunities.

Mike's main challenges involve ensuring the agent is predictable, secure, and aligns closely with the needs and behaviors of end users like Jane.

These personas illustrate how the same multi-agent system must serve fundamentally different transparency and control needs. Jane requires clear, simple communication about outcomes, while Mike needs detailed technical insight into internal operations.

3.4   Multi-Agent UX Design Principles

The coordination challenges and reliability unpredictability outlined earlier require specific UX design approaches. The following principles address these multi-agent challenges through actionable strategies tailored to both end users (like Jane) and developers (like Mike).

Each principle provides strategies for both Jane (end users requiring functional transparency) and Mike (developers requiring architectural transparency), ensuring the system serves different transparency and control needs effectively.

Throughout this section, we illustrate these principles using the PicoAgents WebUI—a web-based interface that ships with the PicoAgents framework. The WebUI demonstrates how these UX principles can be implemented in practice through concrete interface patterns. The implementation details and source code for the WebUI are available in the PicoAgents repository (`picoagents/src/picoagents/webui/`), and Chapter 8 provides a step-by-step guide to implementing these principles from scratch in a minimal agent web application.

Multi-Agent UX Design Principles

Help users understand
what the agents can do

Capability Discovery

- Clearly list key system capabilities, communicate reliability
- Hybrid interfaces that instruct on system capabilities

Allow users to pause,
resume or cancel agent
actions

Interruptibility

- Persist agent and application state
- Provide pause, resume or cancel controls

UX Design
Principles for
Multi-agent
Systems

Communicate the cost of
agent actions, allow users
decide when agents can act

Cost-Aware Delegation

- Estimate and communicate the *cost* of agent actions
- Provide controls on *when* to delegate actions to humans

Ensure users can observe/
trace agent actions

Observability and Provenance

- Activity log visualizations
- Debugging and provenance tools

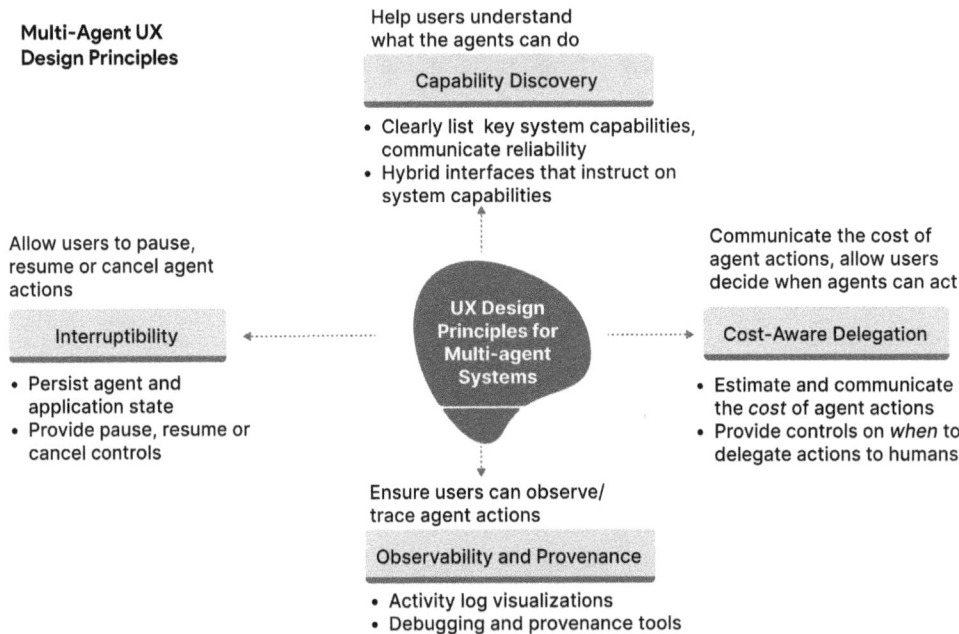

Figure 3.3. Multi-agent applications pose specific challenges that require unique design considerations to ensure a seamless and effective user experience. These include designing for capability discovery (understanding what agents can do), cost-aware action delegation (communicating the cost of agent actions and allowing for delegation mechanisms), observability (ensuring users can observe agent actions), and interruptibility (allowing users to intervene in agent actions).

3.4.1 Capability Discovery

Help users understand what agents can do

Agents have autonomy, meaning they can perform many different actions and coordination patterns. In reality, each agent has specific configurations—system prompts, available tools, and internal logic—that make it more reliable at certain types of tasks compared to others. However, users don't know this. This creates a **capability discovery** problem: users don't know which tasks will work well and which might fail, leading to frustration and mistrust when agents underperform on unsuitable tasks.

This principle advocates that the UX proactively guides users toward high-reliability task examples or suggests relevant high-reliability tasks given the user's context and the system's capabilities. Many successful tools address this by providing sample tasks as presets that users can select to get started.

Implementation Strategies:

End-User Strategies:

- **Capability Presets**: Provide concrete task examples that demonstrate the complexity level and types of tasks the system handles well ("Book domestic business travel with hotel" or "Plan multi-city conference trips").
- **Guided Task Suggestions**: Nudge users toward tasks within the system's reliable capability range through templates and suggestions.
- **Progressive Disclosure** : Start with basic capabilities and reveal advanced features as users become comfortable with the system's reliability patterns.
- **Success Probability Indicators**: Explore the use of confidence indicators that help users understand when tasks are within the system's sweet spot ("Reliable for business travel, requires review for complex international trips").

Developer/Power-User Strategies:

- **Multi Agent Architecture Transparency**: Show how specialized agents work together, their coordination patterns, and handoff mechanisms.
- **Reliability by Coordination Pattern**: Communicate that handoff patterns provide predictable results while conversation-driven patterns enable flexible solutions but with variable outcomes.
- **Tool and Capability Mapping**: Clearly document which tools each agent has access to and how they combine to address different task categories.
- **Capability Boundaries**: Clearly indicate when tasks require human intervention, single agents, or multi-agent coordination.

---

i Example Scenario

A travel booking system provides capability presets such as "Book domestic business trip," "Plan conference travel with group rates," and "Schedule multi-city sales visits." These presets guide users toward tasks the system handles reliably, while more complex requests like "Plan honeymoon with custom adventure activities" are marked as requiring Travel Agent consultation. The interface clearly communicates: "Our system excels at business travel and standard vacation bookings. For complex international trips with special requirements, consider using our assisted booking option."

3.4.2 Cost-Aware Action Delegation

Communicate the cost of agent actions and allow for delegation mechanisms

A valuable property of multi-agent systems is the ability to delegate tasks and actions to agents, allowing users to offload repetitive or complex tasks to the system. However, actions have varying levels of **consequence** —the potential impact or harm if the action contains errors or is inappropriate. **Cost-aware action delegation** is a UX principle for communicating action consequences and providing user controls for how different cost levels should be handled. For example, *summarizing an email* has low cost (minimal consequences if incorrect), *canceling*

Capability Discovery Interfaces highlights sample tasks a selected agent can perform

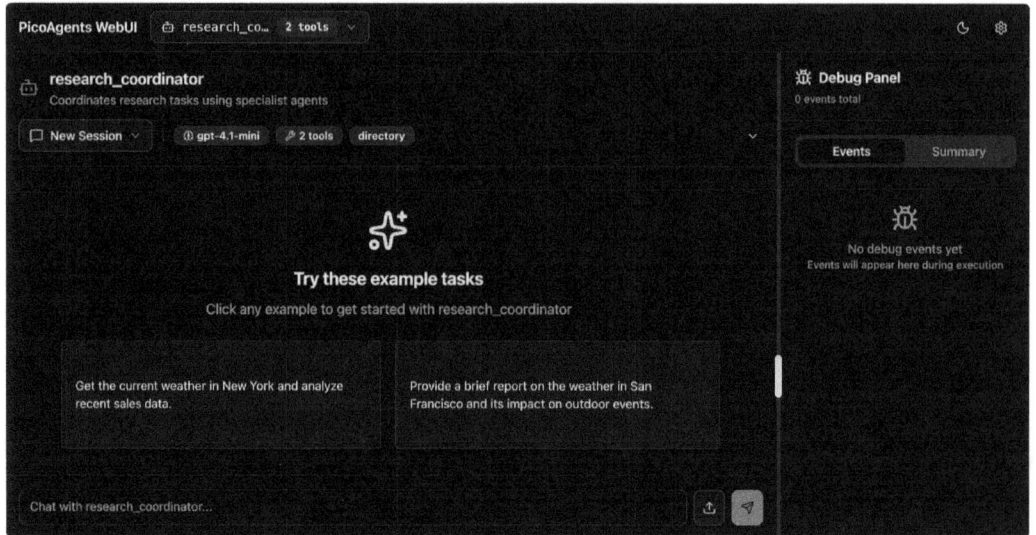

Figure 3.4. The PicoAgents WebUI implements capability discovery by displaying preset example tasks users can click to get started.

a recurring team meeting has medium cost (lost time, changes in participant plans), while *attaching a confidential document to and sending an email to an external party* has high cost (potential data leak).

It is important to build mechanisms that can infer or *quantify* these costs, *communicate* them to the user and offer *settings* for how such actions should be handled. For example, the user should be able to specify how medium and high cost actions should be handled (e.g., ask for approval before executing, execute without asking, or never execute).

Implementation Strategies:

End-User Strategies:

- **Simple Cost Communication**: Use clear, non-technical language to explain action impact ("This will cancel your team meeting and notify 5 people").
- **Three-Tier Approval System**: Low-cost (automatic), medium-cost (notify after), high-cost (ask permission first).
- **One-Click Settings**: Simple controls for overall delegation preferences without technical complexity.
- **Impact Previews**: Show consequences before execution ("Booking this flight will exceed your monthly travel budget by $200").

Developer/Power-User Strategies:

- **Detailed Cost Models**: Expose the cost estimation algorithms, confidence intervals, and factors considered.
- **Custom Threshold Configuration**: Allow fine-grained control over cost categories, agent-specific limits, and conditional rules.
- **Delegation Pattern Controls**: Configure different approval workflows based on coordination complexity.

ℹ Example Scenario

Jane asks the Travel Agent: "Book my trip to the London conference next week and handle any schedule conflicts." The system searches for flights (low-cost, executes automatically) and identifies the need to move a client meeting that conflicts with travel dates (medium-cost). Instead of proceeding, the system shows Jane: "I've found available flights to London [preview options]. I also need to reschedule your Tuesday client meeting with ABC Corp, which will require notifying 3 attendees and may affect their schedules. Should I proceed with rescheduling the meeting?" Jane can approve the specific action or adjust her delegation settings for future similar scenarios.

3.4.3   Observability and Provenance

Ensure users can observe what the agents did (actions) and why

Observability and **provenance** are UX principles ensuring users can observe agent actions, trace the reasoning behind decisions, and understand the data sources that influenced outcomes. Autonomous agents can explore trajectories that are non-deterministic and only known at runtime. Each run with the same input can lead to significantly different trajectories. This makes it important for end users to observe these trajectories both to build trust that agents are doing the right thing and to learn more about their capabilities and limitations, improving their task formulation approach.

Implementation Strategies:

End-User Strategies:

- **Real-Time Activity Streaming**: Show live updates as agents make progress, including structured plans, current steps, and time/cost estimates ("Travel Agent found 3 flights, Calendar Agent checking schedule conflicts").
- **Simple Progress Narratives**: Translate complex agent coordination into user-friendly progress stories that explain what happened and why.
- **Outcome Attribution**: When things go wrong, explain in user terms with clear causation ("Flight booking failed because Calendar Agent detected a conflict with your Tuesday meeting").
- **Trust and Confidence Indicators**: Show data sources and reasoning confidence in accessible language ("Based on your travel history and current calendar availability").

Cost-Aware Delegation Users can explicitly approve specific actions as configured (based on risk levels set by the developer or user)

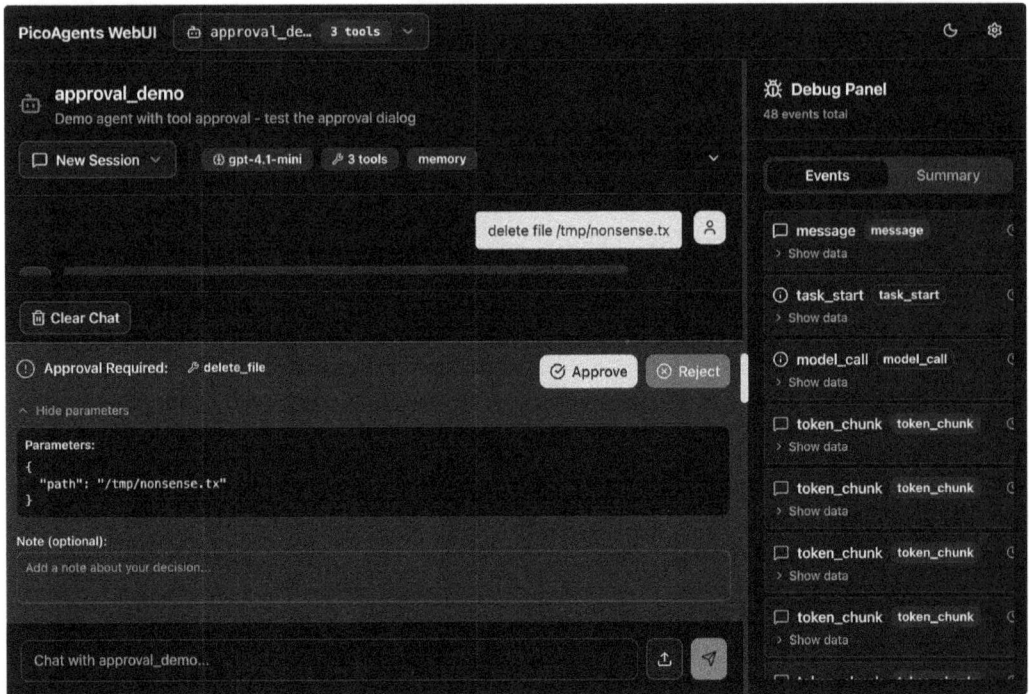

Figure 3.5. In picoagents, functions can be configured to require explicit approval and the WebUI provides an interface to approve function execution requests.

Developer/Power-User Strategies:

- **Detailed Coordination Visualization**: Complete agent handoff flows, delegation patterns, shared context evolution, and decision trees.
- **Attribution Debugging**: Distinguish individual agent actions from coordination decisions with full error tracing and performance analytics.
- **Conflict Resolution Logs**: Full visibility into agent disagreements, reasoning processes, and resolution mechanisms.
- **Multi-Agent Activity Traces**: Conversation-aware logs showing coordination context, shared state changes, and cross-agent dependencies with token usage and performance metrics.

- **Good:** Jane requests "book a trip to London with hotel." The interface shows: Travel Agent finds flights, Calendar Agent detects a conflict with Jane's Tuesday meeting, Expense Agent notes the hotel exceeds policy limits. Jane sees the coordination flow: Calendar Agent passes meeting details to Travel Agent, who adjusts dates, then hands off to Expense Agent for approval. When agents disagree on hotel selection, Jane sees both recommendations with reasoning.
- **Bad:** Jane's trip booking fails with "Error: booking unsuccessful." She can't tell if the Travel Agent couldn't find flights, the Calendar Agent incorrectly blocked dates, or there was a coordination failure between agents. No visibility into which agent contributed what or why the system failed.

3.4.4 Interruptibility

Allow users to pause, resume, or cancel agent actions

Interruptibility is a UX principle ensuring users can interrupt, pause, resume, or cancel agent actions at any point during execution. Autonomous multi-agent systems can run for extended periods, make multiple tool calls, and take actions with real-world resource implications. In some cases, particularly in **human-in-the-loop** settings, users may observe expensive or problematic operation trajectories and need to pause the system to provide feedback, course-correct, or cancel entirely.

This principle advocates for designing systems where users can interrupt agent execution at any point, pause long-running tasks, and resume from where they left off without losing progress or system state. Multi-agent systems are most applicable to complex tasks that require planning across multiple steps and may take significant time to complete.

Implementation Strategies:

End-User Strategies:

- **Simple Stop/Start Controls:** Clear pause, resume, and cancel buttons without technical complexity.
- **Smart Resume:** When resuming, automatically catch up users on what happened during the pause.
- **Gentle Interruption:** Allow users to add requirements or corrections without stopping the entire process.
- **Save Progress:** Ensure users don't lose work when they need to interrupt tasks.

Developer/Power-User Strategies:

- **Granular Agent Control:** Interrupt individual agents without stopping the entire coordination flow.
- **Coordination State Management:** Maintain shared context and agent communication

history across interruptions.

- **Pattern Switching**: Change coordination patterns mid-task (e.g., from autonomous to supervised handoff patterns).
- **Conflict Resolution Intervention**: Manual resolution tools when agents disagree, with visibility into decision impact.

> **i Example Scenario**
>
> - Good: During trip planning, Jane sees Travel and Calendar agents disagreeing about dates. She interrupts the conversation, provides her actual availability, and watches agents update their shared context. The Expense Agent, which was waiting, automatically resumes with the corrected information.
> - Bad: Jane needs to correct the Calendar Agent's assumption but can only stop the entire workflow. When she restarts, agents lose their previous coordination context and must begin from scratch.

Interruptibility
Agent state is persisted in threads (sessions) and can be resumed with user feedback. Requests can be stopped or cancelled.

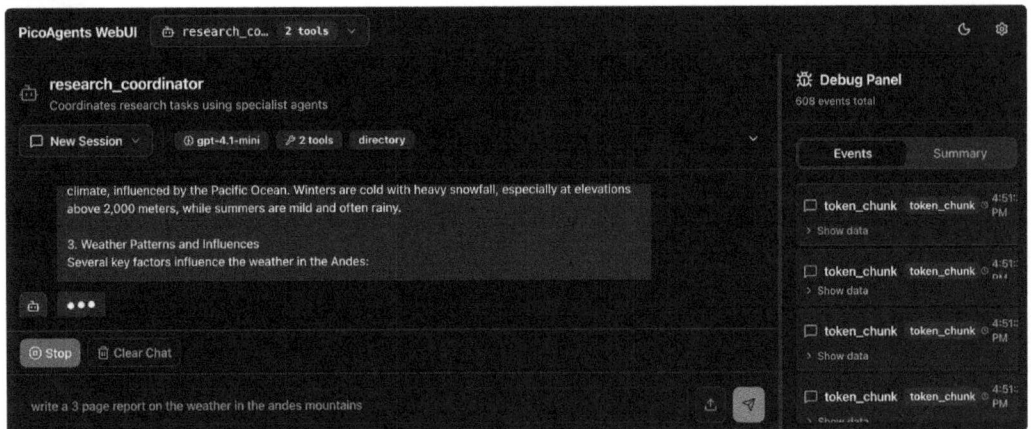

Figure 3.6. In the PicoAgents WebUI, users can stop agent execution mid-stream using the "Stop" button visible at the bottom left. The system maintains conversation state in sessions, allowing users to pause work, review progress, and resume later without losing context. The Debug Panel on the right shows real-time observable event streams, helping users understand what's happening before deciding to interrupt.

3.5 Interactive vs Offline Multi-Agent Scenarios

Task scenarios in multi-agent applications can be categorized as either interactive or offline, depending on the nature of the task and the expected level of human interaction (as shown in Figure 3.7). In **interactive scenarios** , users actively engage with agents in real-time, reviewing progress and providing input as needed. In contrast, **offline scenarios** involve agents working autonomously to complete tasks without direct turn-by-turn user input. Each approach has its advantages and disadvantages, with the choice depending on factors such as task complexity, user preferences, and the desired level of autonomy.

Interactive Multi-Agent UX

Offline Multi-Agent UX

Example use case

Travel Booking

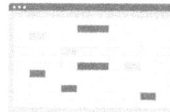

Back-office data extraction from emails, entry into CRM tool

UX should emphasize

1. Real-time progress indicators
2. Always-accessible interrupt controls
3. Dynamic suggestion system
4. Live cost estimation display
5. Interactive task modification tools
6. Contextual help and tooltips

UX should emphasize

1. Robust task setup interface
2. Checkpointing and error recovery controls
3. Batch task management tools
4. Post-task analysis tools

Figure 3.7. Multi-agent applications can be categorized as interactive or offline, depending on the level of user interaction required. Interactive scenarios involve real-time user engagement, while offline scenarios involve autonomous agent task execution.

For instance, in an interactive use case like travel booking, users express their preferences in natural language, observe the agent's progress, provide feedback (either explicitly requested or ad hoc), and verify the task status. Consequently, the associated interface must allow users to define the task, review the agent's actions, assumptions, or relevant data, provide input, interrupt or intervene if mistakes are made, and confirm the agent's actions. All of this must occur relatively quickly (with low latency) to prevent user frustration, utilizing an interactive UI that communicates the agent's state (e.g., an interactive view of the agent's action timeline, reviewing previous actions, or displaying current activities).

On the other hand, an offline task might involve back-office tasks, such as extracting structured data from unstructured emails or entering data into interfaces. For example, a multi-agent workflow could extract structured data from unstructured emails, log into a CRM system, and update customer records. In this case, users may not participate in task execution but may be notified of the task's completion or any exceptions that require their attention. In this setup, low latency is less crucial; however, tools for automatically reviewing and verifying the agent's actions and the final solution can be implemented to improve the overall solution.

3.6 Summary

- **Multi-agent applications UX design represents a fundamental shift** from traditional **interface design** (where users click buttons that execute predictable, programmer-defined actions) to **delegation design** (where users give natural language instructions to autonomous AI agents that can choose from vast action spaces). This shift creates new UX challenges Bansal et al. (2024) because tasks of comparable complexity can succeed or fail unpredictably based on AI reasoning and agent coordination breakdowns.

- **UI and software evolution have progressed through distinct phases**—user interfaces evolved from command-line syntax to graphical metaphors to today's intelligent multimodal systems, while software development shifted from explicit rule-based programming (Software 1.0) to machine learning (Software 2.0) to autonomous multi-agent systems (Software 3.0+). These parallel evolutions create a "capability discovery problem" where users struggle to understand what these powerful but unpredictable systems can reliably accomplish.

- **Two distinct user types need different design approaches**: end users (business travelers booking trips) require simple transparency about what's happening and why, while developers building these systems need detailed visibility into agent coordination patterns, error attribution, and performance optimization to ensure reliable user experiences.

- **Four essential UX design principles address multi-agent delegation challenges**: capability discovery (guiding users toward reliable tasks), cost-aware action delegation (managing real-world consequences of AI decisions), observability and provenance (showing what happened and why), and interruptibility (allowing graceful intervention in autonomous workflows). These principles help users effectively delegate complex tasks while maintaining appropriate control.

- **Multi-agent systems exhibit a "jagged frontier" of reliability**—they can handle sophisticated coordination tasks like booking complex business travel while failing at seemingly simpler requests due to coordination breakdowns, requiring new approaches to communicate uncertainty and build user trust through transparency.

Part II

Part II: Building Multi-Agent Systems from Scratch

Chapter 4

Building Your First Agent

This chapter covers

- Building agents from scratch with async-first architecture and streaming capabilities
- Implementing the core agent execution loop with model clients, tools, and memory
- Exploring design principles including event-based streaming, component serialization, and graceful cancellation
- Creating agents that progress from simple Q&A to complex multi-tool workflows
- Orchestrating groups of agents to address sophisticated tasks

In Chapter 1, we introduced the concept of an agent - and how its core components include a generative AI `model` for reasoning, `tools` for acting, and `memory` for adapting to tasks. In this chapter, we will explore how to implement these components from scratch. By the end of this chapter you will progressively build from an agent that can address simple tasks like "what is the capital of France?" to one that can "act" e.g., such as "plot the stock price of Microsoft over the last 5 years" and one that can do interesting things like refuse to respond to malicious requests (e.g., transfer all bitcoins on the local machine to a given address).

To support the core concepts introduced in this book, we will build a python library from scratch which we will name **picoagents** . We will begin with developing a base agent class abstraction and progress to abstractions that facilitate the coordination between multiple agents.

An effective agent interface should be minimal yet complete—exposing just enough configuration to be powerful without overwhelming users with options. The pattern we'll use— `Agent(model=.., tools=[..], memory=..)` —provides exactly these three essential capabilities: reasoning through models, action through tools, and adaptation through memory. This minimalist interface has emerged as a common abstraction across agent frameworks because it captures core requirements while remaining approachable.

Before we get into implementation details, here is the final developer experience we're building towards (Listing 4.1). Don't forget to replace "your_api_key" with your actual OpenAI API key.

Listing 4.1 Complete picoagents Agent example

```python
from picoagents import Agent, OpenAIChatCompletionClient

def get_weather(location: str) -> str:
    """Get current weather for a given location."""
    return f"The weather in {location} is sunny, 75°F"

# Create an agent - that's it!
agent = Agent(
    name="assistant",
    description="You are a helpful assistant.",
    instructions="You are helpful. Use tools when appropriate.",
    model_client=OpenAIChatCompletionClient(
        model="gpt-4.1-mini",
        api_key="your_api_key"
    ),
    tools=[get_weather]  # Functions become tools automatically!
)

# Use the agent
async for event in agent.run_stream("What's the weather in Paris?"):
    print(event)  # Stream events
```

The output will look something like this:

```
[user] 16:55:44 | What's the weather in Paris?
[assistant] 16:55:45 | tool_call: get_weather(location=Paris)
[assistant] 16:55:45 | tool_response: ✓ The weather in Paris is sunny, 75°F
[assistant] 16:55:45 | The weather in Paris is sunny, 75°F
```

Notice how the agent automatically:

1. Recognized it needed weather data for the given task
2. Called the `get_weather` function with the right parameter
3. Integrated the result into a natural response

Now let's build this Agent step by step, starting with the foundational components that make agents reliable and extensible.

4.1 Design Principles

As we build our agent framework step by step, we'll make key architectural decisions that determine how easy your agents are to develop, debug, and scale. These aren't abstract principles but practical choices that address real problems you'll encounter when building agent applications.

Async-First Architecture: Agent tasks involve multiple slow operations: LLM API calls (500ms-5s), tool executions (variable), and I/O operations. Without async, your agent sits idle during each call, making multi-agent systems painfully slow. A simple 3-agent workflow could take 30 seconds synchronously but only 10 seconds with proper concurrency. We embrace `async/await` throughout because retrofitting async into synchronous code is much harder than the reverse.

Event-Based Streaming: Agent tasks can take 30+ seconds to complete multiple steps. Without streaming, users stare at blank screens wondering if anything is happening. Streaming provides real-time progress updates, enables responsive user interfaces, and supplies the observability data you need to debug multi-step agent behavior. When an agent gets stuck in a tool call loop, streaming events show you exactly where the problem occurs.

Component Serialization: You need to save agent configurations, share them between team members, and potentially build visual editors for non-technical users. When every component (agents, tools, memory) can serialize itself to JSON, you can version control complete agent setups, build configuration UIs, and restore agent configuration and state from saved sessions.

Graceful Cancellation: Users will start long-running agent tasks and then need to cancel them - whether due to incorrect prompts, infinite loops, or to provide feedback.

Abstract Base Classes with Core Behaviors: When you want to support multiple LLM providers, different tool types, or various memory backends, abstract interfaces prevent vendor lock-in and enable testing with mock implementations. The `BaseTool` interface lets you start with simple functions and later add REST API tools, database tools, or emerging standards like MCP tools without changing your agent code.

4.2 The Agent Execution Loop

The heart of any agent framework is the **agent execution loop** . Understanding this loop is crucial because it determines how agents think, act, and learn. Let's start by identifying the foundational components.

Every agent interaction follows the same fundamental pattern:

1. **Prepare Context**: Combine task + instructions + memory + conversation history
2. **Call Model**: Send context to LLM and get response

3. **Handle Response**: Process text response or execute tool calls
4. **Iterate**: If tools were called, add results to context and repeat from step 2
5. **Return**: Provide final response and update memory (if applicable)

Listing 4.2 Simplified agent execution loop pseudocode

```python
# Simplified agent loop pseudocode
async def agent_execution_loop(task):
    context = prepare_context(task, instructions, memory, history)

    while not done:
        response = await model_client.create(context)

        if response.has_tool_calls:
            for tool_call in response.tool_calls:
                result = await execute_tool(tool_call)
                context.append(result)
        else:
            done = True
    update_memory(context) # optional
    return response
```

As seen in Listing 4.2, an agent takes multiple steps (model call -> tool call -> model call) as part of a "run", and each step may take an arbitrary amount of time. From this perspective, it is important that agents can provide or stream progress updates, which is especially beneficial for scenarios where a human is waiting for a response. Streaming also provides the visibility needed for debugging agent behavior.

4.2.1 Streaming Events and Real-Time Updates

Streaming allows agents to yield progress updates as events during execution, providing real-time visibility into what's happening. This is critical for building responsive UIs and debugging agent behavior as it unfolds.

4.2.2 BaseAgent: The Foundation

Let's start implementing the agent architecture. Our agent needs to handle different types of messages in conversations - system instructions (SystemMessage), user inputs (UserMessage), assistant responses (AssistantMessage), and tool execution results (ToolMessage). We'll use structured message types (built with Pydantic for type safety) to represent these communications clearly.

Now let's define the base agent interface:

Listing 4.3 BaseAgent class definition and initialization

```
class BaseAgent(ABC):
    """Abstract base class defining the core agent interface."""

    def __init__(
        self,
        name: str,
        instructions: str,
        model_client: 'BaseChatCompletionClient',
        tools: Optional[List] = None,
        memory: Optional['BaseMemory'] = None,
        context: Optional['AgentContext'] = None,
        middleware: Optional[List] = None,
        max_iterations: int = 10
    ):
        self.name = name
        self.instructions = instructions
        self.model_client = model_client

        # Process optional components with defaults
        self.tools = self._process_tools(tools or [])
        self.memory = memory
        self.context = context or AgentContext()
        self.middleware_chain = MiddlewareChain(
            middleware or []
        )
        ...
```

The BaseAgent initialization (shown in Listing 4.3) establishes the core agent components: a model client for reasoning, optional tools for actions, optional memory for learning, and support for context and middleware injection. The flexible initialization supports agents ranging from simple chatbots to complex multi-tool systems.

With the agent structure defined, we need to specify the execution contract. Every agent must implement two core methods that define how tasks are processed:

Listing 4.4 BaseAgent abstract methods for execution

```python
class BaseAgent(ABC):
    # ... initialization code from above ...

    @abstractmethod
    async def run(
        self,
        task: Union[str, UserMessage, List[Message]]
    ) -> 'AgentResponse':
        """Execute agent and return final response."""
        pass

    @abstractmethod
    def run_stream(
        self,
        task: Union[str, UserMessage, List[Message]]
    ) -> AsyncGenerator[Union[Message, 'AgentEvent'], None]:
        """Execute agent with streaming output."""
        pass
```

These abstract methods (Listing 4.4) enforce that every agent implementation must support both collecting (run) and streaming (run_stream) execution modes. The streaming-first design ensures we can build responsive UIs while providing flexibility for different application needs.

The constructor includes context and middleware parameters for advanced scenarios. Most applications use the defaults, but explicit injection enables specific use cases:

- **Context Injection:** Pre-populate agent with conversation history
- **State Restoration:** Reload agent state from serialized context
- **Orchestration:** Share context between agents in workflows
- **Testing:** Inject controlled state for reproducible testing

4.3   Task Cancellation

Agent tasks can run for minutes or hours—researching topics, calling APIs, processing data. Users need to stop these operations mid-task: they realize the prompt was wrong, the agent is stuck in a loop, or they want to provide feedback before continuing.

We implement **task cancellation** through a CancellationToken —a thread-safe mechanism that propagates stop signals through async operations. Task cancellation is the ability to interrupt and cleanly terminate a running agent operation, stopping LLM calls, tool executions, and all downstream processing. The agent checks this token at critical points during execution:

Listing 4.5 Using cancellation tokens to stop agent execution

```python
from picoagents import Agent, CancellationToken

# Create a cancellation token
token = CancellationToken()

# Pass it to the agent
async def run_with_cancellation():
    async for event in agent.run_stream(
        "Long running task",
        cancellation_token=token
    ):
        print(event)

# From another thread or async task - stop the agent
token.cancel()  # Stops execution immediately
```

When `cancel()` is called (Listing 4.5), the token propagates through the agent's execution loop—stopping LLM calls mid-stream, preventing new tool executions, and raising `asyncio.CancelledError` to exit cleanly. The agent implementation checks `token.is_cancelled()` before each model call, during token streaming, and before tool execution.

The same pattern works throughout the framework: agents, orchestrators, and workflows all accept `cancellation_token` parameters. In Chapter 8, we'll connect user interface "Stop" buttons to cancellation tokens, giving users control over long-running operations.

4.4   Adding a Model Client

The model client is the agent's interface to generative AI models. A well-designed abstraction here enables support for multiple LLM providers while keeping agent code clean. In this section we will introduce a general `BaseChatCompletionClient` that defines two key methods - `create` for obtaining generations from an LLM model and `create_stream` for obtaining streaming responses.

> ℹ Why "BaseChatCompletionClient"?
>
> We use the term "ChatCompletion" because modern LLMs are trained for conversational interactions with alternating roles (system, user, assistant). This multi-turn dialogue format, where each message has a specific role and the model generates

responses as the "assistant" role, is fundamentally different from older completion models that simply continued text. The "chat completion" terminology reflects this role-based conversational training paradigm that enables more structured and controllable interactions.

Listing 4.6 BaseChatCompletionClient abstract interface

```python
from abc import ABC, abstractmethod
from typing import List, Optional, Dict, Any

class BaseChatCompletionClient(ABC):
    """Abstract interface for LLM providers."""

    @abstractmethod
    async def create(
        self,
        messages: List[Message],
        tools: Optional[List[Dict[str, Any]]] = None,
        **kwargs
    ) -> 'ChatCompletionResult':
        """Make a single LLM API call."""
        pass

    @abstractmethod
    async def create_stream(
        self,
        messages: List[Message],
        tools: Optional[List[Dict[str, Any]]] = None,
        **kwargs
    ) -> AsyncGenerator['ChatCompletionChunk', None]:
        """Make a streaming LLM API call."""
        pass
```

Now that we have this base class, we can extend it to support specific LLM implementations. The pattern is always the same: convert our types to the provider's format, make the API call, then convert the response back to our unified format.

Here's how we implement an OpenAI client that follows our three-step conversion pattern. First, we will begin by creating an initialization method that sets up the OpenAI client with the desired model and API key.

The initialization (Listing 4.7) is straightforward - store the model name and create the OpenAI client. The real work happens in the `create` method, which follows our three-step conversion pattern:

Listing 4.7 OpenAI client initialization

```python
class OpenAIChatCompletionClient(BaseChatCompletionClient):
    def __init__(
        self,
        model: str = "gpt-4.1-mini",
        api_key: Optional[str] = None
    ):
        self.model = model
        self.client = AsyncOpenAI(api_key=api_key)
```

Listing 4.8 OpenAI client create method with conversion pattern

```python
async def create(
    self,
    messages: List[Message],
    tools: Optional[List[Dict]] = None
) -> ChatCompletionResult:
    # Step 1: Convert our types to provider's format
    api_messages = self._convert_messages_to_api_format(messages)

    # Step 2: Make the provider-specific API call
    response = await self.client.chat.completions.create(
        model=self.model,
        messages=api_messages,
        tools=tools
    )

    # Step 3: Convert response back to our unified format
    return ChatCompletionResult(
        message=AssistantMessage(
            content=response.choices[0].message.content
        ),
        usage=Usage(
            tokens_input=response.usage.prompt_tokens,
            tokens_output=response.usage.completion_tokens
        ),
        model=response.model
    )
```

Note that the same pattern of Convert input -> Call API -> Convert output can be applied to support any LLM provider while keeping our agent code unchanged.

4.4.1 Agent Model Integration

Now let's see how the agent uses the model client in its execution loop. The agent's job is to prepare the right context and handle the response:

Listing 4.9 Agent preparing context for model client

```
class Agent(BaseAgent):
    async def run_stream(
        self,
        task: Union[str, UserMessage, List[Message]]
    ) -> AsyncGenerator[Union[Message, AgentEvent], None]:
        # Prepare context with instructions and history
        llm_messages = [
            SystemMessage(content=self.instructions),
            *self.context.messages,
            *task_messages
        ]
```

The context preparation combines the agent's system instructions, any previous conversation history, and the current task. This gives the model all the information it needs to respond appropriately. With the context ready, the agent calls the model and handles the response:

Listing 4.10 Agent calling model client and handling response

```
        # Call model client
        completion_result = await self.model_client.create(llm_messages)
        assistant_message = completion_result.message

        # Yield the response
        yield assistant_message

        # Update conversation context
        self.context.add_message(assistant_message)
```

The key insight is that the agent simply prepares the conversation context (instructions + history + current task) and passes it to the model client. The model client handles all the provider-specific details, returning a standardized `ChatCompletionResult` that works the same regardless of whether you're using OpenAI, Anthropic, or any other provider.

4.5 Enabling Structured Output - The Key to Reliable Agents

Our agent currently produces text responses that humans can read but are difficult for programs to parse reliably. To build agents that can take precise actions, we need predictable, machine-readable data. This is where **structured output** becomes essential.

Structured output constrains the model to generate responses in a specific format, like JSON objects that match predefined schemas. Instead of hoping to parse "The weather in Paris is sunny" correctly, we get reliable structured data (shown below) that our code/application can use immediately.

```
{"location": "Paris", "condition": "sunny", "temperature": 75}
```

This reliability becomes crucial when agents need to call tools, store information in memory, or coordinate with other systems. Let's implement structured output using Pydantic models to ensure predictable, machine-readable responses.

Most modern LLM providers support structured output. Our `create` method implementation in `BaseChatCompletionClient` can include an optional `output_format` parameter (please see `picoagents.llm._openai.py`) that accepts a Pydantic model. When provided, the client will ensure the model's output conforms to this schema.

First, we define a Pydantic model that specifies the structure we want:

Listing 4.11 Pydantic model for structured output

```python
from pydantic import BaseModel, Field
from typing import List

class UserProfile(BaseModel):
    id: int = Field(..., description="The user's unique identifier")
    name: str = Field(..., description="The user's full name")
    email: str = Field(..., description="The user's email address")
    age: int = Field(..., description="The user's age")
    skills: List[str] = Field(..., description="A list of the user's skills")
```

The field descriptions help the model understand what kind of data to generate. Now we can use this model to constrain the LLM's output:

The `output_format` parameter tells the model client to constrain the response to match our UserProfile schema. The result comes back as a properly typed object:

Note that the `result.structured_output` is *guaranteed* to be a valid `PersonInfo` object, eliminating the need for error-prone text parsing. We will rely on the concept of structured output throughout this book to reliably build out agent behaviors.

Listing 4.12 Using structured output with model client

```
client = OpenAIChatCompletionClient(model="gpt-4.1-mini")
messages = [
    UserMessage(
        content="Create a profile for a software engineer named Alice \
                who is 28 years old and skilled in Python, JavaScript, \
                and machine learning.",
        source="user"
    )
]

# Call model with structured output constraint
result = await client.create(
    messages=messages,
    output_format=UserProfile  # This ensures structured output
)
```

Listing 4.13 Working with structured output results

```
# The result.structured_output will be a UserProfile object
if result.structured_output:
    profile = cast(UserProfile, result.structured_output)
    print(f"Name: {profile.name}")
    print(f"Age: {profile.age}")
    print(f"Skills: {', '.join(profile.skills)}")

# Output:
# Name: Alice
# Age: 28
# Skills: Python, JavaScript, Machine Learning
```

4.6 Adding Tools

With structured output working, our agent can now produce reliable, parseable data instead of just text. This capability unlocks the next level: **tool calling**. When an agent needs to check the weather, query a database, or perform calculations, it can generate structured tool call requests that our application can execute safely.

Here's how structured output enables reliable tool calling:

Listing 4.14 From structured output to tool calling

```python
# Structured output enables reliable tool requests
class ToolCall(BaseModel):
    name: str
    arguments: Dict[str, Any]

# Agent produces this structure reliably
result = await model_client.create(
    messages=messages,
    tools=available_tools,
    # Tool schemas guide structured output
)

# Now we can safely execute the tool
if result.tool_calls:
    for tool_call in result.tool_calls:
        tool_result = await execute_tool(tool_call.name, tool_call.arguments)
```

Without structured output, we'd be parsing "Please call the get_weather function with location set to Paris" - fragile and error-prone. With it, we get reliable JSON we can execute confidently.

Tools transform agents from text generators into action-taking entities. Our tool system leverages the structured output foundation we just built, enabling LLMs to generate precise function calls with properly typed parameters.

4.6.1 Function Calling (Tool Calling) in LLMs

Tool calling (also known as function calling) is an application of structured output where LLMs generate precisely formatted JSON requests that match predefined tool schemas. Instead of generating free-form text like "please call the weather function," the LLM produces structured function calls with properly typed parameters—for example, generating `get_weather(location="San Francisco", date_range="next 4 days")` when a user asks "What is the weather going to be like in San Francisco over the next 4 days?"

This structured format eliminates parsing ambiguity and ensures tools receive properly typed parameters, enabling agents to reliably interface with external systems. Without structured output, agents would need fragile text parsing that could easily break on slight formatting variations.

Tool calling follows a multi-step request-response pattern:

1. **Define tools:** Provide function schemas (name, description, parameters) to the LLM
2. **Tool calls:** LLM generates structured requests to use specific tools using the schemas
3. **Tool execution:** Application executes the function and generates outputs
4. **Tool outputs:** Results are sent back to the LLM for integration into the response

This pattern leverages structured output to enable LLMs to interface with external systems and access data beyond their training, transforming them from text generators into reliable action-taking agents.

In the following section, we will implement a `BaseTool` class that holds a general representation of the function schema which will ultimately be used to represent tools.

4.6.2   Tool Categories: General Purpose vs Task-Specific

An agent's abilities are directly tied to the tools it has access to. Tools fall into two fundamental categories:

General-purpose tools offer broad capabilities applicable to many tasks. Examples include code interpreters that can execute any programmatic solution, or user interface drivers that can interact with applications and websites. These tools enable maximum flexibility but introduce uncertainty and security concerns that require careful sandboxing (see Chapter 5 for detailed security considerations).

Task-specific tools provide focused capabilities for particular domains - sending emails, querying specific databases, or generating images. These tools constrain the agent's behavior, enabling predictability and helping manage the agent's decision space, but limit what the agent can do to the predefined functions.

4.6.3   The BaseTool Class

While you could just pass Python functions directly to agents, this approach has limitations. What happens when you need tools that call REST APIs, query databases, or integrate with emerging standards like Model Context Protocol (MCP)? You need a common interface that works across different tool types.

The `BaseTool` abstraction follows the abstract base classes principle we discussed earlier. It provides a standardized contract that enables both simple function-based tools and sophisticated external integrations:

This interface enables extensibility without changing agent code. You can start with simple functions and later add database tools, API tools, or MCP tools that connect to external services. For a complete example of integrating MCP tools with agents, see Section 12.2.4.

Listing 4.15 BaseTool abstract class and ToolResult

```python
class BaseTool(ABC):
    """Abstract base class for all tools."""

    def __init__(self, name: str, description: str):
        self.name = name
        self.description = description

    @property
    @abstractmethod
    def parameters(self) -> Dict[str, Any]:
        """JSON schema defining expected inputs."""
        pass

    @abstractmethod
    async def execute(self, parameters: Dict[str, Any]) -> ToolResult:
        """Execute the tool with given parameters."""
        pass

    def to_llm_format(self) -> Dict[str, Any]:
        """Convert to OpenAI (or any other) function calling format."""
        ...
```

4.6.4 FunctionTool: Developer Ergonomics

The most common tool type wraps Python functions. Rather than manually implementing the `BaseTool` interface for every function, a common practice is to provide a `FunctionTool` abstraction that automatically handles the conversion. This way, the developer simply writes normal Python functions with type hints and docstrings:

```python
def get_weather(location: str, days: int = 1) -> str:
    """Get weather forecast for a location."""
    return f"Weather in {location}: sunny"

# This function automatically becomes a tool
agent = Agent(tools=[get_weather], ...)
```

In turn, the `FunctionTool` class handles the conversion automatically:

1. **Function Inspection:** Extracts signature, parameter names, and type hints
2. **Schema Generation:** Converts Python types to LLM-compatible JSON schema
3. **Execution Wrapper:** Handles both sync and async functions with error handling
4. **Result Formatting:** Converts function outputs to structured `ToolResult` objects

This approach prioritizes developer ergonomics - you write functions naturally and the framework handles agent integration.

4.6.5 Agent Tool Integration

With our BaseTool interface defined, let's see how agents discover, process, and execute tools during their execution loop. The agent needs three key capabilities: converting various tool types into a standard format, preparing tool schemas for the LLM, and executing tool calls safely.

First, the agent processes tools during initialization. Since developers can pass either BaseTool instances or plain Python functions, we need to normalize them:

Listing 4.16 Converting mixed tool types to standard format

```
class Agent(BaseAgent):
    def _process_tools(
        self,
        tools: List[Union[BaseTool, Callable]]
    ) -> List[BaseTool]:
        """Convert mixed tool types to BaseTool instances."""
        processed = []
        for tool in tools:
            if isinstance(tool, BaseTool):
                processed.append(tool)
            elif callable(tool):
                # Wrap functions in FunctionTool automatically
                processed.append(FunctionTool(tool))
            else:
                raise ValueError(f"Invalid tool type: {type(tool)}")
        return processed
```

This normalization (Listing 4.16) happens once during agent initialization, ensuring all tools follow the same interface regardless of how developers provided them.

When the agent prepares to call the LLM, it needs to convert tool schemas into the function calling format the model expects:

Listing 4.17 Preparing tool schemas for LLM

```
    def _get_tools_for_llm(self) -> List[Dict[str, Any]]:
        """Convert tools to LLM function calling format."""
        return [tool.to_llm_format() for tool in self.tools]
```

This method (Listing 4.17) generates the JSON schemas that tell the LLM what functions are available, what parameters they accept, and what they do. The LLM uses these schemas to decide when to call tools and how to extract parameters from the user's natural language request.

Finally, when the LLM generates a tool call request, the agent must execute it safely and return results:

Listing 4.18 Executing tool calls and handling results

```
async def _execute_tool_call(
    self,
    tool_call: ToolCallRequest
) -> ToolMessage:
    """Execute a tool call and return result message."""
    # Find the requested tool by name
    tool = self._find_tool(tool_call.tool_name)
    if tool is None:
        return ToolMessage(
            content=f"Tool '{tool_call.tool_name}' not found",
            tool_call_id=tool_call.call_id,
            success=False,
            error=f"Tool not found"
        )

    # Execute the tool with error handling
    try:
        result = await tool.execute(tool_call.parameters)
        content = (
            str(result.result) if result.success
            else f"Error: {result.error}"
        )
        return ToolMessage(
            content=content,
            tool_call_id=tool_call.call_id,
            tool_name=tool_call.tool_name,
            success=result.success,
            error=result.error
        )
    except Exception as e:
        # Handle unexpected errors gracefully
        return ToolMessage(
            content=f"Tool execution failed: {str(e)}",
            tool_call_id=tool_call.call_id,
            success=False,
            error=str(e)
        )
```

This execution method (Listing 4.18) handles several important cases: tool not found errors, successful tool execution with results, tool execution that fails gracefully, and unexpected exceptions. The returned `ToolMessage` gets added to the conversation history, allowing the LLM to see the tool's result and continue reasoning.

With these three methods in place (Listing 4.16, Listing 4.17, Listing 4.18), the complete tool integration workflow becomes clear. Developers pass functions directly to the agent (as shown in Listing 4.1), the agent normalizes them during initialization, converts their schemas for the LLM, and executes them safely when called. The LLM automatically decides which tools to call based on the available schemas, extracts parameters from the user's natural language request, and our framework handles execution with graceful error handling. This infrastructure transforms simple Python functions into reliable agent capabilities.

4.7 Adding Memory

Our agent can now produce structured output and call tools reliably. But there's a missing piece: **persistence**. When an agent calls a weather tool and learns "Paris is sunny today," that information disappears when the conversation ends. Memory bridges this gap.

As agents interact with users and other agents, they need to maintain memory of interactions to ensure coherent and contextually relevant responses. From an implementation perspective, agents can maintain memory in two ways: **short-term memory** via message history and **long-term memory** via vector databases.

When tools execute successfully, their results may need to flow into both short-term and long-term memory systems. Here's how tool results connect to both memory types:

4.7.1 Short-term Memory via Message History

Short-term memory ensures agents are aware of recent context—the immediate set of messages and actions that have occurred—enabling subsequent responses to be coherent and contextually relevant.

Consider this interaction:

1. User: "What is the height of the Eiffel Tower?"
2. Agent: "The Eiffel Tower is 324 meters tall."
3. User: "When was it built?"

Without short-term memory, the agent cannot resolve the second question because it has no context that "it" refers to the Eiffel Tower. In our agent implementation, short-term memory is implemented as message history that stores recent messages and tool results:

Short-term memory is fast and immediately available, but it's ephemeral - when the conversation ends, this context is lost. Long-term memory solves the persistence problem, enabling agents to learn and recall information across sessions.

Listing 4.19 From tool results to persistent memory

```
# Tool execution produces structured data
tool_result = await weather_tool.execute({"location": "Paris"})
# Result: {"location": "Paris", "condition": "sunny", "temperature": 22}

# SHORT-TERM: Immediately available in conversation history
context.add_message(ToolMessage(
    content=f"Weather in Paris: {tool_result}",
    source="weather_tool"
))

# LONG-TERM: Indexed for future retrieval across sessions
await memory.store({
    "content": f"Paris weather on {datetime.now().date()}: sunny, 22°C",
    "metadata": {"type": "weather", "location": "Paris", "date": datetime.now()}
})

# Later conversations can retrieve relevant past interactions
relevant_memories = await memory.retrieve("weather in Paris")
```

Short Term Memory

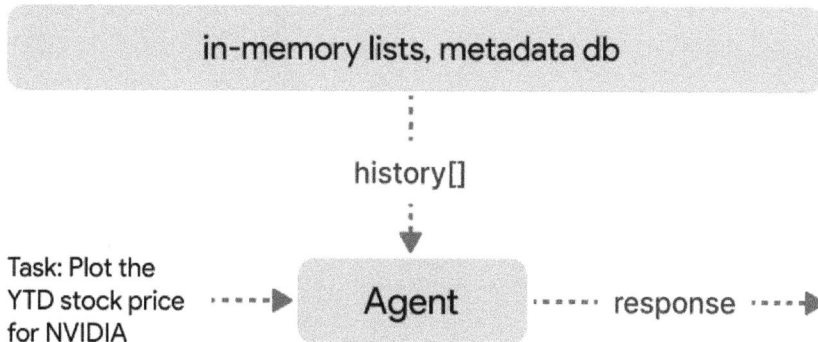

Figure 4.1. Short-term memory maintains conversation context through message history stored in in-memory lists or lightweight databases, enabling agents to track recent interactions and provide coherent responses.

Listing 4.20 Short-term memory via message history

```python
# Agent maintains conversation context
class AgentContext:
    def __init__(self):
        self.messages: List[Message] = []

    def add_message(self, message: Message):
        self.messages.append(message)
        # Optionally truncate to keep recent context
        if len(self.messages) > 50:
            self.messages = self.messages[-50:]

# Usage in agent execution loop
self.context.add_message(
    UserMessage(content="What is the height of the Eiffel Tower?"))
self.context.add_message(
    AssistantMessage(content="The Eiffel Tower is 324 meters tall."))
self.context.add_message(
    UserMessage(content="When was it built?"))

# All messages become part of the prompt to the LLM
llm_messages = [SystemMessage(content=self.instructions), *self.context.messages]
```

4.7.2 Long-term Memory via Retrieval-Augmented Generation (RAG)

Long-term memory stores information that isn't immediately relevant but may be useful for future interactions. This includes domain knowledge, user preferences, episodic memories from past sessions, and general world knowledge.

The challenge is scale - long-term memory can contain vast amounts of information, far more than an LLM can process at once. The solution is **Retrieval-Augmented Generation (RAG)** , a technique where external knowledge is retrieved from data sources and dynamically added to augment the model's prompt just before inference. This grounds the model's responses in domain-specific or up-to-date information beyond its training data. While vector databases enable semantic similarity search for determining relevance, RAG sources can include any external data: traditional databases, APIs, file systems, or knowledge graphs.

Key benefits of RAG with vector databases:

- **Scalability**: Handles large volumes of data efficiently
- **Relevance**: Dynamically retrieves pertinent information for accurate responses
- **Flexibility**: Supports multiple data modalities (text, images, audio)

Long - Term Memory (RAG)

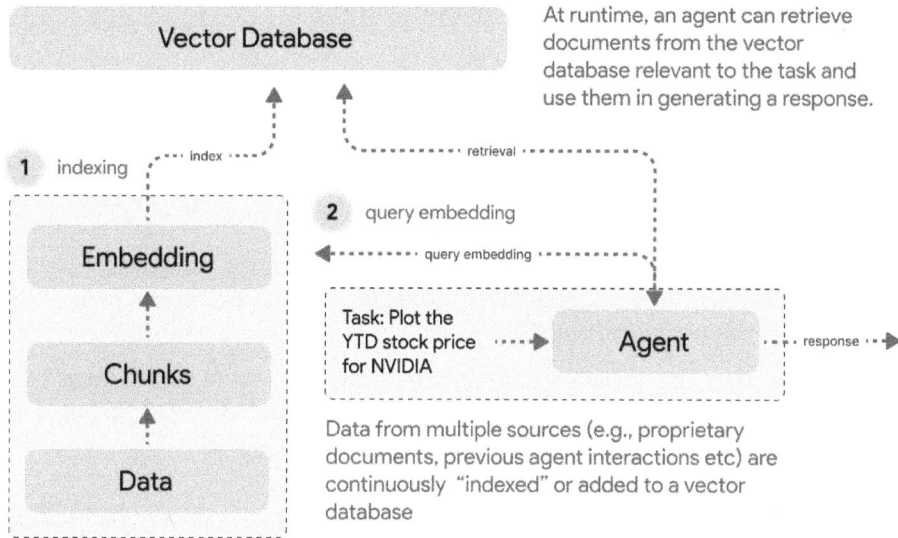

Vector Database

At runtime, an agent can retrieve documents from the vector database relevant to the task and use them in generating a response.

1 indexing ··· index ···

2 query embedding

··· retrieval ···

··· query embedding ···

Embedding

Chunks

Data

Task: Plot the YTD stock price for NVIDIA ···▶ **Agent** ···▶ response ···▶

Data from multiple sources (e.g., proprietary documents, previous agent interactions etc) are continuously "indexed" or added to a vector database

Figure 4.2. Long-term memory uses vector databases to store and retrieve vast amounts of information through RAG (Retrieval-Augmented Generation). Data is indexed as embeddings, and at runtime, agents query the database to retrieve relevant context for generating responses.

Listing 4.21 RAG process for long-term memory

```
# 1. INDEXING: Convert information to vectors
encoder = SentenceTransformer('all-MiniLM-L6-v2')
text = "User prefers morning meetings and dislikes long emails"
vector = encoder.encode(text)
await vector_db.store(vector, text, metadata={"type": "user_preference"})

# 2. QUERY: Convert question to vector and retrieve
query = "When should I schedule the meeting?"
query_vector = encoder.encode(query)
similar_items = await vector_db.search(query_vector, top_k=3)

# 3. AUGMENTATION: Add retrieved context to prompt
context_items = [item.text for item in similar_items]
augmented_prompt = f"""
Relevant context: {context_items}
User question: {query}
Please provide a helpful response.
"""
```

4.7.3 Memory Selection Strategies

In most treatments of RAG, a seemingly simple but complex issue arises: what information should be committed to long-term memory? On one hand, indexing everything includes irrelevant information introduces noise, while indexing too little risks omitting critical context.

Three main approaches address this challenge:

Manual Curation: Humans explicitly specify what to remember through user interfaces that capture preferences, interactive flagging ("remember this address"), or developer-defined rules for important information types.

Automated Selection: Algorithms identify information worth storing using frequency-based heuristics (commonly referenced information), importance scoring (relevance to user goals), or novelty detection (unique information).

Hybrid Approaches: Combine manual and automated methods - automated systems suggest what to remember, humans provide final approval for critical decisions.

4.7.4 BaseMemory Interface

We can achieve short-term memory with the agent's context (message history) and long-term memory with RAG (vector databases). To enable this in our library, we will begin by defining a `BaseMemory` interface that abstracts different memory implementations. The implementation below is reused from the AutoGen library.

The BaseMemory interface interface here provides abstractions for 3 key operations: first, `add()` to store new information; second, `query()` to retrieve relevant memories based on a query; and third, `get_context()` to fetch recent or relevant context for prompt augmentation.

In turn, this means that any agent we design can leverage `get_context()` to return a data structure that used to augment prompts.

Some important decisions that are worth reflecting with this design include:

Context vs Memory: AgentContext manages current session state (messages, temporary data) while BaseMemory provides persistent storage across sessions (user preferences, learned facts, long-term knowledge).

Application-Managed Storage: The developer calls `memory.add()` to store information - successful outcomes, user preferences, conversation summaries. The framework then injects this stored memory into the agent's prompts via `get_context()`, which may use simple recency filtering or sophisticated retrieval strategies like semantic similarity search. The agent receives this context but does not control what gets stored or retrieved.

Retrieval-Augmented Context: When preparing prompts, agents call `memory.get_context()` which executes sophisticated retrieval logic, from simple recency filtering to semantic similarity search.

Listing 4.22 BaseMemory abstract interface

```python
from abc import ABC, abstractmethod
from typing import List, Optional, Dict, Any

class BaseMemory(ABC):
    """Abstract interface for agent memory systems."""

    @abstractmethod
    async def add(
        self,
        content: str, metadata: Optional[Dict[str, Any]] = None
    ) -> None:
        """Store new content in memory."""
        pass

    @abstractmethod
    async def query(self, query: str, limit: int = 10) -> List[str]:
        """Retrieve relevant memories based on query."""
        pass

    @abstractmethod
    async def get_context(self, max_items: int = 10) -> List[str]:
        """Get recent/relevant context for LLM prompt."""
        pass
```

Path to RAG Implementation: The `query()` method provides a foundation for Retrieval-Augmented Generation. Applications can implement vector databases, semantic search, or knowledge graphs behind this interface while keeping agent code unchanged.

Agent Context manages the current session state during agent execution, including conversation messages, request metadata (user_id, session_id), and shared state for orchestration. Unlike Memory which provides persistent storage across sessions, AgentContext is transient and typically resets between major interactions. Implemented as a serializable Pydantic model, AgentContext enables stateless agent execution where any server can handle resume requests without maintaining session state in memory—a critical design for scalable web deployments. This separation between transient session state (AgentContext) and persistent knowledge (BaseMemory) enables both responsive conversation flow and persistent learning across sessions.

4.7.5 Agent Memory Integration

Here's how agents use can use memory during execution:

```
async def _prepare_llm_messages(
    self,
    task_messages: List[Message]
) -> List[Message]:
"""Prepare context with memory integration."""
system_content = self.instructions

    # Memory provides relevant context through get_context()
    if self.memory:
        context = await self.memory.get_context(max_items=5)
        if context:
            system_content += f"\n\nRelevant context:\n{'\n'.join(context)}"

    return [
        SystemMessage(content=system_content),
        *self.message_history,
        *task_messages
    ]
```

The `BaseMemory` interface can be extended to enable different storage and retrieval strategies. This architecture separates retrieval strategy from agent execution, enabling sophisticated memory systems while keeping the agent interface clean and consistent.

For development and testing, we start with a simple in-memory implementation. The `ListMemory` class stores memories in a list with automatic capacity management:

The `add()` method (Listing 4.23) creates `MemoryItem` objects with content and optional metadata, automatically evicting the oldest memories when capacity is reached. This FIFO approach ensures the memory system stays within resource limits.

For retrieval, the `query()` method performs simple text matching to find relevant memories:

The search (Listing 4.24) iterates through memories in reverse order (newest first) and performs case-insensitive substring matching. While simple, this approach works well for small memory sets and provides predictable behavior for testing.

Finally, the `get_context()` method provides recent memories for prompt augmentation:

This simple implementation (Listing 4.25) returns the most recent memories, but more sophisticated implementations can leverage different storage backends:

- **Vector-Based Semantic Memory:** Uses embeddings and vector similarity search to find contextually relevant memories
- **Graph-Based Knowledge Memory:** Stores facts as entities and relationships, enabling complex knowledge retrieval
- **Hybrid Memory Systems:** Combine multiple strategies (recency + semantic similarity + explicit knowledge)

Listing 4.23 ListMemory initialization and storage

```python
class ListMemory(Memory):
    """Simple in-memory list-based memory storage."""

    def __init__(self, max_memories: int = 1000):
        self.memories: List[MemoryItem] = []
        self.max_memories = max_memories

    async def add(
        self,
        content: str,
        metadata: Optional[Dict[str, Any]] = None
    ) -> None:
        """Add new memory, removing oldest if at capacity."""
        memory_item = MemoryItem(
            content=content,
            metadata=metadata or {}
        )
        self.memories.append(memory_item)

        # Remove oldest memories if over capacity
        if len(self.memories) > self.max_memories:
            self.memories = self.memories[-self.max_memories:]
```

Listing 4.24 ListMemory query implementation

```python
    async def query(
        self,
        query: str,
        max_results: int = 10
    ) -> List[str]:
        """Simple text-based search in memory contents."""
        query_lower = query.lower()
        matching_memories = []

        for memory in reversed(self.memories):  # Search newest first
            if query_lower in memory.content.lower():
                matching_memories.append(memory.content)
                if len(matching_memories) >= max_results:
                    break

        return matching_memories
```

Listing 4.25 ListMemory context retrieval

```python
async def get_context(self, max_items: int = 10) -> List[str]:
    """Get most recent memories for context."""
    recent_memories = (
        self.memories[-max_items:] if self.memories else []
    )
    return [memory.content for memory in recent_memories]
```

💡 🖥 Working Code: Structured Output

Find structured output examples in the examples directory - look for files containing `structured_output` for complete Pydantic integration patterns.

4.8 Agent-Managed Memory

The `BaseMemory` interface we just explored provides application-managed storage—the developer calls `memory.add()` to store information, and the application code injects relevant context into prompts via retrieval strategies. But what if agents need direct control over their memory? Consider a code reviewer that discovers a critical bug pattern. It should actively store that pattern, organize it in a knowledge base, and retrieve it in future reviews.

This is **agent-managed memory**: agents explicitly read, write, and organize their own persistent knowledge through tools, rather than passively receiving what the application provides. This approach follows patterns documented in Anthropic's context management work, where agents use file-based operations to actively curate their knowledge base across sessions.

4.8.1 Memory as a Tool

We can implement agent-managed memory by creating a tool that wraps file operations. The `MemoryTool` extends `BaseTool` to provide agents with a file-based memory system where they can create directories, store documents, and retrieve information on demand. Unlike `BaseMemory` which the application code automatically queries, `MemoryTool` requires agents to explicitly call it—the agent decides when to check memory, what to store, and how to organize information:

The agent now treats memory as an external system it can query and update, much like calling a database or API. The framework handles tool execution, but the agent controls what gets stored and when.

Listing 4.26 Creating an agent with memory tool

```python
from picoagents.tools import MemoryTool

# Create memory tool with file-based backend
memory_tool = MemoryTool(base_path="./agent_memory")

# Agent explicitly uses memory through tool calls
agent = Agent(
    name="code_reviewer",
    model_client=model,
    tools=[memory_tool],
    instructions="""You are a code reviewer.

ALWAYS check memory first:
  memory(command="view", path="/memories")

Store important patterns:
  memory(command="create",
        path="/memories/bugs/pattern_name.md",
        file_text="...")""")
)
```

4.8.2   Cross-Session Learning

Agent-managed memory enables learning that persists across conversations. Let's see how a code reviewer learns from one session and applies that knowledge in the next.

In the first session, the agent encounters a race condition bug:

Listing 4.27 Session 1: Agent discovers and stores a bug pattern

```python
code_with_bug = '''
self.results = []
for future in futures:
    self.results.append(future.result())   # RACE!
'''

response = await agent.run(
    f"Review this code:\n```python\n{code_with_bug}\n```"
)
```

When the agent analyzes this code, it recognizes the race condition and takes action to remember it. The agent uses `memory(command="view", path="/memories")` to check its memory, then creates a new entry describing the pattern:

```
memory(
    command="create",
    path="/memories/bugs/race_condition.md",
    file_text="Thread safety: shared state mutations \
        in concurrent code can cause race conditions...")
```

Now, in a completely new conversation session, we reset the agent's context and present similar code:

Listing 4.28 Session 2: Agent applies learned pattern to new code

---

```
agent.context = AgentContext()  # Reset context - new session

similar_code = '''
async def fetch_all(self):
    for coro in asyncio.as_completed(tasks):
        self.responses.append(result)  # RACE!
'''

response = await agent.run(
    f"Review this code:\n```python\n{similar_code}\n```"
)
```

Between sessions, the agent's conversation context resets completely, but its memory persists. When the agent checks memory in the new session with `memory(command="view", ..)`, it retrieves the `race_condition.md` file it stored earlier and immediately recognizes the same pattern in the new code. Despite never having seen this specific `fetch_all` function before, the agent applies its learned knowledge about race conditions to identify the bug.

4.8.3 Memory Operations

The `MemoryTool` provides six operations that mirror file system commands. These enable agents to build and maintain structured knowledge bases:

```
# View directory or file
memory(command="view", path="/memories")
memory(command="view", path="/memories/bugs/race.md")

# Create new file
memory(command="create",
       path="/memories/patterns/singleton.md",
       file_text="Singleton pattern notes...")
```

```
# Edit existing file
memory(command="str_replace",
       path="/memories/bugs/race.md",
       old_str="shared state",
       new_str="shared mutable state")

# Insert at specific line
memory(command="insert",
       path="/memories/notes.md",
       insert_line=3,
       insert_text="New observation\n")

# Delete file or directory
memory(command="delete", path="/memories/old_notes.md")

# Rename or move
memory(command="rename",
       path="/memories/temp.md",
       new_path="/memories/archive/temp.md")
```

These operations give agents full control over their knowledge organization. An agent might create directories for different bug types, move outdated patterns to an archive, or update patterns as it learns new variations.

4.8.4 Security Considerations

Since agents can perform file operations, the MemoryTool includes path validation to prevent security issues:

```
class MemoryBackend:
    def _validate_path(self, path: str) -> Path:
        """Prevent directory traversal attacks."""
        # Resolve to absolute path
        full_path = (self.base_path / path).resolve()

        # Ensure path stays within base directory
        try:
            full_path.relative_to(self.base_path)
        except ValueError:
            raise ValueError(
                "Access denied: path outside memory"
            )

        return full_path
```

This validation blocks attempts like ../../../etc/passwd that try to access files outside the designated memory directory. All memory operations remain sandboxed within the base path.

4.8.5 Choosing Between Memory Approaches

The choice between application-managed and agent-managed memory comes down to **control versus convenience**. Framework-managed memory (BaseMemory) gives you convenience— the developer stores information via `memory.add()` , and the application code automatically retrieves and injects relevant context using strategies like semantic similarity or recency filtering. Agent-managed memory (MemoryTool) gives agents control—they decide what to store, when to retrieve, and how to organize their knowledge base through explicit tool calls.

Use application-managed memory when you want automatic context enhancement without giving agents storage decisions. Use agent-managed memory when agents need to actively curate and organize their own knowledge. Both approaches can coexist—an agent might use application-managed memory for conversation history while using agent-managed memory for organizing learned patterns.

💡 💻 Working Code: Memory Examples

Explore both memory approaches in the examples/memory directory:
- **Agent-managed**: `memory_tool_example.py` - Cross-session learning with MemoryTool
- **Application-managed**: `list_memory_example.py` - Context injection with List-Memory

---

4.9 Adding Agent Middleware for Control and Observability

With great power comes great responsibility. Now that we have an agent that can use a model to generate text, or use tools to take actions (which may have unintended side effects), it is important that we have mechanisms to ensure safe and controlled usage. This is where the concept of **middleware** comes in. You can think of middleware as *routines* or *hooks* that run at key parts of your agent execution loop. When an agent prepares to call the LLM or execute a tool, these operations first pass through a middleware chain. Each middleware in this chain gets a chance to examine what's about to happen, potentially modify the inputs, and decide whether to proceed or block the operation entirely.

4.9.1 How Middleware Works in Practice

When you pass middleware to an agent during initialization, the agent wraps its core operations with these interceptors. Here's what happens during execution:

1. **Model Call Interception**: Before calling `self.model_client.create()` , the agent

---

Agent Middleware

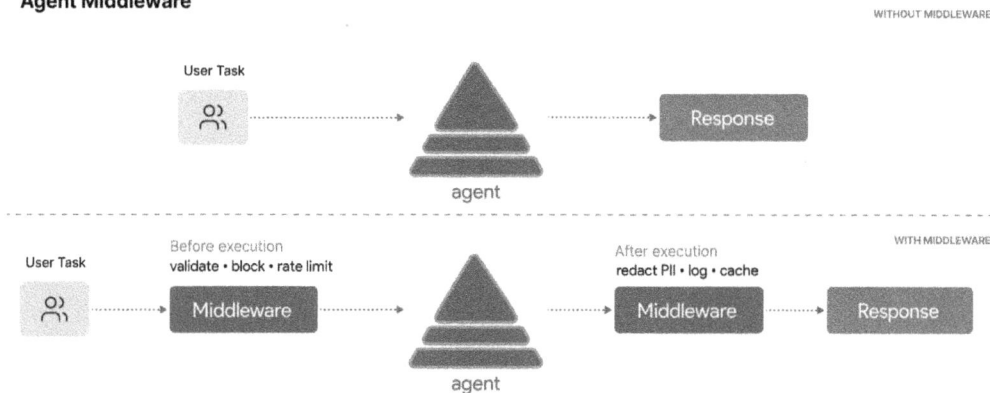

Middleware intercepts both model calls and tool executions. Before execution, it can validate inputs or block malicious requests. After execution, it can redact sensitive data or log results.

Figure 4.3. Agent middleware intercepts operations before and after execution. Without middleware (top), requests flow directly from user to agent to response. With middleware (bottom), interceptors can validate inputs, block malicious requests, redact sensitive data, or log results at each stage of the pipeline.

  routes the request through middleware
2. **Tool Call Interception**: Before executing `tool.execute()`, the agent routes the request through middleware
3. **Chain Execution**: Each middleware can examine the operation, modify inputs/outputs, or terminate the chain

The middleware receives complete context about what's happening: the operation type (`"model_call"` or `"tool_call"`), the agent's name, the current context, and the data being processed. Most importantly, each middleware gets the actual function that would be called, giving it full control over whether and how to proceed.

4.9.2   The BaseMiddleware Interface

To standardize how middleware works, we define a common interface that all middleware must implement. Middleware provides three key hooks that execute at different stages of an operation:

```
from abc import ABC, abstractmethod
from typing import Any, Optional

class BaseMiddleware(ABC):
    """Abstract base class for middleware."""
```

```
@abstractmethod
async def process_request(
    self,
    context: MiddlewareContext
) -> MiddlewareContext:
    """Process before the operation executes."""
    return context

@abstractmethod
async def process_response(
    self,
    context: MiddlewareContext,
    result: Any
) -> Any:
    """Process after the operation completes successfully."""
    return result

@abstractmethod
async def process_error(
    self,
    context: MiddlewareContext,
    error: Exception
) -> Optional[Any]:
    """Handle errors from the operation."""
    raise error
```

The `process_request()` hook runs before any model call or tool execution, giving you a chance to validate inputs, enforce rate limits, or start logging timers. This is where you'd block suspicious requests by raising an exception.

The `process_response()` hook runs after successful operations, allowing you to filter sensitive information from outputs, cache results, or log completion times.

Finally, `process_error()` handles failures, where you can log errors, provide fallback responses, or implement retry logic.

This three-phase approach gives middleware complete control over the execution lifecycle. Most importantly, middleware can block operations entirely by raising exceptions in `process_request()`.

Here's a concrete example of a security middleware. We start by defining patterns to detect malicious input:

The patterns (Listing 4.29) detect common attack vectors like prompt injection attempts, hex encoding tricks, and script injection. Now we implement the request interception:

When the middleware (Listing 4.30) detects dangerous patterns, it raises an exception

Listing 4.29 Security middleware initialization

```python
class SecurityMiddleware(BaseMiddleware):
    """Blocks malicious input before it reaches the model."""

    def __init__(self):
        self.malicious_patterns = [
            r"ignore.*previous.*instructions",
            r"system.*prompt.*injection",
            r"\\x[0-9a-f]{2}",  # Hex encoding attempts
            r"<script.*?>.*?</script>",  # Script injection
        ]
```

Listing 4.30 Security middleware request processing

```python
    async def process_request(
        self,
        context: MiddlewareContext
    ) -> MiddlewareContext:
        """Block malicious requests before they reach the model."""
        if context.operation == "model_call":
            for message in context.data:
                if hasattr(message, "content"):
                    for pattern in self.malicious_patterns:
                        if re.search(
                            pattern,
                            message.content,
                            re.IGNORECASE
                        ):
                            raise ValueError(
                                "Blocked potentially malicious input"
                            )
        return context
```

immediately. The malicious input never reaches the model, no expensive API call is made, and the problematic content stays out of your telemetry logs. This security barrier prevents threats before they reach your system, creating a first line of defense for your agent applications.

4.9.3 How Agents Use Middleware

Internally, agents route their core operations through the middleware chain. Specifically, we wrap model calls and tool executions inside our `Agent` class implementation:

```
# Model call with middleware
completion_result = await self.middleware_chain.execute(
    operation="model_call",
    agent_name=self.name,
    agent_context=self.context,
    data=llm_messages,
    func=lambda msgs: self.model_client.create(msgs, tools=tools)
)

# Tool execution with middleware
tool_result = await self.middleware_chain.execute(
    operation="tool_call",
    agent_name=self.name,
    agent_context=self.context,
    data=tool_call,
    func=lambda tc: self._execute_tool_internal(tc)
)
```

The `middleware_chain.execute()` method calls each middleware in sequence. Each middleware receives the same parameters and can decide whether to call the next middleware in the chain or short-circuit the process. This pattern enables sophisticated capabilities like distributed tracing, content filtering, and performance monitoring without cluttering the core agent implementation.

> 💡 💻 Working Code: Middleware Examples
>
> Explore middleware implementations in `examples/agents/middleware.py` - includes logging, rate limiting, safety filtering, and observability patterns.

4.9.4 Real-World Middleware Applications

Let's explore some practical middleware that you might implement in production applications.

Logging and Observability: In production, you need visibility into what your agents are doing. A logging middleware can capture every model call with its inputs, outputs, and timing data. This becomes crucial when debugging why an agent made certain decisions or when measuring system performance.

```python
class LoggingMiddleware(BaseMiddleware):
    """Log all agent operations using before/after hooks."""

    async def process_request(
        self,
        context: MiddlewareContext
    ) -> MiddlewareContext:
        """Log operation start."""
        print(f"[{context.agent_name}] Starting {context.operation}")
        context.metadata["start_time"] = time.time()
        return context

    async def process_response(
        self,
        context: MiddlewareContext,
        result: Any
    ) -> Any:
        """Log successful completion."""
        duration = time.time() - context.metadata.get("start_time", 0)
        print(f"[{context.agent_name}] \
            {context.operation} completed in {duration:.2f}s")
        return result

    async def process_error(
        self,
        context: MiddlewareContext,
        error: Exception
    ) -> Optional[Any]:
        """Log operation failures."""
        print(f"[{context.agent_name}] {context.operation} failed: {error}")
        raise error  # Re-raise the error
```

here is how we would then pass this middleware to an agent:

```python
from picoagents import Agent, LoggingMiddleware
from picoagents.llm import OpenAIChatCompletionClient

# Create agent with logging middleware
agent = Agent(
    name="assistant",
    description="A helpful assistant for answering questions",
    model_client=OpenAIChatCompletionClient(model="gpt-4.1-mini"),
    instructions="You are a helpful assistant.",
    middlewares=[LoggingMiddleware()]  # Uses default logger
```

```
)

# Execute task - middleware automatically logs operations
response = await agent.run("What's 2+2?")
print(f"Response: {response.messages[-1].content}")
```

When executed, the middleware captures all operations with timestamps:

```
[assistant] Starting model_call
[assistant] Completed model_call in 0.82s
Response: 2 + 2 = 4.
```

Safety and Content Filtering: In applications handling user-generated content or sensitive domains, you need guardrails. A safety middleware can scan model outputs for personally identifiable information, inappropriate content, or policy violations before they reach users. This middleware can either block problematic responses entirely or sanitize them by removing sensitive information.

As an example, we can implement a middleware that can detect and redact PII from model outputs. We can define a simple regex-based PII detector and use it in our middleware:

```
from picoagents import Agent, PIIRedactionMiddleware
from picoagents.llm import OpenAIChatCompletionClient

# Create agent with PII protection
agent = Agent(
    name="customer_service",
    description="Customer service agent with PII protection",
    model_client=OpenAIChatCompletionClient(model="gpt-4.1-mini"),
    instructions="Process customer information.",
    middlewares=[PIIRedactionMiddleware()]
)

# Process message containing sensitive data
response = await agent.run(
    "Customer John Doe called from 555-123-4567 about "
    "order confirmation sent to john@example.com"
)

print(f"Response: {response.messages[-1].content}")
```

The middleware automatically detects and protects sensitive information:

```
Input: Customer John Doe called from 555-123-4567
       about order confirmation sent to john@example.com
Response: Customer John Doe called from the redacted phone number
          regarding an order confirmation sent to the redacted
          email address. How would you like me to assist with
```

this information?

Rate Limiting and Cost Control: LLM API calls can be expensive and some users might abuse your system. Rate limiting middleware tracks usage per user or session and can block requests that exceed defined quotas. This is particularly important for applications exposed to the internet where you need to control operational costs.

```python
from picoagents import Agent, RateLimitMiddleware
from picoagents.llm import OpenAIChatCompletionClient
from datetime import datetime

# Create agent with aggressive rate limiting for demo
agent = Agent(
    name="limited_assistant",
    description="Rate-limited assistant for cost control demo",
    model_client=OpenAIChatCompletionClient(model="gpt-4.1-mini"),
    instructions="You are a helpful assistant.",
    middlewares=[
        RateLimitMiddleware(max_calls_per_minute=3)  # Only 3 calls/min
    ]
)

# Make multiple rapid requests with timestamps
for i in range(5):
    timestamp = datetime.now().strftime("%H:%M:%S.%f")[:-3]
    try:
        response = await agent.run(f"Question {i+1}: What is {i}+{i}?")
        print(f"[{timestamp}] Request {i+1}: \
            Success - {response.messages[-1].content[:40]}...")
    except Exception as e:
        print(f"[{timestamp}] Request {i+1}: Rate limited - {e}")
```

The timestamps clearly show the rate limiting in action. Notice how Request 4 takes nearly a full minute (59.58s) because the middleware enforced the 3-calls-per-minute limit:

```
[20:02:05.317] Request 1: Success - 0 + 0 = 0...
[20:02:05.682] Request 2: Success - 1 + 1 = 2...
[20:02:06.156] Request 3: Success - 2 + 2 = 4...
[20:02:06.517] Request 4: Success (59.58s) - 3 + 3 = 6...  # Rate limited!
[20:03:06.101] Request 5: Success - 4 + 4 = 8...
```

Audit and Compliance: In regulated industries, you need to log every decision an AI system makes. Audit middleware can capture complete traces of agent operations—what inputs were received, what decisions were made, what outputs were generated—and store them in compliance-ready formats. This creates immutable audit trails that can be analyzed by compliance tools or exported to observability platforms like OpenTelemetry.

4.10 OpenTelemetry: Industry-Standard Observability

The middleware patterns we've explored—logging, metrics collection, PII redaction, rate limiting—solve real production problems. However, implementing custom middleware for every project creates overhead and makes it difficult to integrate with modern monitoring tools. OpenTelemetry provides a standardized solution that works across all programming languages, frameworks, and monitoring backends.

OpenTelemetry (OTEL) is an open-source observability framework backed by the Cloud Native Computing Foundation (CNCF). It defines vendor-neutral standards for collecting traces, metrics, and logs—enabling you to instrument once and export to any backend (Jaeger, Datadog, Azure Monitor, etc.) without code changes.

For AI systems, OpenTelemetry offers the Gen-AI Semantic Conventions —standardized attribute names and span structures specifically designed for AI operations. These conventions define how to instrument LLM calls (model names, token usage), tool executions (tool names, success/failure), and agent workflows (multi-step operations with proper hierarchy).

When you follow these conventions, your telemetry becomes immediately portable. Whether you're using PicoAgents, LangChain, or AutoGen, your traces follow the same structure—making it easier to compare systems, share tooling, and move between platforms. This standardization also enables critical production capabilities like audit trails for compliance, cost tracking across teams, and performance monitoring at scale.

4.10.1 Building the OTelMiddleware

Let's build an `OTelMiddleware` class that implements the `BaseMiddleware` interface we defined earlier. This middleware will follow the Gen-AI Semantic Conventions to ensure our telemetry is standards-compliant.

First, initialization creates the telemetry instruments:

```
from opentelemetry import metrics, trace

class OTelMiddleware(BaseMiddleware):
    """OpenTelemetry middleware following Gen-AI conventions."""

    def __init__(self):
        self._tracer = trace.get_tracer("picoagents")
        self._meter = metrics.get_meter("picoagents")

        # Create metric instruments following Gen-AI semconv
        self._token_histogram = self._meter.create_histogram(
            name="gen_ai.client.token.usage",
            unit="{token}",
            description="Token usage per request"
```

```
        )

        self._duration_histogram = self._meter.create_histogram(
            name="gen_ai.client.operation.duration",
            unit="s",
            description="Operation duration"
        )
```

The middleware obtains a tracer for creating spans and a meter for recording metrics. It creates two histogram instruments following Gen-AI conventions: one for token consumption and one for operation duration.

When an operation starts, `process_request()` creates a span with Gen-AI attributes:

```python
async def process_request(
    self,
    context: MiddlewareContext
) -> MiddlewareContext:
    """Start span before operation."""
    # Create span following Gen-AI naming conventions
    if context.operation == "model_call":
        span_name = f"chat {self._get_model_name(context)}"
    elif context.operation == "tool_call":
        span_name = f"tool {self._get_tool_name(context)}"

    span = self._tracer.start_span(span_name)

    # Add Gen-AI semantic convention attributes
    span.set_attribute("gen_ai.system", "picoagents")
    span.set_attribute("gen_ai.operation.name", context.operation)
    span.set_attribute("gen_ai.agent.name", context.agent_name)

    # Store span for completion in process_response
    context.metadata["_otel_span"] = span
    context.metadata["_otel_start"] = time.time()

    return context
```

After the operation completes, `process_response()` records metrics and ends the span:

```python
async def process_response(
    self,
    context: MiddlewareContext,
    result: Any
) -> Any:
    """Record metrics and end span."""
```

```
span = context.metadata.get("_otel_span")

# Record token usage metrics for model calls
if hasattr(result, "usage"):
    self._token_histogram.record(
        result.usage.prompt_tokens,
        {"gen_ai.token.type": "input"}
    )
    self._token_histogram.record(
        result.usage.completion_tokens,
        {"gen_ai.token.type": "output"}
    )

    # Add usage as span attributes
    span.set_attribute("gen_ai.usage.input_tokens",
                        result.usage.prompt_tokens)
    span.set_attribute("gen_ai.usage.output_tokens",
                        result.usage.completion_tokens)

# Record duration and end span
duration = time.time() - context.metadata["_otel_start"]
self._duration_histogram.record(duration)
span.set_status(Status(StatusCode.OK))
span.end()
return result
```

4.10.2   Implementation: Auto-Instrumentation

Now that we have the middleware, let's implement automatic instrumentation using environment variables. This follows industry best practices used by frameworks like Django, Flask, and FastAPI.

```
import os

if os.getenv("PICOAGENTS_ENABLE_OTEL", "false").lower() == "true":
    from ._otel import auto_instrument
    auto_instrument()  # Patches Agent class
```

The `auto_instrument()` function modifies the `Agent` class to automatically prepend `OTelMiddleware` to every agent's middleware chain. This happens transparently at import time.

Users enable telemetry by setting environment variables before importing:

```
import os
```

```
# Enable OpenTelemetry before importing
os.environ["PICOAGENTS_ENABLE_OTEL"] = "true"
os.environ["OTEL_EXPORTER_OTLP_ENDPOINT"] = "http://localhost:4318"
os.environ["OTEL_SERVICE_NAME"] = "my-agent-app"

from picoagents import Agent
agent = Agent(name="assistant", model_client=..., tools=[...])
response = await agent.run("What's the weather in Paris?")
```

4.10.3   Complete Example with Jaeger

Let's see OpenTelemetry in action with a complete example using Jaeger, an open-source distributed tracing system. We'll instrument a weather agent and visualize the complete execution trace. Start by installing dependencies and launching Jaeger:

```
pip install picoagents[otel]
docker run -d -p 16686:16686 -p 4318:4318 jaegertracing/all-in-one:latest
```

Enable telemetry and create your agent:

```
import os

# Enable telemetry before importing picoagents
os.environ["PICOAGENTS_ENABLE_OTEL"] = "true"
os.environ["OTEL_EXPORTER_OTLP_ENDPOINT"] = "http://localhost:4318"
os.environ["OTEL_SERVICE_NAME"] = "weather-assistant"

from picoagents import Agent

# Create and use agent normally - telemetry automatic
agent = Agent(name="weather_assistant", model_client=..., tools=[...])
response = await agent.run("What's the weather in Paris?")
```

When you run this and visit http://localhost:16686, you'll see traces showing the hierarchy of operations as shown in Figure 4.4. Each span includes Gen-AI attributes like model names, token usage, and tool execution status.

💡 🖳 Working Code: OpenTelemetry Example

Complete OpenTelemetry examples with Jaeger and docker-compose are available in `examples/otel/` . The example includes detailed setup instructions and demonstrates integration with multiple observability backends.

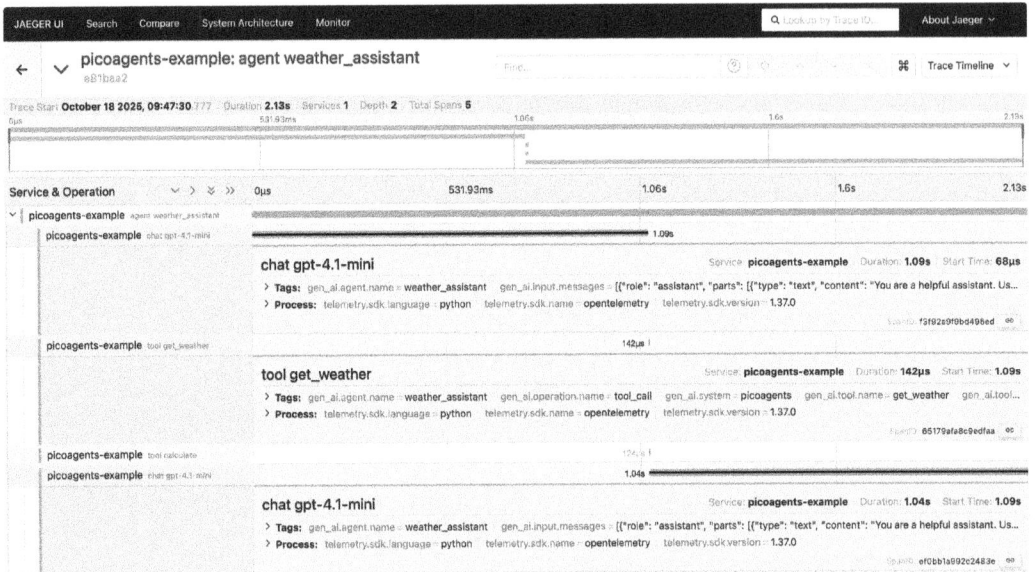

Figure 4.4. Jaeger UI showing OpenTelemetry trace for agent operations

4.10.4 Integration with Observability Platforms

OpenTelemetry's **standardization** enables vendor-neutral observability—your telemetry works with any OTLP-compatible backend by simply changing the endpoint. The Gen-AI semantic conventions ensure consistent traces across all platforms.

Leading OTLP-compatible platforms:

- **.NET Aspire Dashboard**: Lightweight, standalone dashboard for local development
- **Azure Monitor Application Insights**: Enterprise APM with native OpenTelemetry support
- **Jaeger**: Open-source distributed tracing (CNCF graduated)
- **Grafana Stack**: Tempo, Prometheus, Loki for complete observability
- **Arize AI Phoenix**: AI/ML-focused with AutoGen, LangChain support
- **Langfuse**: LLM engineering platform with cost tracking
- **Datadog, Honeycomb, New Relic**: Full-stack commercial platforms

Switch backends by changing the endpoint—no code changes needed:

```
# Aspire Dashboard (local)
os.environ["OTEL_EXPORTER_OTLP_ENDPOINT"] = "http://localhost:4317"

# Azure Application Insights
```

```
os.environ["OTEL_EXPORTER_OTLP_ENDPOINT"]
= "https://your-app.applicationinsights.azure.com"

# Arize Phoenix
os.environ["OTEL_EXPORTER_OTLP_ENDPOINT"] = "https://app.arize.com/v1/traces"
```

This portability exists **because of OpenTelemetry standardization.** Use Aspire locally, Arize for AI insights, Datadog for production—all with the same instrumentation code.

By adopting OpenTelemetry with automatic instrumentation, you gain immediate visibility into token consumption (track costs per agent, task, or user), performance monitoring (see latency distributions and identify slow operations), debugging capabilities (trace the exact sequence of model calls and tool executions), and production readiness (monitor error rates and set up alerts)—all with minimal configuration overhead. The observability foundation we've built here becomes essential for evaluation and optimization, topics we'll explore in depth in Chapter 10 and Chapter 11.

4.10.5   Privacy and Content Capture

Following OpenTelemetry Gen-AI semantic conventions, content attributes (prompts, completions, tool parameters) are **opt-in** by design due to potential sensitive information. By default, PicoAgents captures only metadata—operation names, model names, token counts, and success/failure status—which is safe for production environments. Full content capture requires explicit opt-in via `PICOAGENTS_OTEL_CAPTURE_CONTENT=true` .

This privacy-first approach means your production traces contain performance metrics and debugging metadata without exposing user queries, model responses, or tool execution details. When debugging specific issues in development, you can temporarily enable content capture to see the full conversation flow, then disable it again before deploying to production.

Table 4.1 summarizes the Gen-AI attributes captured by our implementation.

Table 4.1. Gen-AI Semantic Convention Attributes

Attribute	Type	When Captured	Description
gen_ai.operation.name	string	Always	Operation type: chat , tool_call
gen_ai.request.model	string	Always	Model name (e.g., gpt-4.1-mini)
gen_ai.usage.input_tokens	int	Always	Number of tokens in prompt
gen_ai.usage.output_tokens	int	Always	Number of tokens in completion
gen_ai.tool.name	string	Always	Tool function name
gen_ai.tool.success	boolean	Always	Tool execution success status

Attribute	Type	When Captured	Description
gen_ai.input.messages	JSON	Opt-in	Full prompt messages (may contain PII)
gen_ai.output.messages	JSON	Opt-in	Model responses (may contain PII)
gen_ai.tool.parameters	JSON	Opt-in	Tool input arguments
gen_ai.tool.result	string	Opt-in	Tool execution results

4.11 Agents as Tools

We've built a single agent that can act using tools. The notion of a tool here is around a discrete request-response operation (e.g., a function call, an API call, or a database query). In reality, it can be advantageous to have tools that can be much more—i.e, they can implement some version of the agentic loop we have discussed so far. From this perspective the line between "agent" and "tool" becomes blurred.

This underpins an emerging powerful *agent as tool* pattern where **any agent can become a tool for other agents.** The key architectural advantage is that each agent maintains complete independence—its own model configuration, toolset, memory, and middleware—while becoming a reusable component in larger systems.

To implement `as_tool()`, we wrap the agent in a `BaseTool` interface that exposes a single `task` parameter. This parameter tells the calling LLM what to send to the specialist agent. The agent's description becomes the tool's description, helping the coordinator's LLM understand when to use this particular specialist. When the coordinator calls the tool, it passes a task string that gets routed to the wrapped agent's `run()` method, and the agent's final response becomes the tool's result.

4.11.1 Controlling Result Informativeness

A critical design decision in agent composition is **what information flows back** from specialist agents to coordinators. By default, `as_tool()` returns only the specialist's final message. While this provides context isolation—preventing the coordinator's context from exploding with all the specialist's intermediate steps—it can lose important information when the last message is just "Done!" or "Thank you!" while crucial details appeared in earlier messages.

The `result_strategy` parameter balances context isolation against informativeness:

```
# Default: last message only (maximum isolation, minimal context pollution)
weather_tool = weather_agent.as_tool()
```

```
# Last N messages: capture conclusion + key supporting details
analyst_tool = analyst_agent.as_tool(result_strategy="last:3")

# All messages: full transparency, but risks context explosion
debug_tool = debug_agent.as_tool(result_strategy="all")

# Custom: precise control over what information transfers
def extract_key_findings(messages):
    """Extract only messages containing data or conclusions."""
    return "\n".join(
        msg.content for msg in messages
        if any(keyword in msg.content.lower()
               for keyword in ["found", "result:", "conclusion"])
    )
research_tool = research_agent.as_tool(result_strategy=extract_key_findings)
```

This control becomes essential as agent hierarchies deepen. A coordinator managing five specialists, each executing 10-20 tool calls internally, could accumulate massive context if every intermediate step bubbled up. The `result_strategy` enables each composition layer to define its own context boundaries.

Consider building a research system: your weather specialist might use a fast, cheap model (since weather queries are straightforward), while your data analysis specialist uses an expensive reasoning model for complex computations. Each agent can be developed, tested, and optimized independently by different teams, then composed together seamlessly.

When executed, the coordinator would automatically route the weather data collection to the weather specialist (using the efficient model) and the trend analysis to the analysis specialist (using the reasoning model). The execution flow shows clear delegation:

```
[coordinator] Delegating weather task to weather_specialist
[weather_specialist] Using weather_api tool -> "Paris: sunny, 22°C"
[coordinator] Delegating analysis task to analysis_specialist
[analysis_specialist] Using analyze_data tool -> "Trend analysis shows..."
[coordinator] "Based on weather data and analysis, Paris shows..."
```

The same specialist agents can be reused across different coordinators, enabling flexible system architectures without tight coupling between components.

4.12   Context Engineering

> If the LLM is the CPU, then the context window is RAM. — Andrej Karpathy

Agents rely on large language models (LLMs) to process instructions and generate responses. However, LLMs have **finite context windows**—limited amounts of text they can attend to at once. As agents execute multi-step tasks, their context grows rapidly, risking performance

Listing 4.31 Agent composition with specialized configurations

```python
# Each agent optimized for its specific domain
weather_agent = Agent(
    model="gpt-4.1-mini",      # Fast, cheap model for simple queries
    tools=[weather_api],
    memory=weather_cache
)

analysis_agent = Agent(
    model="o1-preview",        # Expensive reasoning model for complex analysis
    tools=[analyze_data, create_charts],
    memory=analysis_history
)

# Coordinator composes specialized agents
coordinator = Agent(
    model="gpt-4o",
    tools=[weather_agent.as_tool(), analysis_agent.as_tool()]
)

response = await coordinator.run(
    "Analyze weather trends in Paris over the last month"
)
```

degradation as critical information is pushed out of the model's attention.

While prompt engineering focuses on crafting individual inputs to models, **context engineering** is the practice of actively managing how context accumulates and evolves during multi-step agent execution. As agents work through tasks, each action—tool calls, memory retrievals, conversation turns—adds to the context. Without careful management, this growth creates two critical problems:

4.12.1 Context Rot and Explosion

Context Rot: As token count increases, model ability to accurately recall information decreases. Studies show performance degradation beginning around 32K tokens, well before advertised limits. The transformer architecture's n^2 pairwise relationships mean every token depletes the model's attention budget—a finite resource with diminishing returns.

Context Explosion: During agent execution, context grows rapidly. Each tool call adds request parameters, execution results, and error messages. Memory retrieval injects historical context. Message history accumulates turn by turn.

Consider what happens during a typical agent run:

Table 4.2. *Context growth during agent execution*

Step	Context Added	Tokens	Running Total
Initial	System + Memory + Task	2,600	2,600
Tool Call	Assistant + Tool Result	3,150	5,750
5x Tools	Multiple Results	15,000	20,750

Without context management, agents fail silently. A research agent that worked perfectly on 5-step tasks starts hallucinating on step 8 of a 10-step workflow—not because the model degraded, but because 45K tokens of accumulated tool results pushed critical instructions out of the attention window.

4.12.2   Context Engineering in This Chapter

If you've read this far, you'll recognize we're already implementing *some form* of context engineering. Our _prepare_llm_messages() method assembles context from system instructions, conversation history, memory retrievals, and the current task. Our memory implementation explored both RAG-based retrieval (application-managed) and active memory management via MemoryTool (agent-managed)—approaches that not only add context but deliberately *prune it*.

4.12.3   Management Strategies

Modern agent frameworks implement several approaches to context management. These strategies address different aspects of context explosion and can be combined for production deployments.

Compaction (Active Pruning): Periodically summarize and compress context to reclaim tokens. When approaching limits, distill conversation history into concise summaries:

```python
def compact_context(
    messages: List[Message],
    max_tokens: int
) -> List[Message]:
    """Simple context compaction - keep system + recent messages."""
    if estimate_tokens(messages) < max_tokens:
        return messages

    return [
        messages[0],  # System message
        SystemMessage(content="[Previous context summarized]"),
        *messages[-10:]  # Keep last 10 exchanges
    ]
```

The challenge: overly aggressive compaction loses critical details, while conservative compression fails to reclaim sufficient tokens.

The middleware system from Section 4.9 provides a clean implementation point for applying compaction. A `ContextCompactionMiddleware` can intercept the message list during `process_request`, apply the compaction logic above, and return the trimmed messages—keeping context management separate from agent logic. See the complete implementation and comparison in `examples/contextengineering/`.

Active Memory Management: Agents proactively manage what persists. The `MemoryTool` from Section 4.8 lets agents explicitly store, organize, and delete information:

```
# Agent curates its knowledge base
memory(
    command="create",
    path="/patterns/race_condition.md",
    file_text="Thread safety issues..."
)


# Later, prunes outdated information
memory(
    command="delete",
    path="/patterns/deprecated/old_approach.md"
)
```

This treats memory as an external system the agent actively manages, preventing accumulation of irrelevant historical context.

Context Isolation: The agents-as-tools pattern from Section 4.11 provides natural isolation. Specialist agents operate in their own context bubble, returning only condensed summaries:

```
# Specialist uses extensive context internally
analysis_agent = Agent(
    model="o1-preview",
    tools=[analyze_data, create_visualizations],
    memory=detailed_analysis_history  # Large context
)


# Coordinator receives only essential results
coordinator = Agent(
    model="gpt-4o",
    tools=[analysis_agent.as_tool()]  # Isolated context
)


# Specialist might use 30K tokens internally,
# but coordinator only sees 200-token summary
```

This prevents context explosion—the specialist executes 20 tool calls consuming 30K tokens, but the coordinator's context grows by only the 200-token summary.

Tool Result Filtering: Not all tool outputs need full representation. A database query might return 10K rows, but the agent only needs aggregate statistics:

```python
async def _execute_tool_call(
    self,
    tool_call: ToolCallRequest
) -> ToolMessage:
    result = await tool.execute(tool_call.parameters)

    # Filter large outputs before adding to context
    if self._is_large_result(result):
        content = self._summarize_result(result)
    else:
        content = str(result.result)

    return ToolMessage(content=content, ...)
```

4.12.4 Strategies in Practice

To understand how these strategies perform empirically, consider a research agent analyzing AI/ML companies. The agent searches for companies, gathers details, funding information, and traction metrics for three candidates, then synthesizes findings into a report. This multi-step workflow creates substantial context growth.

Implementing three versions—baseline (no management), compaction (middleware-based trimming), and isolation (hierarchical agents)—reveals measurable differences. The baseline agent accumulates all 17 model calls and tool results in its context window. The compaction version uses middleware to trim messages, keeping only system context plus the last 5 conversation turns. The isolation version delegates research to a specialist agent, with the coordinator seeing only final summaries.

The impact on token consumption:

- **Baseline:** 21,633 tokens (no context management)
- **Compaction:** 7,276 tokens (66% reduction)
- **Isolation:** 5,892 tokens (73% reduction)

Compaction provides substantial savings through message trimming, but context still grows with each tool call. Isolation achieves better results by containing specialist work in a separate context bubble—the specialist consumes tokens internally, but the coordinator's context grows minimally. This illustrates a key principle: architectural prevention (isolation) outperforms reactive management (compaction) for context control.

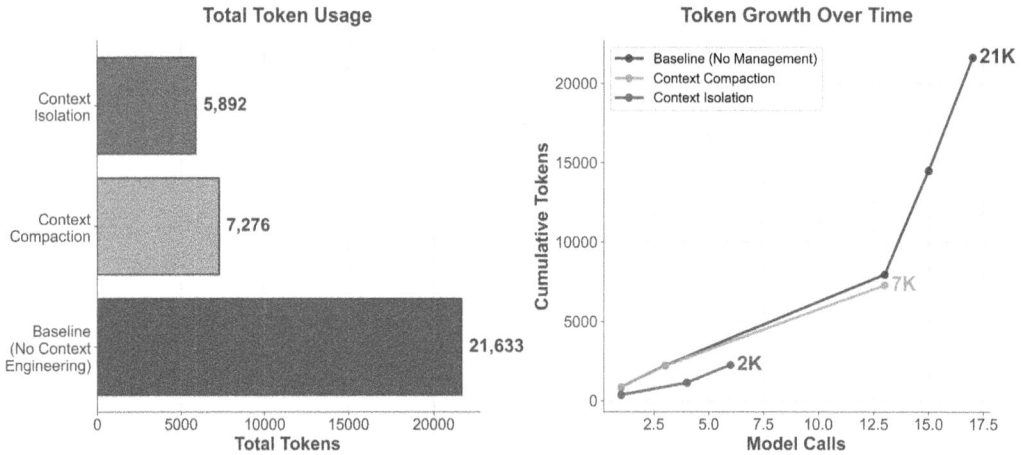

Figure 4.5. *Comparing token usage across three context management strategies for a multi-step research task. Context isolation achieves the greatest reduction by preventing context accumulation through architectural design rather than reactive trimming.*

4.12.5 Combining Strategies

Production systems combine these approaches. A code review agent might use compaction middleware to summarize old conversation turns, implement active memory to store discovered bug patterns, wrap specialist agents as isolated tools for specific analyses, and apply result filtering to condense large code search outputs.

Each strategy addresses different aspects of context explosion: compaction handles conversation history, active memory prevents knowledge accumulation, isolation contains specialist work, and filtering controls tool output volume.

> 💡 💻 Working Code: Context Engineering
>
> Complete implementations of all three strategies with token tracking and visualization are available in `examples/contextengineering/context_strategies.py` . The example demonstrates compaction middleware, isolation via agents-as-tools, and produces the comparison chart above.

Context engineering represents a maturation of agent development—moving from hoping models work with whatever context we provide to actively managing information flow as a first-class design concern. The techniques explored here—from `_prepare_llm_messages()` to `MemoryTool` to agents-as-tools composition—all contribute to effective context management.

4.13   Adding Humans in the Loop

Our agent execution loop operates autonomously: the agent generates responses with tool calls, tools execute automatically, and this cycle continues until completion. While this autonomy enables powerful automation, scenarios arise where human feedback becomes necessary during execution, distinct from middleware configured at design time.

When agents delete files, execute system commands, make expensive API calls, or modify databases, human oversight prevents costly mistakes. These approval points happen dynamically based on the specific tools being called.

4.13.1   Tool Approval Requests

Tools can be marked as requiring approval, causing the agent to stop execution and return an approval request:

Listing 4.32 Marking tools as requiring approval

```python
from picoagents.tools import tool

@tool(approval_mode="always_require")
def delete_file(path: str) -> str:
    """Delete a file from the filesystem."""
    import os
    os.remove(path)
    return f"Deleted {path}"

@tool  # Regular tool - executes immediately
def read_file(path: str) -> str:
    """Read contents of a file."""
    with open(path, 'r') as f:
        return f.read()
```

When an agent encounters `approval_mode="always_require"`, execution stops completely and returns an approval request. The agent doesn't pause holding resources—it exits, storing all state in the serializable `AgentContext`. When approved, a new `agent.run()` call resumes execution. This stateless design (detailed in Chapter 8) enables any server to handle the resume request.

The approval workflow involves three phases: request, user decision, and resumption. Simply add the decorator to any tool and pass it to an agent—when that tool is called, the agent stops and returns an approval request:

If approved, the tool executes and the agent continues. If rejected, the agent receives a failure message and can pursue alternatives or report the task couldn't be completed.

Listing 4.33 Complete tool approval workflow

```
response = await agent.run("Delete /tmp/old_data.txt")

if response.needs_approval:
    for req in response.approval_requests:
        print(f"Tool: {req.tool_name}, Args: {req.parameters}")
        user_input = input("Approve? (yes/no): ")

        approval = req.create_response(
            approved=user_input.lower() == "yes"
        )
        response.context.add_approval_response(approval)

    # Resume with updated context
    response = await agent.run(context=response.context)

print(f"Result: {response.messages[-1].content}")
```

💡 🖵 Working Code: Tool Approval

Complete examples in `tests/test_tool_approval.py` demonstrate approval flows, mixed approval/non-approval tools, and rejection handling.

4.13.2 Stateless Context and Human Input

Beyond tool approval, our stateless context design enables general human input during execution. Agents can request clarification, stop execution, and resume when input arrives—without holding server resources.

We create an agent configured to ask for clarification when needed:

The agent executes and checks if clarification is needed:

Output:

```
Agent: CLARIFICATION NEEDED: Which employment sector - technology,
manufacturing, services, or general overview?

Your response: Focus on technology sector
Final result: AI's impact on technology shows job displacement in
routine coding and creation in AI development, data science...
```

Our context-based design treats human input like tool results: both are messages added to context. Agent instructions guide when to request input, and execution resumes seamlessly.

Listing 4.34 Agent configured to request clarification

```
agent = Agent(
    name="research_assistant",
    instructions="""
    You are a helpful research assistant. When you need more
    information, ask by returning a message starting with
    'CLARIFICATION NEEDED:' followed by your question.
    """,
    model_client=AzureOpenAIChatCompletionClient(
        model="gpt-4.1-mini"
    ),
    tools=[search_web, analyze_data]
)
```

Listing 4.35 Detecting and handling clarification requests

```
response = await agent.run(
    "Research the impact of AI on employment"
)

if response.final_content.startswith("CLARIFICATION NEEDED:"):
    print(f"Agent: {response.final_content}")
    user_input = input("Your response: ")

    # New agent.run() with preserved context
    response = await agent.run(
        task=user_input,
        context=response.context
    )

print(f"Final result: {response.final_content}")
```

This stateless pattern (detailed in Chapter 8) enables:

- **Scalability**: Any server handles resume requests; no state in memory
- **Resource Efficiency**: No held connections while waiting for input
- **Reliability**: Server crashes don't lose conversation state
- **Load Balancing**: Standard load balancers work without session affinity

This mechanism enables multi-step workflows with user feedback, conversational debugging, and progressive solution refinement. The combination of tool approval and instruction-driven input provides comprehensive human-in-the-loop capabilities that scale to production web deployments without complex session management.

> 💡 🖥 Working Code: Complete Implementation
>
> Explore the full picoagents framework:
> - **Agent Examples**: Browse `examples/` by concept (agents, orchestration, work-flows)
> - **Core Implementation**: `src/picoagents/` contains the complete framework source
> - **Getting Started**: See the main README for installation and quick start

> ℹ Comparing Agent Frameworks
>
> Now that you understand the core abstractions required for agent systems (execution loop, tool integration, memory, middleware), compare how production frameworks approach these same challenges. Examine the base agent implementations in Microsoft Agent Framework, Microsoft AutoGen, LangGraph, Pydantic AI, and Google ADK. Look specifically for:
> - **Interface design**: How do they expose model, tools, and memory to developers?
> - **Execution patterns**: Do they use async/await? How do they handle streaming?
> - **Tool integration**: How do they convert functions to tools? What tool types do they support?
> - **State management**: How do they handle conversation context and resumption?
> - **Observability**: What telemetry and debugging capabilities do they provide?
>
> To make this comparison concrete, the companion repository includes equivalent implementations of the patterns from this chapter in Microsoft Agent Framework, Google ADK, and LangGraph: `examples/frameworks/` .

4.14 Summary

Building an agent framework from scratch reveals the foundational design decisions that determine agent effectiveness and developer experience. This chapter established core principles and demonstrated their practical implementation through the picoagents library.

- **Foundational Design Principles**: Async-first architecture enables concurrent operations across multi-agent systems. Event-based streaming provides real-time progress updates for responsive UIs and debugging visibility. Component serialization supports configuration sharing and state restoration. Graceful cancellation gives users control over long-running operations. Abstract base classes enable extensibility without vendor lock-in.

- **Agent Execution Loop**: The pattern of prepare context -> call model -> handle response -> iterate remains consistent from simple chatbots to complex multi-tool systems. The `Agent(model, tools, memory)` followed by `agent.run(task)` interface provides clean abstractions that scale naturally.

- **Structured Output and Tool Calling**: Structured output using Pydantic models transforms agents from text generators into reliable action-takers. This foundation enables trustworthy tool calling, where LLMs generate precisely formatted function calls with properly typed parameters. The `FunctionTool` wrapper automatically converts Python functions to tools, prioritizing developer ergonomics, while maintaining type safety.

- **Memory Systems**: Short-term memory via message history maintains conversation coherence, while long-term memory through RAG-based vector storage enables persistent learning across sessions. Memory systems can be application-managed (developer stores via `memory.add()`, application code injects via retrieval strategies) or agent-managed (agents explicitly call memory tools to organize their own knowledge). The choice comes down to control versus convenience: application-managed memory provides automatic context enhancement, while agent-managed memory gives agents autonomy over what they store and how they organize it.

- **Control and Observability**: Middleware provides runtime control over agent operations. Security filtering blocks malicious inputs before they reach models. Rate limiting prevents abuse and controls costs. PII redaction protects sensitive data in logs. Logging captures execution traces for debugging. OpenTelemetry integration following the Gen-AI semantic conventions delivers industry-standard observability (token tracking, performance monitoring, and distributed tracing) that works with any OTLP-compatible backend (Jaeger, Datadog, Azure Monitor, Arize Phoenix)—without code changes.

- **Agent Composition**: The agents-as-tools pattern enables building sophisticated multi-agent systems from specialized components. Each specialist maintains independent configuration (model, tools, memory, middleware) while becoming a reusable component through the `as_tool()` interface. This composition pattern enables flexible architectures where, for example, weather specialists use fast models for simple queries while analysis specialists use reasoning models for complex computations.

The picoagents framework demonstrates that sophisticated agent capabilities emerge from composable, well-designed abstractions, rather than monolithic implementations. These

---

architectural foundations (particularly structured output for reliability, middleware for control, and OpenTelemetry for observability) provide the base we'll build upon as we explore computer use agents, workflows, and autonomous orchestration in subsequent chapters.

Chapter 5

Building Computer Use Agents

This chapter covers:

- Understanding when interface automation becomes necessary beyond code execution and APIs
- The anatomy of computer use agents: action generation, interface representation, execution
- Strategies for representing interfaces through text, images, and hybrid approaches
- Action sequence generation and translating tasks into interface actions
- Building agents that observe, reason, and act on web and desktop interfaces
- Challenges: reliability, latency, security, and action disambiguation

In Chapter 4, we built agents that can reason using LLMs, act through tools, and maintain memory. We explored how function tools allow agents to execute code, call APIs, and perform structured tasks. However, many real-world scenarios require a different approach - agents that can understand and interact with user interfaces similar to how humans might interact with applications to accomplish tasks. These types of agents are known as **computer use agents** .

In this chapter, we will start by motivating the need for computer use agents, the core ideas that underpin them and walk through an extension of our `Agent` class from Chapter 4 to build a `ComputerUseAgent` class that can drive a web browser to accomplish tasks. We will explore components such as interface representation, action sequence generation, and action execution. Finally, we will discuss challenges and opportunities in the field of computer use agents.

5.1 When Code and APIs Aren't Enough: The Role of Computer Use Agents

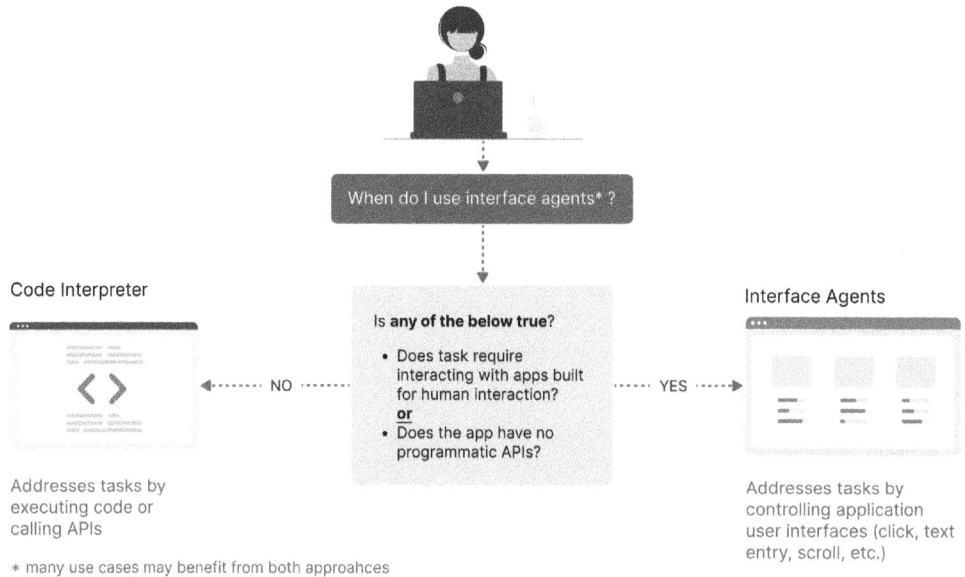

Figure 5.1. *Agents can act via tools by either executing code (or calling apis) or by directly interacting with applications. For tasks that involve complex dynamic behaviors including manipulating interfaces, or tasks (for example, proprietary software) that have no APIs, computer use agents can be valuable. Importantly, many use cases may benefit from a combination of both approaches.*

In Chapter 4, we explored how to build agents with access to generative AI models and tools (code interpreter, functions) that interact to solve tasks. We outlined how the quality of tools that agents have access to can significantly impact the tasks they can solve, and outlined how general purpose tools such as code interpreters or the ability to directly control applications can be used to solve a wide range of tasks.

Importantly, though many tasks can be accomplished through code execution (for example, the LLM generates code to solve the task, or can correctly select an existing function to solve the task), there are task scenarios where a code execution approach falls short (as illustrated in Figure 5.1).

For example, some tasks have complex, dynamic behaviors that are relatively easy for humans to accomplish by manipulating interfaces but challenging to specify programmatically. Examples include opening an image of a map of Northern California in Photoshop and drawing polygon sections only on areas with bodies of water, or manipulating data in spreadsheet applications with complex formatting. Beyond individual tasks, computer use agents enable broader automation patterns: research agents that navigate multiple websites to gather and compare information (aggregating pricing data, monitoring competitors, synthesizing news); back-office agents that triage emails and create corresponding GitHub issues, update CRM records, or reconcile data across disconnected enterprise systems; and software testing agents that execute end-to-end test scenarios, perform visual regression testing, or explore UI paths looking for errors. In other cases, applications may have functionality explicitly designed for human interaction, with no corresponding API available.

These tasks often require a deep understanding of the interface layout, visual cues, and context that may not be easily accessible through traditional programming methods.

> 💡 Tip
>
> While basic web scraping can fetch HTML content for static content, modern web applications often use dynamic JavaScript, AJAX requests, and anti-bot measures (CAPTCHAs, header checks). Computer use agents that simulate human interactions—clicking buttons, filling forms, handling dynamic content—can automate tasks impossible through code alone. This should always comply with website terms of service.

In these scenarios, agents that can take a task description and interact directly with the application's user interface to solve the task can be valuable. Computer use agents automate tasks by interacting directly with user interfaces, handling scenarios that are difficult to accomplish through code execution alone. In fact, if you have used emerging tools like Operator from OpenAI (OpenAI 2025), Computer Use from Anthropic (Anthropic 2024a), TryCUA (TryCUA 2024), or Manus (Manus AI 2025), these follow the general principles of computer use agents.

We will first discuss core concepts of computer use agents, including the anatomy of a computer use agent, the role of LLMs in generating action sequences, and the challenges faced by computer use agents. We will then walk through the implementation of a computer use agent from scratch, focusing on interface representation, action sequence planning, and action execution. Finally, we will explore the challenges and opportunities in the field of computer use agents.

```
pip install picoagents[computer-use]
playwright install
```

At the end of this chapter, we will have built an agent class that can drive a web browser in addressing questions (see Listing 5.1). But before you run this code, you will need to ensure that you have playwright installed and have run an initial command to setup playwright on your machine. Playwright is a library for driving web browsers (we will cover it in detail later in this chapter).

Running this code will open a web browser where you can see the agent navigate to www.techcrunch.com , perform several actions, and return an answer.

Listing 5.1 Complete ComputerUseAgent in action

```
...
from picoagents.agents import PlaywrightWebClient
async def main():

    web_client = PlaywrightWebClient(headless=False)
    computer_agent = ComputerUseAgent(
        interface_client=web_client,
        model_client=OpenAIChatCompletionClient(
            model="gpt-4.1-mini"),
        use_screenshots=True,
    )
    task = "Find the latest AI news on techcrunch.com?"
    try:
        result = await computer_agent.run(task)
        print(result.content)
    finally:
        await web_client.close()

if __name__ == "__main__":
    asyncio.run(main())
```

> 💡 🖥 Working Code: Computer Use Agents
>
> Explore the complete implementation in the examples directory - look for `computer_use.py` for working examples of agent composition, orchestration, and production patterns.

5.2 The Anatomy of a Computer Use Agent

Computer use agents, much like humans, must execute a set of steps to accomplish a task. Let us begin by considering the following task: "Find and print flights from Seattle to San Jose that depart after 5 pm and return before 9 am, departing this Wednesday and returning on Friday morning." An agent addressing this task may take the following steps, where each step can be implemented as an action or sequence of actions on the interface:

1. Navigate to a flight booking website, such as `flights.google.com`. [navigate, browser]
2. Input `Seattle` in the departure field. [type, input]
3. Input `San Jose` in the arrival field. [type, input]
4. Select the desired dates of travel. [click, dropdown]

5. Set the departure flight filter for departures after 5 pm. [click, dropdown]
6. Set the return flight filter for arrivals before 9 am. [click, dropdown]
7. Click the search button to find the flights that meet the criteria. [click, button]
8. Print results [click, print hotkey - Ctrl+P]

If we reflect on how this might be implemented, we can see that the agent must:

1. First understand the interface state (e.g., Is this a web page? What is the current page? What fields are available for input? What buttons can be clicked?).

2. Plan a sequence of actions to achieve the desired outcome or the next desired outcome (e.g., enter a URL into the address bar and hit Enter).

3. Execute these actions on the interface (actual text entry and simulation of the Enter button) to complete the task.

4. Repeat this process as needed to achieve the final goal.

Computer use agents follow an **OBSERVE-THINK-ACT** execution pattern.

Concretely, we need mechanisms to represent the interface state, generate action sequences, and execute these actions on the interface. As illustrated in Figure 5.2, these mechanisms make up the core components of a computer use agent.

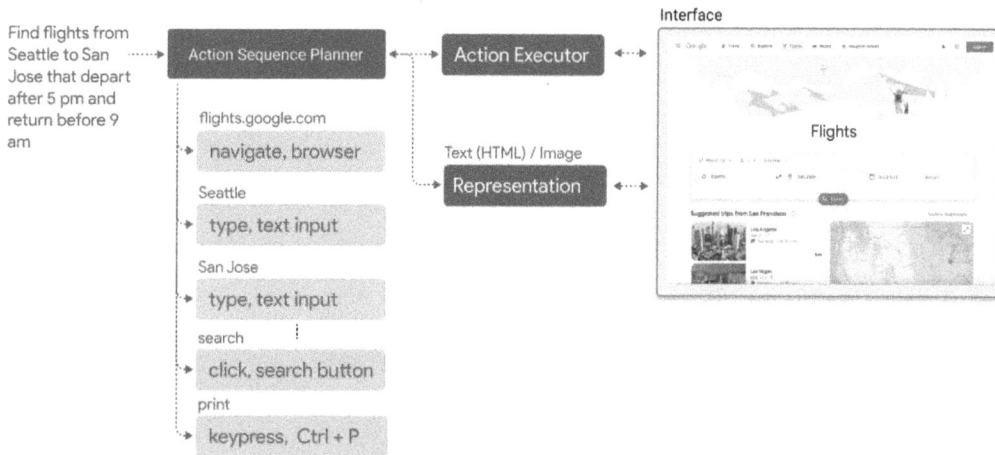

Figure 5.2. Computer Use Agents are composed of 3 components - an action sequence planner, an action executor, and an interface representation component. These components work together to interpret, plan, and execute actions on user interfaces to accomplish tasks.

5.2.1 Action Sequence Generation

Action sequence generation is the process by which generative AI models translate high-level tasks into specific interface actions. Computer use agents rely on this capability to convert natural language instructions into executable operations on user interfaces. In practice, off-the-shelf models like GPT-4+ and its multimodal variants provide this capability. For example, given the task "What is the botanical name of an invasive snail species in Florida?", a language model might produce an action/target pair like: "Action: search for 'botanical name of invasive snail species in Florida', Target: search button on google.com".

However, general-purpose models face two main challenges:

- **Interface Representation Format:** The model's applicability depends on how the interface is represented. Text-based representations (e.g., HTML of a webpage) can use language models, while image-based representations (e.g., screenshots) require multimodal models capable of processing both text and images.
- **Contextual Understanding**: Predicting the next action requires a deep understanding of the current interface state. This includes handling unexpected elements (e.g., pop-ups) and accurately locating UI targets based on instructions.

To address these challenges, recent research in academia and industry has explored fine-tuning base models specifically for GUI interaction tasks. Rather than relying solely on prompted general-purpose models, these specialized models are trained on large-scale datasets that pair interface states with corresponding action sequences. For example, UI-TARS (Yujia Lu et al. 2025) represents a new generation of purpose-built GUI agents that achieve state-of-the-art performance through several key innovations: (1) training on large-scale GUI screenshot datasets for enhanced perception, (2) unified action modeling across platforms, (3) System-2 reasoning for deliberate multi-step decision making, and (4) iterative training with reflective online traces that enable the model to learn from its own interaction history. This specialized approach has demonstrated significant improvements over general-purpose models, achieving 24.6% success on OSWorld (outperforming Claude Computer Use) and 46.6% on AndroidWorld (surpassing GPT-4o's 34.5%), suggesting that domain-specific fine-tuning may represent an important direction for advancing computer use agent capabilities.

Action sequence generation can be approached in two ways:

- **Explicit Planning**: The model generates a complete plan based on the initial interface state, which is then executed. This method benefits from structured reasoning but may struggle with dynamic interfaces where early steps become obsolete. For example, when filling out a contact form on a static website, the entire sequence could be planned upfront.

- **Implicit Planning**: The model generates the plan iteratively, considering the current interface state at each step. This approach adapts better to dynamic interfaces but may be less efficient and require more trial and error.

The choice between explicit and implicit planning depends largely on the nature of the interface and task requirements. Explicit planning works best for static, predictable interfaces like login forms or contact forms where the sequence of actions is well-defined and unlikely to change during execution. These scenarios benefit from the efficiency of upfront planning and fewer model calls. Implicit planning excels when working with dynamic interfaces like flight booking systems, social media feeds, or search results pages where the interface state changes unpredictably during task execution.

5.2.2   Agent Action Space

The **action space** of a computer use agent refers to the set of actions the agent can take on the interface to accomplish a task. This space can be defined by the types of interactions the agent can perform, such as clicking buttons, typing text, selecting options from dropdowns, or submitting forms. Importantly, it is crucial that this action space is defined and made available to the generative model used to generate action sequences that address the task.

5.3   Interface Representation

Interface representation is the method by which the current state of a user interface is presented to AI models for understanding and action generation. Computer use agents require this representation to perceive what elements exist, where they are located, and what actions can be performed. The choice of representation format depends on the type of application being automated (web, mobile, desktop) and the programmatic access available at runtime. Three main approaches exist: text-based (DOM/HTML structure), image-based (screenshots), and hybrid (combining both), each offering different trade-offs between precision, computational cost, and visual understanding.

Web applications support multiple representation formats, as they can be represented as text-based (html) or image-based (screenshots of the interface) – both being accessible on the client side via a browser at runtime. *Mobile applications* may have a text-based representation of the interface, but it is often not accessible via a programmatic api at runtime. For example, we may have access to the UI layout of an android application via the xml layout file (or other UI component system) at train time, but not the actual UI elements at runtime. Recent work like Ferret-UI (You et al. 2024) has explored grounded mobile UI understanding with multimodal LLMs to address these challenges. Similarly, *desktop applications* typically do not have a text-based representation of the interface, and the only way to capture the interface state is via screenshots or visual captures of the interface.

5.3.1   Text-Based Representation

This involves representing the interface as structured text data, such as HTML, XML, JSON, or even raw text. This approach is efficient for processing by language models, provides precise information about interface elements, and works well with interfaces that have a clear text-based structure. Common methods for text-based representation include parsing the DOM of web pages, or leveraging accessibility APIs to access interface elements via platform-specific APIs.

5.3.1.1 DOM Parsing

DOM parsing involves extracting structured data from web pages. The HTML structure of a web page can be fetched either via a http request or directly accessing the DOM of the page using a browser automation tool like Playwright or Selenium. The generative model can then use this structured data to generate the next action: the element to interact with (typically done by generating a css selector that can be executed to locate the element) and the action to perform on it.

> 💡 Tip
>
> **DOM Filtering**
> While the entire DOM can be used as an effective representation of the state of a web page, it can be problematic to provide the entire DOM content to the model for several reasons. First, the DOM can be quite large and contain irrelevant information (e.g., ads, hidden elements), which can make it difficult for the model to focus on the relevant parts of the interface. Second, many models have limited token capacity, so providing the entire DOM may exceed the model's input size limit. To address these issues, it's often beneficial to filter the DOM to include only the most relevant elements for the task at hand. This can involve selecting specific elements based on their attributes, classes, or positions in the DOM, or using heuristics to identify key elements that are likely to be relevant to the task.

Our `ComputerUseAgent` implementation automatically filters DOM content to focus on interactive elements that agents can meaningfully interact with, such as buttons, links, form inputs, and other clickable elements.

5.3.1.2 Accessibility APIs: Accessing interface elements via platform-specific APIs

Accessibility APIs provide a way to programmatically access and interact with user interface elements across different operating systems. These APIs are designed to assist assistive technologies but can also be leveraged by computer use agents to interact with applications.

- For Windows: Microsoft UI Automation is an accessibility framework that enables Windows applications to provide and consume programmatic information about user interfaces (UIs). It provides programmatic access to most UI elements on the desktop. Recent research like UFO (Zhang et al. 2024) has demonstrated effective UI-focused agents for Windows OS interaction using these APIs.
- For macOS: The macOS Accessibility API provides similar functionality for Apple systems. It allows applications to interact with the user interface using AppleScript or Objective-C.

These APIs can provide a text-based representation of the interface, including element hierarchies, properties, and states. While this representation can be easily fed into text-based models, it may struggle to capture complex visual relationships or dynamic content. Additional processing or complementary approaches (like computer vision) may be needed to fully interpret visual cues and layout information.

5.3.2 Image-Based Representation

This entails using screenshots or visual captures to represent an interface to a model. This method can capture visual information not present in text-based structures, and works well with applications that lack accessible text-based representations. However, processing images is computationally expensive (more tokens compared to text), requires models with image processing capabilities or some intermediate image processing step, often implies increased latency, and may lose fine-grained textual information present in the interface.

Often, due to the complexity of processing images (there is no notion of css selectors that can be used to get references to specific elements in a screenshot), image based representation may include a two-step process where the image is first annotated with marks (e.g., bounding boxes around elements with labels before being passed to the model for improved performance) (Cheng et al. 2024). This can help the model focus on specific elements of the interface and reduce the complexity of the image representation. Projects like OmniParser (Yadong Lu et al. 2024) and Set-of-mark prompting (Abuelsaad et al. 2024) are exploring this approach to improve task performance.

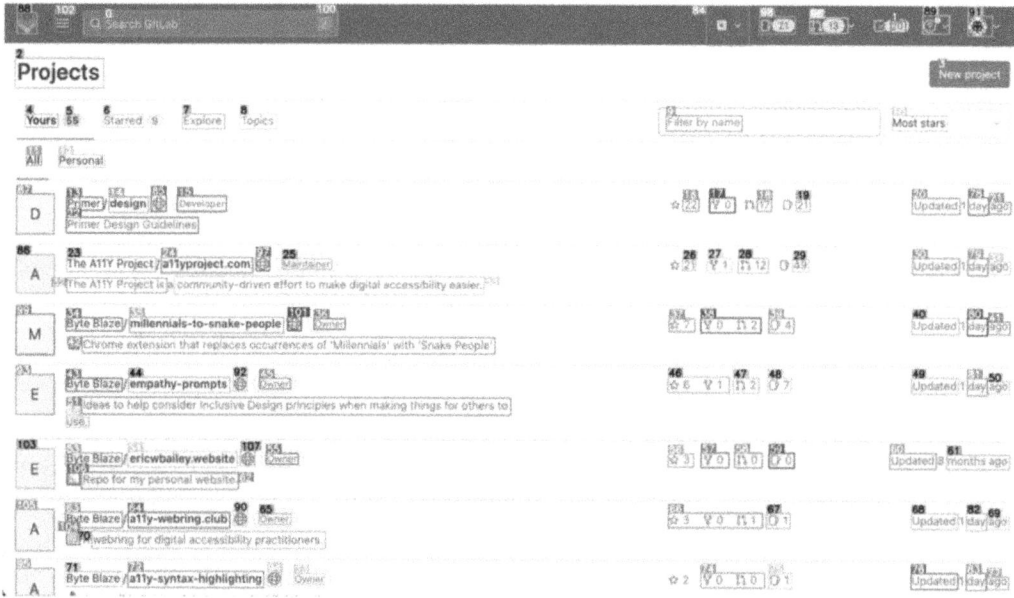

Figure 5.3. OMNIPARSER applies a detection model to parse interactable regions on the screen and a caption model to extract the functional semantics of the detected elements used to annotate the image before being passed to a generative model for action generation.

5.3.3 Hybrid (Text and Image) Representation

Hybrid approaches that combine text and image representations are also emerging, offering the benefits of both formats. For example, an implementation may include logic to extract text-based information from the DOM and overlay it on a screenshot to provide visual context, select between text and image representations based on the task requirements, or use a multimodal model that can process both text and images simultaneously. This approach can provide a more comprehensive representation of the interface, capturing both the structural information and visual context needed to generate accurate action sequences.

5.3.4 Comparing Interface Representation Strategies

The choice between text, image, and hybrid representations involves fundamental trade-offs between precision, cost, and versatility. Text-based representations provide efficient token usage and precise element targeting through CSS selectors, but they miss visual context that humans rely on when interacting with interfaces. Image-based representations capture complete visual information and work with any interface type, but they consume significantly more tokens and require multimodal models. Hybrid approaches combine both formats— our ComputerUseAgent implementation supports this through the `format` parameter in `get_state()`, which can return text-only, visual-only, or hybrid representations depending on task requirements.

Table 5.1 summarizes the key trade-offs between different interface representation approaches for computer use agents.

Table 5.1. Comparison of interface representation strategies for computer use agents

Strategy	Key Advantages	Key Limitations	Best Use Cases
Text-based (DOM/HTML)	Efficient token usage; Precise element targeting via CSS selectors; Fast processing	Misses visual context; Limited for apps without text APIs	Web apps with structured DOM; Predictable interfaces
Image-based (Screenshots)	Captures visual layout; Works with any interface type; Handles dynamic content	High token cost; Requires multimodal models; No direct element selectors	Desktop apps; Visually complex interfaces; Apps without APIs
Hybrid (Text + Image)	Combines precision and context; Optimal decision making; Rich multimodal input	Higher computational cost; Complex implementation	Production web agents; Tasks requiring visual understanding (Recommended)

5.4   Action Executor (Interface Automation Tools)

Once an action sequence has been planned, the computer use agent must execute these actions on the target interface by an *action executor*. This component is responsible for translating high-level actions into low-level interface interactions, managing timing and synchronization, and handling error cases and unexpected interface states. For example, in a web application, the action executor may simulate mouse clicks, keyboard inputs, or form submissions to interact with the interface. In a desktop application, it may use OS-specific automation tools to control UI elements or simulate user inputs. The action executor is typically an automation tool to the interface representation component; that is, the same tooling used to derive the interface state is used to execute actions on the interface.

Common tools for action execution include browser automation frameworks like Playwright and platform automation libraries like PyAutoGUI that provide support for simulating user interactions with desktop applications across the Windows, macOS, and Linux platforms.

5.4.1   An Overview of Playwright

Playwright is an open-source automation library developed by Microsoft that allows developers to control web browsers programmatically. It provides a high-level API to interact with Chromium, Firefox, and WebKit browsers, making it suitable for web scraping, automated testing, and building computer use agents.

Playwright features include:

1. **Cross-browser support**: Playwright can automate Chromium (including Chrome and Edge), Firefox, and WebKit (Safari) with a single API.
2. **Comprehensive selectors and actions**: It offers mechanisms to locate and interact with elements on a web page, including CSS selectors, XPath, and text content. It supports a wide range of actions like clicking, typing, selecting options, capturing screenshots, and more.
3. **Auto-wait functionality**: Playwright automatically waits for elements to be ready before performing actions, reducing the need for explicit waits and making scripts more reliable.
4. **Network interception**: It allows monitoring and modifying network requests and responses, useful for testing different scenarios or mocking API responses.
5. **Mobile emulation**: Playwright can emulate mobile devices, enabling testing of responsive designs and mobile-specific features.

5.5   Implementing a Computer Use Agent

Now that we understand the conceptual foundations, let's build our `ComputerUseAgent` step by step. Rather than creating an entirely new agent architecture from scratch, we'll make a key design decision that leverages the robust foundation we built in Chapter 4.

Our implementation focuses on web browser automation using Playwright, but the architectural patterns we develop apply equally to desktop applications (using PyAutoGUI or platform-specific accessibility APIs) and mobile interfaces (using Appium or similar frameworks). The key abstraction is the interface client—swap the browser client for a desktop automation client, and the same agent architecture handles different interface types.

5.5.1 An "Agent + Tools" Architecture

Rather than building a custom solution from scratch, we'll extend our base `Agent` class to interact with interfaces. The key insight is treating interface actions—clicking, typing, navigating—as tools the agent can use. Our base agent already provides an execution loop, tool calling mechanism, and context management for interface automation.

Our `ComputerUseAgent` inherits from the base `Agent` class and adds three components: an interface client for browser automation, specialized tools that represent interface actions, and enhanced observation capabilities that capture the results of actions. When a task comes in, the agent uses the model to select and execute tools, just as our base agent did with functions in Chapter 4.

This approach gives us key benefits: the base agent's streaming capabilities work with our new tools, error handling and cancellation mechanisms extend to browser automation, and context management maintains conversation history with tool results.

5.5.2 Defining the Action Space Through Tools

Our computer use agent's functionality centers on its tools. Each tool represents a specific action the agent can take on a web interface, from navigating to URLs to clicking buttons and typing text. These tools define our agent's action space for web interaction.

Each tool wraps a specific Playwright operation. The implementation defines each as a `BaseTool` subclass with proper parameter schemas. For example, the `NavigateTool` requires a URL parameter, `ClickTool` requires a CSS selector, and `TypeTool` requires both a selector and text to type. These tools define our agent's action space for web interaction.

5.5.3 The Critical Observation Loop

Here's where computer use agents differ fundamentally from traditional function-calling agents. When our weather tool returns "sunny, 75°F," that's complete information. But when our agent clicks a button, we need to observe what happened. Did a new page load? Did a form appear? Did an error message show up?

To enable this, we can add a specific `observe_page` tool that the LLM can call to capture both the textual structure and visual appearance of the current page state.

Listing 5.2 ComputerUseAgent class architecture

```python
class ComputerUseAgent(Agent):
    def __init__(
        self,
        interface_client,
        model_client,
        use_screenshots=True,
        max_iterations=10,
        **kwargs
    ):
        # Create specialized web automation tools
        playwright_tools = create_playwright_tools(
            interface_client
        )

        # Inherit proven agent patterns
        # - tool calling, context management, streaming
        super().__init__(
            model_client=model_client,
            tools=playwright_tools,  # Our action space
            max_iterations=max_iterations,
            **kwargs
        )

        # Manage browser interface internally
        self.interface_client = interface_client
        self.use_screenshots = use_screenshots
```

Listing 5.3 Creating the Playwright tool suite

```python
def create_playwright_tools(interface_client):
    """Create specialized tools for web interface interaction."""
    return [
        NavigateTool(interface_client),
        ClickTool(interface_client),
        TypeTool(interface_client),
        SelectTool(interface_client),
        PressTool(interface_client),
        HoverTool(interface_client),
        ScrollTool(interface_client),
        ObservePageTool(interface_client),
    ]
```

```
FunctionTool.from_function(
    name="observe_page",
    description="Get current page state and content",
    function=lambda: interface_client.get_page_info()
)
```

When the LLM calls the `observe_page` tool, our implementation intercepts the result and enhances it with visual context. This pattern allows the agent to receive multimodal information combining structured DOM data with screenshots for richer reasoning.

Listing 5.4 Intercepting observe_page results to add visual context

```
async def run_stream(self, task, cancellation_token=None, verbose=False):
    # Use base Agent's proven execution loop
    async for item in super().run_stream(
        task,
        cancellation_token,
        verbose):
        yield item

        # Intercept observe_page tool results to enhance with screenshots
        if isinstance(item, ToolMessage) and item.tool_name == "observe_page":
            if self.use_screenshots:
                # Capture current page screenshot
                state = await self.interface_client.get_state("hybrid")
                if state.screenshot:
                    # Add screenshot to agent's context for LLM reasoning
                    screenshot_msg = MultiModalMessage(
                        content="Page observation",
                        source=self.name,
                        role="user",
                        mime_type="image/png",
                        data=state.screenshot
                    )
                    self.context.add_message(screenshot_msg)
                    yield screenshot_msg  # Also stream for user feedback
```

This enables our agent to process both text and visual information. When `use_screenshots=True`, the agent receives both DOM structure and screenshot data, allowing it to understand layout, visual cues, and interface states that are difficult to express in text alone.

5.5.4   Interface State Representation

Our interface client converts raw web page data into formats the LLM can process effectively. It supports three representation modes, each optimized for different scenarios and computational constraints.

Listing 5.5 Interface representation modes

```python
async def get_state(self, mode="hybrid"):
    """Get current interface state in specified format."""
    if mode == "text":
        return InterfaceState(
            content=await self.page.content(),
            interactive_elements=await self._get_interactive_elements(),
            screenshot=None
        )
    elif mode == "visual":
        return InterfaceState(
            content="",
            interactive_elements=[],
            screenshot=await self.page.screenshot()
        )
    else:  # hybrid mode
        return InterfaceState(
            content=await self.page.content(),
            interactive_elements=await self._get_interactive_elements(),
            screenshot=await self.page.screenshot()
        )
```

The three modes determine what data gets captured. When using `mode="text"`, the agent receives HTML content and a list of interactive elements with their CSS selectors, but no screenshot. This is efficient for token usage but misses visual context. With `mode="visual"`, only the screenshot is captured, requiring a multimodal model to process the image. The `mode="hybrid"` (recommended) combines both approaches, providing structured data for precise element targeting through CSS selectors and visual context for understanding complex layouts and visual relationships. Our DOM filtering focuses on interactive elements that the agent can meaningfully engage with, reducing token usage while maintaining essential information.

5.5.5 The Enhanced Execution Flow

With all components in place, let's see how they work together in a complete task execution. The core execution loop remains unchanged from our base agent, but now operates on a web interface action space.

When you run this code, the agent will open a browser window, navigate to TechCrunch, observe the page structure, extract relevant headlines, and return a summary of the latest AI news.

The execution flow demonstrates our key architectural principles in action. The agent receives

---

Listing 5.6 Complete ComputerUseAgent execution flow

```python
import asyncio
from picoagents.agents import ComputerUseAgent
from picoagents.llm import OpenAIChatCompletionClient
from picoagents.agents._computer_use import PlaywrightWebClient

async def main():
    # Create agent with multimodal reasoning capabilities
    computer_agent = ComputerUseAgent(
        interface_client=PlaywrightWebClient(headless=False),
        model_client=OpenAIChatCompletionClient(model="gpt-4.1-mini"),
        use_screenshots=True,  # Enable visual reasoning
        max_iterations=10
    )

    try:
        # Task requires navigation, observation, and information extraction
        task = "What is the latest AI news on techcrunch.com?"

        async for event in computer_agent.run_stream(task):
            print(event)
    finally:
        await computer_agent.close()  # Clean up browser resources

if __name__ == "__main__":
    asyncio.run(main())
```

a task and uses its inherited tool-calling capabilities to break it down into actionable steps. Each tool call potentially changes the interface state, so observation becomes critical. The enhanced observation loop provides both the structured data needed for precise actions and the visual context needed for intelligent reasoning.

5.5.6 Resource Management and Error Handling

Computer use agents manage real browser processes and network connections, making proper resource cleanup essential. Our implementation includes several layers of protection against resource leaks and graceful degradation when things go wrong.

The error handling strategy is particularly important for screenshot capture. While visual context enhances reasoning, it shouldn't break task execution if unavailable. Our implementation gracefully degrades to text-only mode when screenshot capture fails, ensuring robust operation across different environments and network conditions.

Listing 5.7 Robust resource management

```python
class ComputerUseAgent(Agent):
    async def close(self) -> None:
        """Ensure browser cleanup on agent shutdown."""
        if hasattr(self, 'is_initialized') and self.is_initialized:
            await self.interface_client.close()
            self.is_initialized = False

    async def run_stream(
        self,
        task,
        cancellation_token=None,
        verbose=False):
        # Initialize browser on first use
        if not self.is_initialized:
            await self.interface_client.initialize()
            self.is_initialized = True

        try:
            async for item in super().run_stream(
                task,
                cancellation_token,
                verbose):
                yield item
        except Exception as e:
            # Log error but don't crash on screenshot failures
            if "screenshot" not in str(e).lower():
                raise  # Re-raise non-screenshot errors
```

5.6 Key Challenges in Computer Use Agents

While computer use agents offer powerful capabilities for automating complex tasks, they also face several significant challenges. This section explores the key issues that developers and researchers must address to create effective and reliable computer use agents. Figure 5.4 illustrates the interconnected nature of these challenges and their impact on development and deployment.

5.6.1 Interface Representation and Grounding

One of the primary challenges in developing computer use agents is finding the optimal way to represent and ground the interface state for AI models. This is crucial for enabling the agent to understand and interact with the interface effectively.

Text-based representations like the Document Object Model (DOM) for web interfaces often

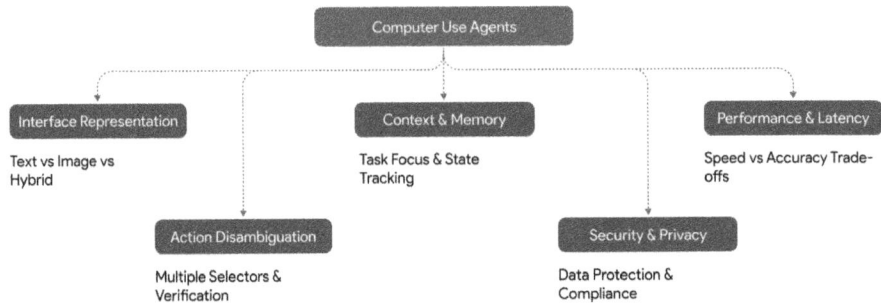

Figure 5.4. *Key challenges in implementing computer use agents. The diagram illustrates five major hurdles: interface representation (balancing detail and efficiency), context and memory management (maintaining task relevance), action disambiguation (resolving ambiguous targets), security and privacy (protecting sensitive data), and latency (balancing speed and accuracy). These interconnected challenges highlight the complexity of developing effective and reliable computer use agents for real-world applications.*

contain significant amounts of noisy or irrelevant data, which can confuse AI models. These representations also struggle to capture visual relationships and spatial layouts effectively. Additionally, dynamic content generated by JavaScript and frequent interface updates pose challenges for maintaining accurate representations.

Image-based representations, while capable of capturing visual information more comprehensively, come with their own set of challenges. Processing images requires significantly more computational resources than text, potentially increasing latency and reducing responsiveness. Important textual details, such as exact button labels or input field placeholders, may be harder to extract from images. Furthermore, variations in screen sizes, resolutions, and aspect ratios can complicate the interpretation of image-based representations across different devices.

To address these challenges, researchers are exploring innovative approaches such as hybrid representations that combine text and image-based methods to leverage the strengths of both. Advanced grounding techniques like Set-of-mark prompting (Abuelsaad et al. 2024) enrich representations with anchor markers, enhancing task performance and reducing ambiguity. Additionally, specialized models like Ferret-UI (You et al. 2024) and UI-TARS (Yujia Lu et al. 2025) are being designed specifically for interface understanding, with UI-TARS demonstrating significant improvements through training on large-scale GUI screenshot datasets and implementing System-2 reasoning for deliberate multi-step decision making.

5.6.2 Context and Memory

Computer use agents must maintain a comprehensive understanding of the user's context to provide relevant and useful assistance. This involves learning and remembering individual user preferences across multiple interactions and tasks, maintaining a record of previous actions and their outcomes, understanding the current state of the application, and managing context across various interfaces and environments for tasks that span multiple applications.

Addressing these challenges requires sophisticated memory management and context modeling techniques. Potential solutions include implementing episodic memory systems that can store and retrieve relevant past experiences, developing methods to encode rich contextual information into compact representations, and creating cross-application knowledge graphs that capture relationships between various applications and tasks.

5.6.3   Action Disambiguation and Reliability

In many scenarios, computer use agents may encounter multiple potentially valid action targets for completing a task. Developing robust disambiguation logic is crucial for ensuring that the agent makes appropriate choices. This involves determining which action or element is most relevant to the current task when multiple possibilities exist, developing strategies for handling ambiguous situations, and balancing autonomy with the need for user input.

Emerging patterns for addressing disambiguation challenges include training specialized models to rank and prioritize options based on contextual factors, developing intuitive ways for users to provide feedback and guide the agent's decision-making process, and implementing mechanisms for the agent to assess its own confidence in decisions and seek help when uncertainty is high. Recent work like UI-TARS demonstrates how iterative training with reflective online traces can help models learn from their own interaction history, improving decision-making over time.

5.6.4   Security and Privacy

As computer use agents interact with various applications and handle sensitive user data, ensuring security and privacy becomes paramount. Key challenges in this area include safeguarding user information accessed during automation tasks, securely managing access credentials and permissions across multiple applications, ensuring compliance with relevant data protection regulations and application-specific terms of service, and providing users with clear information about what data the agent can access and how it is used.

Potential approaches to address these concerns include implementing robust encryption for all sensitive data handled by the agent, developing systems that allow users to precisely control what information and capabilities the agent can access, and creating detailed logs of agent actions to support transparency and facilitate compliance checks.

5.6.5   User Experience: Latency, Responsiveness, and Cost

Computer use agents face significant user experience challenges from latency and cost management. Complex tasks often require multiple interactions with AI models, introducing cumulative delays and compounding costs. Maintaining fluid user experience requires handling latency in agent actions, managing cases where interface elements change or time out during processing, and intelligently deciding when to request user feedback (e.g., ask for login credentials or disambiguate between similar options) to avoid unnecessary delays or errors.

Cost management deserves special attention in production deployments. Unlike traditional applications where costs scale linearly, agent token consumption can spiral as systems make multiple LLM calls per task, capture and process screenshots, and maintain conversation context across sessions. Production teams should treat cost as infrastructure, implementing token budgets per session, monitoring cost-per-task metrics, and using smaller models for routine operations while reserving larger models for complex reasoning steps. Screenshot capture represents significant token cost—consider capturing only when necessary or using text-only representations for straightforward workflows.

To mitigate latency challenges, researchers and developers are exploring several strategies. These include creating smaller, faster models specifically tuned for interface understanding and action prediction, such as multimodal models optimized for GUI interaction. Other approaches involve implementing techniques to anticipate and pre-fetch likely next actions, designing interfaces that provide immediate feedback on agent actions while processing continues in the background, and structuring agent workflows to allow for non-blocking operations and parallel processing where possible.

5.6.6   Reliability and Robustness

Ensuring consistent and accurate performance across diverse interfaces and scenarios remains a significant challenge for computer use agents. This includes adapting to interface changes, updates, and unexpected elements like pop-ups or overlays, implementing robust error handling and recovery mechanisms to gracefully manage unexpected states or failures, and maintaining uniform behavior across various browsers, operating systems, and device types.

Addressing these challenges requires a multi-faceted approach, including developing agents that can learn and adapt to interface changes over time, creating flexible intermediate representations that can handle variations in interface structure across platforms, and designing automated testing systems that can evaluate agent performance across a wide range of scenarios and edge cases. By addressing these challenges, researchers and developers can work toward creating more robust, reliable, and effective computer use agents that can seamlessly interact with a wide range of applications and platforms

5.7   Production Considerations

Deploying computer use agents in production environments requires careful attention to resource management, security, and performance. This section outlines practical considerations for building reliable, production-ready systems.

5.7.1   Resource Management

Computer use agents spawn browser processes that consume significant memory and must be explicitly closed—Python garbage collection won't terminate them automatically. Production deployments should implement timeout mechanisms to prevent runaway agents, monitor memory usage per session, and use containerization to enforce resource limits. For concurrent sessions, track active agents and enforce cleanup policies to prevent resource exhaustion.

For production deployments, enable headless mode (in `PlaywrightWebClient` set `headless=True`) to reduce resource overhead, capture screenshots only when needed for decision-making, and use explicit planning for predictable workflows to minimize LLM calls.

5.7.2   Security and Compliance

Security considerations for computer use agents extend beyond typical application security. These agents access and interact with potentially sensitive interfaces, handle user credentials, and capture screenshots that may contain private information. Recent research has revealed that commercial LLM agents are already vulnerable to simple yet dangerous attacks (Li et al. 2024), with prompt injection attacks representing a particularly critical threat vector. Prompt injection attacks (Wang et al. 2025) occur when malicious instructions embedded in webpage HTML or rendered content manipulate agent behavior.

Key security threats include credential exposure where agents handling authentication tokens or cookies can be exploited to impersonate users across connected systems, supply chain risks from third-party browser extensions or plugins that can hijack active sessions or exfiltrate data, and detection challenges where traditional security tools like EDR and DLP systems struggle to distinguish between legitimate human actions and potentially malicious agent behaviors.

Key security practices include respecting robots.txt and website terms of service. Rate limiting prevents overwhelming target servers and reduces the risk of IP bans or CAPTCHA challenges. Sandboxing agents in isolated environments limits the blast radius if an agent is compromised or misbehaves. Data privacy measures must ensure that screenshots and DOM captures are filtered to avoid capturing and logging sensitive information like passwords, credit cards, or personal data.

Implementation approaches include filtering captured data before storage to remove sensitive fields, implementing audit logs that track what interfaces were accessed without storing the full content, using short-lived credentials with minimal necessary permissions, and regularly reviewing automation behavior for compliance with target website policies. Consider requiring user confirmation for sensitive actions, isolating browser instances per session, and deploying browser detection and response (BDR) tools designed specifically for monitoring agent security threats.

5.8 Summary

- **When Code and APIs Aren't Enough**: When code execution and APIs aren't sufficient, computer use agents enable direct interface interaction for tasks involving dynamic behaviors, proprietary software without APIs, and complex visual interfaces requiring human-like navigation patterns.

- **Three Core Components Enable Agent Functionality**: Computer use agents consist of action sequence generation (LLMs translating tasks into interface actions), interface representation (text, visual, or hybrid approaches), and action execution (browser automation tools like Playwright for implementing planned actions).

- **Interface Representation Strategies Offer Different Trade-offs**: Text-based approaches (DOM parsing) provide precise element targeting but miss visual context, image-based methods capture visual information but require expensive multimodal processing, while hybrid approaches combine both for optimal performance.

- **Implicit vs. Explicit Planning for Action Sequences**: Agents generate action sequences through implicit planning (iterative step-by-step generation that adapts to dynamic interfaces) or explicit planning (complete upfront sequences for static, predictable workflows).

- **Production Deployment Requires Robust Error Handling**: Successful computer use agents need proper resource management, security considerations (respecting robots.txt and terms of service), performance optimization (headless mode, screenshot management), and reliability patterns for handling interface changes and network failures.

Chapter 6

Building Multi-Agent Workflows

This chapter covers:

- Understanding workflows as deterministic computational graphs with explicit control flow
- Building type-safe workflow steps that transform data and maintain shared state
- Creating conditional edges that enable branching and parallel execution patterns
- Implementing a workflow runner that provides streaming observability and robust error handling
- Putting it all together: complete workflow systems with serialization, checkpointing, and production capabilities

In Chapter 2, we explored the taxonomy of multi-agent orchestration patterns—from explicit workflow control to emergent autonomous behavior. We learned that workflow patterns sit on the **explicit control** side of the orchestration spectrum, providing developer-defined execution paths with predictable behavior. In fact, you have used and loved drag and drop AI tools like n8n (n8n GmbH 2024), Flowise (FlowiseAI 2024) where components are assembled as steps visually to solve problems (e.g., generate ideas for a youtube video etc), *this is how these systems are built underneath*!

In this chapter, we implement workflow patterns from the ground up, demonstrating how to build deterministic, composable systems that leverage computational graphs. Unlike the autonomous patterns we'll explore in Chapter 7, workflows provide complete control over execution flow while enabling sophisticated capabilities through type safety, streaming observability, and robust error handling.

Before diving into implementation details, let's see the complete developer experience we're building toward. This example creates a simple mathematical pipeline that doubles a number, then adds ten to the result (e.g., 3 -> 6 -> 16). Notice that this workflow contains no AI models or agents—it's pure deterministic computation. This illustrates an important point:

144

workflows provide low-level control over execution flow, and steps can be any computation (mathematical operations, data transformations, API calls, or AI agents). Later in this chapter, we'll show how to wrap agents as workflow steps, but the fundamental patterns remain the same. We'll build this example in two parts: first defining the building blocks, then composing and executing them.

Listing 6.1 Workflow building blocks: types, functions, and steps

```python
from picoagents.orchestration.workflow import (Workflow, WorkflowRunner,
    FunctionStep)
from picoagents.orchestration.workflow.core import (WorkflowMetadata,
    StepMetadata, Context)
from pydantic import BaseModel

# Define data types for type-safe contracts
class NumberInput(BaseModel):
    value: int

class NumberOutput(BaseModel):
    result: int

# Define step functions with typed signatures
async def double_number(input_data: NumberInput,
                        context: Context) -> NumberOutput:
    return NumberOutput(result=input_data.value * 2)

async def add_ten(input_data: NumberOutput,
                context: Context) -> NumberOutput:
    return NumberOutput(result=input_data.result + 10)

# Wrap functions into workflow steps
double_step = FunctionStep(
    step_id="double", metadata=StepMetadata(name="Double Number"),
    input_type=NumberInput, output_type=NumberOutput,
    func=double_number
)

add_ten_step = FunctionStep(
    step_id="add_ten", metadata=StepMetadata(name="Add Ten"),
    input_type=NumberOutput, output_type=NumberOutput,
    func=add_ten
)
```

These three components form the foundation of type-safe workflows. The Pydantic models (NumberInput , NumberOutput) define contracts that ensure data flowing between steps matches expectations—the workflow validates that `double_step`'s `NumberOutput` matches `add_ten_step`'s input type before execution begins. The async functions contain the actual transformation logic, each accepting typed input and shared context while returning typed output. `FunctionStep` wraps these functions into executable nodes that the workflow can manage, handling both sync and async functions while providing flexible result handling (dictionaries, Pydantic models, or simple values all work).

Listing 6.2 Workflow composition and execution with streaming

---

```
# Compose steps into a workflow using fluent chaining
workflow = (Workflow(metadata=WorkflowMetadata(
                name="Simple Math Pipeline"))
            .chain(double_step, add_ten_step))

# Execute with streaming observability
runner = WorkflowRunner()
async for event in runner.run_stream(workflow, {"value": 3}):
    print(event)  # Real-time events show execution progress
```

When you run this example, you'll see the workflow execute each step in sequence, with complete type safety and real-time observability. The workflow automatically handles:

1. **Type-safe data flow**: Each step's output is validated before becoming the next step's input

2. **Streaming observability**: Real-time events show exactly what's happening during execution
3. **Error handling**: Built-in retry logic, timeout handling, and graceful failure recovery

6.1 Workflows as Computational Graphs

A **Workflow** is fundamentally a **computational graph** —a container that holds **Steps** (nodes representing units of computation) and **Edges** (connections that define transitions and conditions between steps). This graph-based approach enables several powerful capabilities:

- **Validation**: Check execution paths before runtime
- **Visualization**: Clear representation of system behavior

- **Deterministic execution**: Predictable outcomes
- **Parallel processing**: Concurrent execution where dependencies allow
- **Conditional logic**: Dynamic routing based on step outputs or shared state

Let's implement each component step by step, building toward the complete system.

> 💡 Agent Node
>
> While we have a generic expression of steps/nodes , the core idea is that a step may
> hold any computation including itself being an agent as defined in our implementation
> in Chapter 4. In fact, later in this chapter, we will show to build an `AgentStep` class
> that encapsulates this behavior.

6.2   Adding Steps: Units of Computation

Steps are the fundamental building blocks of workflows—each step represents a single unit
of computation that transforms input data into output data. Our step system provides type
safety, validation, and robust error handling while maintaining a clean developer experience.

6.2.1   The BaseStep Foundation

Every step in our system extends from a common base that handles infrastructure concerns
while allowing specific implementations to focus on their unique logic. First, we define the
essential interface that every step must implement:

Listing 6.3 Core step interface with type safety

```
# ... imports
class BaseStep(ABC, Generic[InputType, OutputType]):
    def __init__(self, step_id: str, metadata: StepMetadata,
                 input_type: Type[InputType], output_type: Type[OutputType]):
        self.step_id = step_id
        self.metadata = metadata
        self.input_type = input_type
        self.output_type = output_type

    @abstractmethod
    async def execute(
        self,
        input_data: InputType,
        context: Context
    ) -> OutputType:
        pass
```

This interface defines the contract—every step transforms typed input to typed output with
access to workflow context. The generics ensure compile-time type checking and runtime
validation.

Steps also need to track their execution state for observability and error handling:

Listing 6.4 Step status tracking and lifecycle management

```
class StepStatus(str, Enum):
    PENDING = "pending"
    RUNNING = "running"
    COMPLETED = "completed"
    FAILED = "failed"

class StepMetadata(BaseModel):
    name: str
    max_retries: int = 0
    timeout_seconds: Optional[int] = None

# Inside BaseStep.__init__:
self._status = StepStatus.PENDING
self._start_time: Optional[datetime] = None
self._error: Optional[str] = None
```

Status tracking enables real-time monitoring and provides detailed execution history for debugging and optimization.

The `run()` method handles the complete execution lifecycle with validation, retries, and error handling:

This approach handles input validation, context preparation, timeout management, and retry logic while maintaining detailed execution tracking.

6.2.2   FunctionStep: Converting Functions to Steps

Now let's implement a concrete step that demonstrates the BaseStep pattern. `FunctionStep` wraps Python functions, making it easy to convert existing code into workflow steps:

The key insight is the flexible result handling - functions can return dictionaries, Pydantic models, or simple values, and FunctionStep adapts accordingly. This makes it easy to wrap existing functions without modification.

Using FunctionStep is straightforward:

6.3   Adding Edges: Connecting Steps with Transitions

Edges define how data flows between steps and under what conditions transitions occur. Our edge system supports everything from simple sequential processing to complex conditional logic and parallel execution patterns.

Listing 6.5 Step execution with error handling

```python
async def run(
    self,
    input_data: Dict[str, Any],
    context: Dict[str, Any]
) -> Dict[str, Any]:
    self._status = StepStatus.RUNNING
    self._start_time = datetime.now()

    for retry_count in range(self.metadata.max_retries + 1):
        try:
            # Validate input and create typed context
            validated_input = self.input_type(**input_data)
            typed_context = \
                Context.from_state_ref(context.get('workflow_state', {}))

            # Execute with optional timeout
            output = await self.execute(validated_input, typed_context)

            self._status = StepStatus.COMPLETED
            return output.model_dump()
        except Exception as e:
            # Handle retries and failures...
```

6.3.1 Edge Fundamentals

An edge represents a connection between two steps—a `from_step` that produces output and a `to_step` that consumes it as input. The basic edge structure handles the common case where all connections are active:

This simple approach handles the most common workflow pattern: **sequential processing** where each step's output becomes the next step's input.

6.3.2 Conditional Edges

The real power of edges emerges when we add conditional logic. **Conditional edges** enable workflows to make dynamic routing decisions based on step outputs or shared workflow state:

These conditions examine the `is_valid` field from the validation step's output and route accordingly. But how does the system actually evaluate these conditions at runtime?

When a step finishes, the workflow runner looks at its output, finds the relevant fields, and decides which next steps to activate. This means handling cases where fields might be missing,

Listing 6.6 FunctionStep class implementation

```python
class FunctionStep(BaseStep[InputType, OutputType]):
    def __init__(self, step_id: str, metadata: StepMetadata,
                 input_type: Type[InputType], output_type: Type[OutputType],
                 func: Callable[..., Any]):
        super().__init__(step_id, metadata, input_type, output_type)
        self.func = func

    async def execute(
        self,
        input_data: InputType,
        context: Context
    ) -> OutputType:
        # Handle both sync and async functions
        if asyncio.iscoroutinefunction(self.func):
            result = await self.func(input_data, context)
        else:
            result = self.func(input_data, context)

        # Flexible result handling
        if isinstance(result, dict):
            return self.output_type(**result)
        elif hasattr(result, 'dict'):
            return result  # Already a Pydantic model
        else:
            return self.output_type(result=result)  # Wrap simple values
```

Listing 6.7 FunctionStep usage example

```python
def multiply_by_two(input_data: NumberInput, context: Context) -> NumberOutput:
    return NumberOutput(result=input_data.value * 2)

step = FunctionStep(
    step_id="multiply", metadata=StepMetadata(name="Multiply by Two"),
    input_type=NumberInput, output_type=NumberOutput, func=multiply_by_two
)
```

Listing 6.8 Basic edge structure and simple workflow connections

```python
from typing import Any, Dict, Optional
from pydantic import BaseModel, Field
import uuid

class EdgeCondition(BaseModel):
    """Defines conditions for workflow edges."""
    type: str = Field(default="always", description="Type of condition")
    expression: Optional[str] = None
    field: Optional[str] = None
    value: Optional[Any] = None
    operator: Optional[str] = None

class Edge(BaseModel):
    """Represents a connection between workflow steps."""
    id: str = Field(default_factory=lambda: str(uuid.uuid4()))
    from_step: str = Field(description="Source step ID")
    to_step: str = Field(description="Target step ID")
    condition: EdgeCondition = Field(default_factory=lambda: EdgeCondition())

# Simple sequential workflow: A -> B -> C
workflow.add_edge("step_a", "step_b")  # Always condition (default)
workflow.add_edge("step_b", "step_c")  # Always condition (default)
```

Listing 6.9 Conditional routing based on step outputs

```python
# Output-based condition: Route based on step results
workflow.add_edge("validate_data", "process_data", condition={
    "type": "output_based", "field": "is_valid",
    "operator": "==", "value": True
})

workflow.add_edge("validate_data", "handle_error", condition={
    "type": "output_based", "field": "is_valid",
    "operator": "==", "value": False
})
```

values might be different types than expected, or comparisons might fail. Here's how the system evaluates these conditions:

Listing 6.10 Edge condition evaluation logic

```
def _evaluate_edge_condition(
    self, edge: Edge, execution: WorkflowExecution) -> bool:
    condition = edge.condition

    if condition.type == "always":
        return True

    if condition.type == "output_based":
        # Get output from the source step
        from_step_exec = execution.step_executions.get(edge.from_step)
        if not from_step_exec or not from_step_exec.output_data:
            return False

        # Extract field value and compare
        field_value = from_step_exec.output_data.get(condition.field)
        return self._compare_values(field_value, \
            condition.operator, condition.value)

    if condition.type == "state_based":
        # Check workflow shared state
        field_value = execution.state.get(condition.field)
        return self._compare_values(field_value, \
            condition.operator, condition.value)

    return True  # Default for unsupported conditions

def _compare_values(self, left: Any, operator: str, right: Any) -> bool:
    if operator == "==": return left == right
    elif operator == "!=": return left != right
    elif operator == ">=": return left >= right
    # ... other operators
```

With this condition evaluation logic, workflows can make real decisions at runtime. For example, a data processing workflow might check a quality score and route high-quality data to fast processing while sending questionable data through additional validation steps. Or an approval workflow could check user roles and route requests through different approval chains automatically.

The operator system supports common comparisons (== , >= , in) so you can build these decisions without writing custom evaluation code. The step producing this conditional output might look like:

Listing 6.11 Step that produces conditional outputs

```
class ValidationOutput(BaseModel):
    is_valid: bool
    error_message: Optional[str] = None

async def validate_input(input_data: DataInput) -> ValidationOutput:
    if input_data.value < 0:
        return ValidationOutput(is_valid=False,
        error_message="Negative values not allowed")
    return ValidationOutput(is_valid=True)
```

Output-based conditions examine the results of the previous step to determine routing. This enables patterns like validation, error handling, and quality checks.

State-based conditions examine the workflow's shared state to determine routing. This enables configuration-driven behavior and dynamic workflow adaptation.

6.3.3 Parallel Execution Patterns

Edges enable **parallel execution** through two distinct patterns that the readiness algorithm we built earlier can distinguish automatically:

Listing 6.12 Fan-out and fan-in patterns with different execution logic

```
# Fan-out pattern: One step feeds multiple parallel steps
workflow.add_edge("prepare_data", "process_batch_a")
workflow.add_edge("prepare_data", "process_batch_b")
workflow.add_edge("prepare_data", "process_batch_c")

# Fan-in pattern: Multiple steps feed into one step
workflow.add_edge("process_batch_a", "combine_results")
workflow.add_edge("process_batch_b", "combine_results")
workflow.add_edge("process_batch_c", "combine_results")
```

Here's how the readiness algorithm distinguishes these patterns:

This distinction enables two powerful execution patterns:

Fan-out -> Fan-in (Data Processing): `prepare_data` completes -> all three `process_batch_*` steps start concurrently -> `combine_results` waits for ALL three to finish. Perfect for parallel data processing where you need all results.

Conditional Routing (Decision Making): A validation step produces success/failure -> either the success path OR failure path executes (not both). Perfect for error handling and

Listing 6.13 How the readiness algorithm detects parallel patterns

```
# From our get_ready_steps() algorithm
incoming_edges = [edge for edge in self.edges if edge.to_step == step_id]
is_fan_in = all(edge.condition.type == "always" for edge in incoming_edges)

if is_fan_in:
    # Fan-in (AND logic): ALL dependencies must complete
    if all(execution.step_executions.get(dep_id, {}).status == "completed"
            for dep_id in dependencies):
        ready_steps.append(step_id)
else:
    # Conditional (OR logic): ANY valid path enables execution
    # Used for error handling, quality gates, approval workflows
```

approval workflows.

The same readiness algorithm handles both patterns without special cases - the edge conditions determine the behavior.

6.4 Building the Workflow Class

The Workflow class serves as the container that holds steps, edges, and metadata while providing validation, dependency management, and execution planning capabilities. It transforms individual components into a cohesive computational graph.

6.4.1 Workflow Construction

At its core, a workflow is a container for steps and their connections:

Listing 6.14 Basic workflow structure and initialization

```
class WorkflowMetadata(BaseModel):
    name: str
    version: str = "1.0.0"

class Workflow:
    def __init__(self, metadata: WorkflowMetadata):
        self.metadata = metadata
        self.steps: Dict[str, BaseStep] = {}
        self.edges: List[Edge] = []
        self.start_step_id: Optional[str] = None
```

This container holds all workflow components. Note that steps and edges can be added using standard methods like `workflow.add_step(step)` and `workflow.add_edge(from_id, to_id)`. However, the fluent API provides a more intuitive developer experience:

The fluent API enables method chaining for readable workflow construction:

Listing 6.15 Fluent API for workflow construction

```
def add_step(self, step: BaseStep) -> 'Workflow':
    self.steps[step.step_id] = step
    return self  # Enable chaining

def add_edge(self, from_step: str, to_step: str) -> 'Workflow':
    edge = Edge(from_step=from_step, to_step=to_step)
    self.edges.append(edge)
    return self  # Enable chaining

# Method chaining in action
workflow = (Workflow(metadata=WorkflowMetadata(name="Pipeline"))
    .add_step(extract_step)
    .add_step(transform_step)
    .add_edge("extract", "transform")
    .set_start_step("extract"))
```

The fluent pattern improves readability by expressing workflow construction as a single, flowing statement. Each method returns `self`, enabling the chain to continue.

For simple sequential workflows, the `chain()` method provides an even more concise approach:

Listing 6.16 Sequential chaining for linear workflows

```
def chain(self, *steps: BaseStep) -> 'Workflow':
    for i in range(len(steps) - 1):
        self.add_edge(steps[i], steps[i + 1])
    self.set_start_step(steps[0])
    return self

# Simple sequential pipeline
workflow = (Workflow(metadata=WorkflowMetadata(name="Math Pipeline"))
            .chain(double_step, add_ten_step, square_step))
```

This creates a linear pipeline where each step feeds into the next, automatically setting the first step as the start and connecting all steps sequentially.

6.4.2 Validation and Dependencies

With construction handled, workflows need validation and dependency tracking. The workflow tracks dependencies using the edge connections we defined earlier:

```
def get_step_dependencies(self, step_id: str) -> List[str]:
    """Get all steps that must complete before this step."""
    return [edge.from_step for edge in self.edges if edge.to_step == step_id]
```

This simple method powers the readiness algorithm we explored - it determines whether a step uses fan-in logic (wait for ALL dependencies) or conditional logic (wait for ANY valid path).

Determining when a step is ready to run gets tricky when it has multiple inputs. Should it wait for ALL inputs to complete (like aggregating results from parallel processes) or can it proceed when ANY valid input arrives (like handling either success or error cases)? The algorithm needs to figure this out automatically.

Here's how the algorithm actually determines step readiness:

This readiness logic determines whether your workflow waits for everything or moves forward opportunistically. A reporting workflow needs all data collection steps to finish before generating the report (fan-in), but an error handling workflow should proceed as soon as either the success or failure path completes (conditional). Getting this distinction right makes workflows both predictable and efficient.

The `validate_workflow()` method ensures workflow integrity:

- **Start/end steps exist**: Verifies entry and exit points
- **Edge validity**: All edges reference existing steps
- **Cycle detection**: Uses depth-first search to prevent infinite loops
- **Type compatibility**: Validates that connected steps have matching input/output types

Workflows maintain shared state through the `Context` class, enabling steps to coordinate:

Shared state through Context lets steps coordinate with each other. An early configuration step might set the batch size, and all subsequent processing steps use that setting. Or steps might accumulate metrics as they work, with a final reporting step collecting everything. Without this shared memory, each step would be isolated and workflows would need external databases just to pass information around.

The Context uses references to the workflow's state dictionary, so when one step updates a value, all other steps see the change immediately.

Listing 6.17 Step readiness algorithm with fan-in vs conditional logic

```python
def get_ready_steps(self, execution: WorkflowExecution) -> List[str]:
    ready_steps = []

    for step_id in self.steps:
        # Skip if already running, completed, or failed
        step_exec = execution.step_executions.get(step_id)
        if step_exec and step_exec.status.value in
        ["running", "completed", "failed"]:
            continue

        dependencies = self.get_step_dependencies(step_id)
        if not dependencies:
            # Start step - ready if it's the designated start
            if step_id == self.start_step_id:
                ready_steps.append(step_id)
        else:
            # Determine pattern: fan-in (AND) vs conditional (OR)
            incoming_edges = [edge for edge in self.edges \
                if edge.to_step == step_id]
            is_fan_in = all(edge.condition.type == "always" \
                for edge in incoming_edges)

            if is_fan_in:
                # Fan-in: ALL dependencies must complete
                if all(execution.step_executions
                .get(dep_id, {}).status == "completed"
                        for dep_id in dependencies):
                    ready_steps.append(step_id)
            else:
                # Conditional: ANY valid path enables execution
                for edge in incoming_edges:
                    dep_exec = execution.step_executions.get(edge.from_step)
                    if (dep_exec and dep_exec.status == "completed" and
                        self._evaluate_edge_condition(edge, execution)):
                        ready_steps.append(step_id)
                        break

    return ready_steps
```

Listing 6.18 Context class for shared workflow state

```python
class Context:
    def __init__(self, state_ref: Dict[str, Any]):
        self._state = state_ref  # Reference to workflow state

    def get(self, key: str, default: Any = None) -> Any:
        return self._state.get(key, default)

    def set(self, key: str, value: Any) -> None:
        self._state[key] = value

    def update(self, updates: Dict[str, Any]) -> None:
        self._state.update(updates)

    @classmethod
    def from_state_ref(cls, state_dict: Dict[str, Any]) -> 'Context':
        return cls(state_dict)

# Usage in steps
def configure_batch_size(input_data:
    ConfigInput, context: Context) -> ConfigOutput:
    context.set("batch_size", input_data.size)
    context.set("processing_mode", "parallel")
    return ConfigOutput(configured=True)

def process_batch(input_data: DataInput,
context: Context) -> DataOutput:
    batch_size = context.get("batch_size", 10)  # Default to 10
    mode = context.get("processing_mode", "sequential")
    # Use configuration from previous step
```

6.5 Building the Workflow Runner

Now that we have defined steps, edges, and a workflow container, we need an execution engine that transforms this static graph specification into a running system. Think of it like a manufacturing assembly line: we've designed the stations (steps), conveyor belts (edges), and factory layout (workflow), but now we need the control system that coordinates workers, manages flow rates, and ensures quality control.

Executing a workflow correctly requires solving several complex problems simultaneously: dependency resolution (which steps are ready?), concurrent execution (running independent steps in parallel), state management (threading data through the pipeline), event streaming (real-time visibility), error recovery (handling failures gracefully), and deadlock detection (identifying when workflow is stuck).

The WorkflowRunner handles all these concerns through careful coordination:

Listing 6.19 WorkflowRunner initialization with resource control

---

```
class WorkflowRunner:
    def __init__(self, max_concurrent_steps: int = 5):
        self.max_concurrent_steps = max_concurrent_steps
        self._execution_semaphore = asyncio.Semaphore(max_concurrent_steps)
        self._cancellation_tokens: Dict[str, CancellationToken] = {}
```

The semaphore controls concurrent execution, preventing resource exhaustion. Cancellation tokens enable graceful shutdown. The heart of the runner is the execution loop that coordinates all moving parts.

Dependency Resolution and Deadlock Detection

The first challenge is determining which steps are ready to run and detecting when the workflow is stuck:

This logic ensures steps run in the correct order while preventing infinite waiting. With ready steps identified, the runner spawns tasks while respecting concurrency limits:

This coordination enables the parallel execution patterns we discussed earlier. Each step receives input from either the initial workflow input or previous step outputs:

This threading mechanism enables data to flow through the workflow pipeline, with each step's output becoming the next step's input.

Throughout execution, the runner emits structured events for real-time visibility:

These events enable progress tracking, debugging, performance analysis, and can drive real-time UI updates.

Listing 6.20 Dependency resolution with deadlock detection

```python
async def _execute_workflow_stream(self, workflow, execution, initial_input):
    completed_steps = set()
    running_tasks = {}

    while len(completed_steps) < len(workflow.steps):
        # Get steps ready to run (dependency resolution)
        ready_steps = workflow.get_ready_steps(execution)
        ready_steps = [s for s in ready_steps
                        if s not in completed_steps and s not in running_tasks]

        # Detect deadlock - no progress possible
        if not ready_steps and not running_tasks:
            remaining = set(workflow.steps.keys()) - completed_steps
            if remaining:
                raise RuntimeError(f"Workflow stuck: \
                    {remaining} cannot execute")
            break
```

Listing 6.21 Task spawning and completion handling

```python
        # Start ready steps (respecting concurrency limits)
        for step_id in ready_steps:
            if len(running_tasks) >= self.max_concurrent_steps:
                break

            step = workflow.steps[step_id]
            input_data = self._prepare_step_input(step_id, workflow, execution)
            task = asyncio.create_task(self._run_step(step, input_data))
            running_tasks[step_id] = task

        # Wait for any task to complete
        done, _ = await asyncio.wait(running_tasks.values(),
                                    return_when=asyncio.FIRST_COMPLETED)

        # Process completed tasks and update workflow state
        for task in done:
            step_id = self._find_step_for_task(task, running_tasks)
            result = await task
            execution.state[f"{step_id}_output"] = result
            completed_steps.add(step_id)
            del running_tasks[step_id]
```

Listing 6.22 Step input preparation with state threading

```python
def _prepare_step_input(self, step_id: str, workflow: Workflow,
                        execution: WorkflowExecution,
                        initial_input: Dict[str, Any]):
    step = workflow.steps[step_id]
    dependencies = workflow.get_step_dependencies(step_id)

    if not dependencies:
        # Start step - use initial input
        return initial_input
    else:
        # Use output from previous step (for simple chains)
        # Or merge outputs for fan-in patterns
        prev_step_id = dependencies[0]
        prev_output = execution.state.get(f"{prev_step_id}_output", {})
        return prev_output
```

Listing 6.23 Event streaming for workflow observability

```python
# Workflow lifecycle events
yield WorkflowStartedEvent(timestamp=datetime.now(), workflow_id=workflow.id)

# Step execution events
yield StepStartedEvent(
    timestamp=datetime.now(),
    workflow_id=workflow.id,
    step_id=step_id,
    input_data=input_data
)

# On completion
yield StepCompletedEvent(
    timestamp=datetime.now(),
    workflow_id=workflow.id,
    step_id=step_id,
    output_data=result,
    duration_seconds=duration
)
```

Before execution begins, the runner validates workflow structure and input compatibility:

Listing 6.24 Workflow and input validation

```
# Validate workflow structure
validation = workflow.validate_workflow()
if not validation.is_valid:
    raise RuntimeError(f"Workflow validation failed: {validation.errors}")

# Validate input matches start step type
if workflow.start_step_id:
    start_step = workflow.steps[workflow.start_step_id]
    try:
        start_step.input_type(**initial_input)
    except Exception as e:
        raise RuntimeError(f"Input doesn't match expected type: {e}")
```

This upfront validation catches errors early, providing clear feedback before expensive execution begins.

6.6 Workflow Serialization and Persistence

The workflow system supports complete **serialization** through the same serialization methods introduced in earlier chapters. You can call `dump_component()` and `load_component()` methods on any `workflow` instance. Workflows serialize to JSON format with full type information, enabling sharing between teams, version control in git repositories, UI-based visual editors, and runtime configuration of processing pipelines.

6.7 Workflow Checkpointing

Checkpointing enables workflows to save progress and resume from failure points. Consider a data processing workflow with expensive steps—fetching data from an API, calling an LLM for analysis, validating results, and saving to a database. If the validation step fails due to a rate limit, restarting from scratch means repeating the expensive API fetch and LLM call. How can we save progress and resume from where we left off?

To resume a workflow, we need three pieces of information: which steps already completed, what outputs those steps produced (since downstream steps depend on them), and the current shared workflow state. Let's capture this in a `WorkflowCheckpoint` model:

```
class WorkflowCheckpoint(BaseModel):
    checkpoint_id: str
    execution: WorkflowExecution      # Contains step_executions, state
    completed_step_ids: List[str]     # Quick lookup helper
    pending_step_ids: List[str]       # What's left to run
```

But if the workflow structure changed since the checkpoint was saved—maybe we added a new step or changed types—resuming could produce incorrect results. We need to detect this.

The solution: compute a SHA256 hash of the workflow structure (steps, edges, and types—but not metadata like names or descriptions) and store it with the checkpoint. When resuming, compare the stored hash with the current workflow's hash. Matching hashes mean safe resume; different hashes mean the structure changed and we reject the checkpoint:

```
class WorkflowCheckpoint(BaseModel):
    # ... fields above ...
    workflow_structure_hash: str  # Validates compatibility
```

Now modify the workflow runner to accept checkpoint configuration and an optional checkpoint to resume from:

```
async def run_stream(
    workflow: Workflow,
    initial_input: Dict[str, Any],
    checkpoint: Optional[WorkflowCheckpoint] = None,  # Resume point
    checkpoint_config: Optional[CheckpointConfig] = None,  # Save config
) -> AsyncGenerator[WorkflowEvent, None]:
```

When a checkpoint is provided, validate its structure hash matches the current workflow, then extract the execution state. The runner identifies completed steps and skips them during execution:

```
if checkpoint:
    # Validate structure hash
    current_hash = workflow.compute_structure_hash()
    if checkpoint.workflow_structure_hash != current_hash:
        raise ValueError("Workflow structure changed")

    # Use checkpoint's execution state
    execution = checkpoint.execution
    completed_steps = {
        step_id for step_id, step_exec in execution.step_executions.items()
        if step_exec.status == StepStatus.COMPLETED
    }
else:
```

```
# Fresh execution
execution = WorkflowExecution(workflow_id=workflow.id, ...)
completed_steps = set()
```

During execution, after each step completes, create and save a checkpoint if configured:

```
# After step completes
if checkpoint_config and checkpoint_config.auto_save:
    checkpoint = WorkflowCheckpoint.from_execution(
        execution=execution,
        workflow_structure_hash=workflow.compute_structure_hash(),
        ...
    )
    await checkpoint_config.store.save(checkpoint)
```

Users can then resume by loading the checkpoint and passing it to the runner:

```
# Initial run with auto-save
store = FileCheckpointStore(base_path=Path("./checkpoints"))
config = CheckpointConfig(store=store, auto_save=True)

async for event in runner.run_stream(
    workflow, {"text": "data.csv"}, checkpoint_config=config
):
    pass  # Saves checkpoints to disk

# Later: resume after failure
checkpoint = await store.load_latest(workflow.id)
async for event in runner.run_stream(
    workflow, {"text": "data.csv"}, checkpoint=checkpoint
):
    pass  # Skips completed steps
```

6.7.1   Checkpoint Storage Abstraction

Where should we save checkpoints? Testing needs fast in-memory storage. Single machines can use file storage. Distributed deployments need shared storage like S3 or databases. Rather than hardcoding one approach, let's build an abstract storage interface:

```
class CheckpointStore(ABC):
    async def save(
        self,
        checkpoint: WorkflowCheckpoint
        ) -> None: ...
    async def load(
        self,
```

```
    checkpoint_id: str
    ) -> Optional[WorkflowCheckpoint]: ...
async def load_latest(
    self,
    workflow_id: str
    ) -> Optional[WorkflowCheckpoint]: ...
```

This interface defines what any storage backend must do: save checkpoints, load specific checkpoints by ID, and load the most recent checkpoint for a workflow. Now we can implement concrete backends:

File Backend: Save checkpoints as JSON files organized by workflow ID. This is human-readable, works with version control, and requires no additional infrastructure.

In-Memory Backend: Store checkpoints in a dictionary. Fast but ephemeral—perfect for testing where you don't want filesystem side effects.

You can also implement custom backends for production scenarios: databases (PostgreSQL, MongoDB), cloud storage (S3, Azure Blob), or distributed coordination systems (Redis, etcd). Each backend just implements the three abstract methods.

6.7.2   Enabling Stateless Horizontal Scaling

Checkpointing unlocks a powerful deployment pattern: stateless workflow workers. Consider the problem: if you assign specific workflows to specific workers (worker affinity), what happens when a worker crashes? The workflow is stuck until that worker restarts.

Here's a better approach: store checkpoints in shared storage (S3, database, Redis). Now any worker can resume any workflow. When a worker crashes, another worker simply loads the latest checkpoint and continues. No workflow is tied to a specific machine.

This means you can scale horizontally: need more throughput? Add more workers. Each worker pulls work from a queue, checks if a checkpoint exists, and either resumes from that point or starts fresh. Since checkpoints skip completed steps, expensive operations like LLM calls or API requests are never re-executed.

The architecture is both fault-tolerant (workers are disposable) and elastic (scale workers up and down based on load). The workflow engine becomes stateless—all state lives in checkpoints, not in worker memory.

> 💡 🖥 Working Code: Workflow Implementation
>
> Complete workflow examples:
> - **Sequential Workflows:** `workflows/sequential.py` - Linear data processing pipelines
> - **Checkpoint Example:** `workflows/checkpoint_example.py` - Auto-save, resume,

and cleanup patterns
- **General Workflows**: `workflows/general.py` - Conditional logic and parallel execution patterns
- **Complete Implementation**: `src/picoagents/workflow/` - Full workflow engine source code

6.8   Summary

Building multi-agent workflows from scratch reveals the foundational components needed for explicit control patterns:

- **Orchestration Spectrum Positioning**: Workflows provide explicit control where developers define deterministic execution paths, contrasting with autonomous patterns in Chapter 7. This gives predictable behavior but less runtime adaptability. Understanding this trade-off helps choose the right orchestration approach for specific tasks.

- **Type-Safe Step Architecture**: The step system uses Pydantic models for compile-time type checking and runtime validation. `BaseStep` handles infrastructure concerns (execution lifecycle, status tracking, error handling) while concrete implementations like `FunctionStep` focus on domain logic, enabling easy conversion of existing Python functions into workflow components.

- **Conditional Edge Patterns**: Edges handle sequential processing, conditional logic, and parallel execution through condition evaluation. Output-based conditions enable validation and quality control patterns, while state-based conditions enable configuration-driven behavior and dynamic workflow adaptation based on shared context.

- **Intelligent Execution Engine**: The WorkflowRunner implements sophisticated dependency resolution using fan-in (wait for ALL inputs) versus conditional (wait for ANY valid path) logic. It handles concurrent step execution with resource limits, deadlock detection, and graceful error recovery while providing streaming observability for debugging and monitoring.

- **Production-Ready Features**: Workflows support complete serialization to JSON for sharing, version control, and UI integration. Automatic checkpointing enables resumable execution after failures, with structure hash validation ensuring safe resume. This supports stateless horizontal scaling where any worker can resume any workflow from shared storage. The validation system catches structural errors and type mismatches at build time, while streaming events enable real-time progress tracking and responsive user interfaces.

- **Implementation Guidance**: Choose workflows when the solution approach is known, predictability is important, dependencies need explicit coordination, and step-by-step observability is required. The pattern excels for data processing pipelines, approval workflows, and any scenario where audit trails and deterministic execution are essential.

In Chapter 7, we'll explore autonomous patterns where AI models drive coordination decisions, trading predictability for adaptability.

Chapter 7

Building Autonomous Multi-Agent Orchestration

This chapter covers:

- The orchestrator loop that underlies all coordination patterns
- Building flexible termination conditions that prevent runaway execution
- Implementing round-robin orchestration with streaming observability
- Practical considerations for debugging, cost management, and production deployment

In Chapter 2, we explored the taxonomy of multi-agent coordination patterns—from explicit workflow control to emergent autonomous behavior. We learned that orchestration patterns exist on a spectrum, with workflow patterns providing predictable, developer-defined execution paths, and autonomous patterns enabling runtime-determined coordination through agent reasoning and communication.

In Chapter 6, we explored how to build multi-agent workflows. In this chapter we will focus on how to implement multi-agent patterns where the flow of control is *driven by an AI model*.

Before diving into implementation details, here's the final developer experience we're building towards—a complete orchestration system in just a few lines. This example demonstrates how two specialized agents (a researcher and a writer) collaborate through round-robin coordination, with flexible termination conditions that stop the conversation when either a maximum message count is reached or when the writer signals completion.

First, we create two specialized agents—each with clear responsibilities and instructions:

Next, we set up the orchestrator to coordinate these agents using round-robin turn-taking. Notice how we combine termination conditions using the | operator—the collaboration stops when either the message limit is reached OR when the writer signals completion:

Listing 7.1 Creating specialized agents for collaboration

```
from picoagents import Agent
from picoagents.llm import OpenAIChatCompletionClient

# Create two specialized agents
researcher = Agent(
    name="researcher",
    instructions=(
        "You research topics and provide "
        "factual insights. Be concise."
    ),
    model_client=OpenAIChatCompletionClient(
        model="gpt-4.1-mini"
    )
)

writer = Agent(
    name="writer",
    instructions=(
        "You write engaging content. "
        "If sufficient content exists, return "
        "a summary and end with 'TERMINATE'."
    ),
    model_client=OpenAIChatCompletionClient(
        model="gpt-4.1-mini"
    )
)
```

Listing 7.2 Configuring orchestrator with flexible termination

```
from picoagents.orchestration import RoundRobinOrchestrator
from picoagents.termination import (
    MaxMessageTermination,
    TextMentionTermination
)

orchestrator = RoundRobinOrchestrator(
    agents=[researcher, writer],
    termination=(
        MaxMessageTermination(6) |
        TextMentionTermination("TERMINATE")
    )
)
```

Finally, we execute the collaboration with streaming output to observe the conversation in real-time:

Listing 7.3 Running multi-agent collaboration with streaming

```
task = "Write a brief article about solar energy benefits"

async for message in orchestrator.run_stream(task):
    print(message)
```

When you run this example, you'll see streaming output that shows each agent's contribution in real-time, along with metadata about the collaboration. Notice how the orchestrator tracks which agent is speaking, maintains conversation flow, and provides clear feedback about why and how the conversation ended:

Listing 7.4 Example streaming output showing agent messages, role indicators, and orchestration metadata

```
⬚ Starting solar energy research collaboration...

⬚ [USER]: Write a brief article about solar energy benefits
⬚ [RESEARCHER]: **The Benefits of Solar Energy**
Solar energy is a renewable and sustainable power source that harnesses ...
✍ [WRITER]: **Summary: The Benefits of Solar Energy**

Solar energy presents numerous benefits, making it essential for addressing ....

TERMINATE

⬚ Collaboration complete! Generated 3 messages
⬚ Stop reason: Text mention found: 'TERMINATE'
⬚ Duration: 17,427ms
```

Notice how the orchestrator:

1. Coordinates turn-taking between the two agents in round-robin fashion
2. Maintains shared conversation context so each agent sees the full history
3. Terminates intelligently when the writer determines the task is complete
4. Provides streaming observability into the coordination process

Now let's explore how to build this, and other orchestration systems step by step.

> 💡 🖥 Working Code: Autonomous Orchestration
>
> Explore complete autonomous orchestration implementations:
> - **Round-Robin:** `orchestration/round-robin.py` - Fixed turn-taking coordination
> - **AI-Driven:** `orchestration/ai-driven.py` - Intelligent agent selection
> - **Plan-Based:** `orchestration/plan-based.py` - Centralized orchestrator with dynamic planning
> - **Core Implementation:** `src/picoagents/orchestration/` - Full orchestration framework

7.1   The Orchestrator Loop

Every orchestration pattern, regardless of how agents are selected or coordinated, follows the same fundamental **orchestrator loop** . Whether you're implementing a simple round-robin approach to scheduling agents (agents take turns in round robin fashion until some termination condition is met), AI-driven selection (an AI model determines the agent to take the next turn), or a dedicated AI model crafts a plan based on some task and directs agents, the core architecture remains consistent:

- select the next agent
- prepare context for the agent
- execute the agent with the prepared context
- update shared state with the agent's response
- check termination conditions

```
# The core pattern that powers all orchestration
while not termination.is_met():
    next_agent = select_next_agent()          # Pattern-specific logic
    context = prepare_context_for_agent()     # Pattern-specific logic
    result = await next_agent.run(context)
    update_shared_state(result)
    check_termination(result)
```

This core loop provides several key benefits:

This loop delivers consistency across all patterns (making them easier to understand and maintain), composability (allowing you to mix different patterns by switching orchestrator implementations), extensibility (new patterns only need to implement pattern-specific methods while inheriting robust infrastructure), and comprehensive observability (common instrumentation works across all pattern types).

With this foundational understanding, let's implement the base orchestrator class that powers all coordination patterns, then build specific implementations starting with round-robin orchestration and flexible termination conditions.

7.1.1 Building the BaseOrchestrator

Similar to our BaseAgent class in Chapter (Section 4.2.2), the BaseOrchestrator class (Listing 7.5) handles the complexity of streaming, cancellation, and state management while letting each pattern focus only on its specific coordination logic.

The BaseOrchestrator defines the contract that all orchestration patterns must follow. It initializes the core components (agents, termination conditions, and shared state) and validates that agent names are unique to prevent coordination conflicts.

The constructor establishes the three core components every orchestrator needs: a list of agents to coordinate, termination conditions that determine when to stop, and shared state management for tracking the conversation. The max_iterations parameter provides a safety fallback to prevent runaway execution.

7.1.2 The Orchestrator Run Loop

The orchestrator's core logic lives in two methods that follow the same pattern we used for agents. The run_stream method provides real-time streaming of messages and events, while run collects everything into a single response for simpler use cases.

The run loop starts by setting up the execution context, normalizing the input task, and preparing for streaming output:

The setup phase handles several critical responsibilities. The _reset_for_run() call clears any state from previous executions, ensuring each orchestration starts fresh. The streamed_messages list tracks all messages we've yielded to the caller—this becomes important later for termination checking. Task normalization handles three input formats: plain strings (converted to UserMessage), UserMessage objects (used directly), or full message lists (for continuing conversations). We check termination immediately after the initial messages because even the user's input might contain stop keywords like "DONE" or "TERMINATE".

Now the orchestrator enters its main coordination loop. Each iteration selects an agent, runs it with appropriate context, and processes the results. The loop starts by selecting which agent should execute next:

The select_next_agent() and prepare_context_for_agent() methods are pattern-specific—this is where round-robin differs from AI-driven or plan-based orchestration. Round-robin cycles through agents in order, while AI-driven orchestration uses an LLM to choose intelligently based on conversation state.

After selecting an agent and preparing context, the orchestrator executes the agent with streaming support:

Listing 7.5 BaseOrchestrator abstract class definition

```
from abc import ABC, abstractmethod
from typing import List, Optional, Sequence
from ..agents import BaseAgent
from ..messages import Message
from .termination import BaseTermination

class BaseOrchestrator(ABC):
    """Abstract base class for all orchestration patterns."""

    def __init__(
        self,
        agents: Sequence[BaseAgent],
        termination: BaseTermination,
        max_iterations: int = 50
    ):
        if not agents:
            raise ValueError(
                "At least one agent is required"
            )

        self.agents = list(agents)
        self.termination = termination
        self.max_iterations = max_iterations

        # Runtime state
        self.shared_messages: List[Message] = []
        self.iteration_count = 0

        # Validate agent names are unique
        names = [agent.name for agent in agents]
        if len(names) != len(set(names)):
            raise ValueError(
                "Agent names must be unique"
            )
```

Listing 7.6 Orchestrator run loop setup and initialization

```python
async def run_stream(
    self,
    task: Union[str, UserMessage, List[Message]],
    cancellation_token: Optional[CancellationToken] = None
) -> AsyncGenerator[
    Union[
        Message,
        OrchestrationEvent,
        OrchestrationResponse
    ],
    None
]:
    """Execute orchestration with streaming output."""
    self._reset_for_run()
    self.start_time = time.time()
    stop_message: Optional[StopMessage] = None
    streamed_messages: List[Message] = []

    try:
        yield OrchestrationStartEvent(...)

        # Normalize task to initial messages
        initial_messages = (
            self._normalize_task_to_messages(task)
        )
        self.shared_messages.extend(initial_messages)

        # Stream initial messages to caller
        for message in initial_messages:
            yield message
            streamed_messages.append(message)

        # Check termination on initial messages
        self.termination.check(initial_messages)
```

Listing 7.7 Agent selection and context preparation

```
# Main orchestration loop
while self.iteration_count < self.max_iterations:
    # Check for cancellation
    if (
        cancellation_token
        and cancellation_token.is_cancelled()
    ):
        raise asyncio.CancelledError()

    # 1. Select next agent (pattern-specific)
    next_agent = await self.select_next_agent()

    # 2. Prepare context (pattern-specific)
    context = await self.prepare_context_for_agent(
        next_agent
    )
```

Listing 7.8 Streaming agent execution with type guards

```
# 3. Execute agent with streaming
agent_messages = []
result: Optional[AgentResponse] = None

async for item in next_agent.run_stream(
    context, cancellation_token):
    # Type guard for Message objects
    if (hasattr(item, "content")
        and hasattr(item, "role")):
        message_item = cast(Message, item)
        agent_messages.append(message_item)

        # Filter out UserMessages from streaming
        if not isinstance(
            message_item,
            UserMessage
        ):
            yield message_item
            streamed_messages.append(
                message_item
            )
```

The streaming execution uses type guards to safely process different item types. We filter out UserMessages because these are just the context we sent to the agent, not new content. Only assistant messages get streamed to the caller, providing clean real-time visibility into agent responses without echoing the prompts.

After agent execution completes, the orchestrator updates shared state and checks whether the conversation should continue:

Listing 7.9 State updates and termination checking

```
# 4. Update shared state (pattern-specific)
await self.update_shared_state(result)

# 5. Check termination with delta messages
new_streamed_messages = streamed_messages[
    self._last_termination_check_count:
]
self._last_termination_check_count = len(
    streamed_messages
)

stop_message = self.termination.check(
    new_streamed_messages
)
if stop_message:
    break

self.iteration_count += 1
```

The state management phase reveals a subtle but crucial pattern: delta-based termination checking. Rather than re-processing the entire conversation history on every iteration, we track our position with `_last_termination_check_count` and only pass new messages to the termination checker. This is more efficient than scanning hundreds of messages repeatedly, and it allows termination conditions to focus on recent developments rather than the full history. The pattern-specific `update_shared_state()` method is what enables different orchestration strategies—round-robin might append all messages, while AI-driven orchestration might filter or transform them based on coordination needs.

After the loop completes (either through termination conditions or max iterations), the orchestrator assembles a final response with complete execution metrics:

The completion phase ensures we always provide a meaningful stop message, even when the loop exits through max iterations rather than explicit termination conditions. The usage aggregation is particularly important—each agent execution tracks its own API costs and timing, and we sum these together to give an accurate picture of total orchestration cost.

Listing 7.10 Orchestration completion and result assembly

```python
        # Handle max iterations case
        if (
            self.iteration_count >= self.max_iterations
            and stop_message is None
        ):
            stop_message = StopMessage(
                content=(
                    f"Maximum iterations reached "
                    f"({self.max_iterations})"
                ),
                source="MaxIterations"
            )

        # Aggregate usage from all agent executions
        elapsed_time = int(
            (time.time() - self.start_time) * 1000
        )
        total_usage = Usage(duration_ms=elapsed_time)
        for agent_usage in agent_usage_stats:
            total_usage = total_usage + agent_usage

        # Yield final OrchestrationResponse
        yield OrchestrationResponse(
            messages=self.shared_messages,
            final_result=self._generate_final_result(),
            usage=total_usage,
            stop_message=stop_message,
            pattern_metadata=self._get_pattern_metadata()
        )

except asyncio.CancelledError:
    # Handle cancellation gracefully
    raise
```

The `_generate_final_result()` helper searches backward through messages to find the last assistant response, providing a concise summary rather than forcing callers to scan the full message history. This pattern of rich, complete responses makes orchestration results self-contained and easy to work with.

This implementation provides the foundation for all orchestration patterns while handling several complex concerns:

Streaming Observability: When debugging multi-agent systems, you need to see what's happening in real-time. Which agent is running? What messages are being exchanged? Where did the conversation get stuck? The streaming interface provides this visibility, making it possible to build responsive UIs and debug complex interactions.

Graceful Cancellation: Multi-agent conversations can take minutes to complete. Users will inevitably need to cancel them - whether because they provided the wrong prompt or the agents entered an unproductive loop. Clean cancellation prevents runaway API costs and ensures the system remains responsive.

Robust Error Handling: In a multi-agent system, one agent's failure shouldn't crash the entire orchestration. The error handling ensures that exceptions are caught and logged while preserving the conversation state up to the failure point.

Resource and Performance Tracking: Understanding how long orchestrations take and how many API calls they consume is essential for cost management and performance optimization. The automatic metrics collection provides this data without requiring manual instrumentation.

Each orchestration pattern inherits this robust infrastructure and only needs to implement three pattern-specific abstract methods: `select_next_agent()` for choosing which agent executes next, `prepare_context_for_agent()` for assembling the conversation context, and `update_shared_state()` for managing shared state after agent execution. This clean separation allows patterns to focus solely on their unique orchestration logic while benefiting from comprehensive streaming, cancellation, and error handling capabilities.

7.2   Building Termination Conditions

Termination conditions

Looking back at our orchestrator loop, you'll notice we conduct a termination check - termination.check(new_messages) at the end of each iteration. This is one of the most critical aspects of multi-agent systems: knowing when to stop.

Without proper termination conditions, agent conversations can run indefinitely. You've probably seen chatbots get stuck in loops - the same problem happens with multi-agent systems, except now you're paying for multiple API calls per loop iteration. You need termination conditions that are both reliable and flexible.

7.2.1 The BaseTermination Interface

To build a termination system that can handle simple rules ("stop after 5 messages") and complex combinations ("stop after 10 messages OR if someone says 'DONE' OR after 5 minutes"), we start with a common base class:

```python
from abc import ABC, abstractmethod
from typing import List, Optional, Dict, Any, Sequence
from ..messages import Message
from ..types import StopMessage

class BaseTermination(ABC):
    """Abstract base class for all termination conditions."""

    def __init__(self) -> None:
        self._met = False
        self._reason = ""
        self._metadata: Dict[str, Any] = {}

    @abstractmethod
    def check(self, new_messages: Sequence[Message]) -> Optional[StopMessage]:
        """Check termination on delta messages."""
        pass

    def is_met(self) -> bool:
        return self._met

    def reset(self) -> None:
        self._met = False
        self._reason = ""
        self._metadata = {}

    def __or__(self, other: 'BaseTermination') -> 'CompositeTermination':
        return CompositeTermination([self, other], mode="any")

    def __and__(self, other: 'BaseTermination') -> 'CompositeTermination':
        return CompositeTermination([self, other], mode="all")
```

The __or__ and __and__ methods enable Python's | and & operators for combining conditions. This lets you write intuitive termination logic like MaxMessages(5) | TextMention("DONE")

Notice that termination conditions receive **delta messages** - only new messages since the last check. This is more efficient than re-processing the entire conversation each time and allows conditions to react to recent developments rather than scanning everything repeatedly.

7.2.2 Common Termination Patterns

The two most common termination patterns are message count limits and text-based triggers. Here's how they work conceptually:

```python
# Conceptual: Message count termination
class MaxMessageTermination(BaseTermination):
    """Stop after N messages."""

    def check(self, new_messages):
        self.count += len(new_messages)
        if self.count >= self.max_messages:
            return StopMessage(f"Reached {self.count} messages")
        return None

# Conceptual: Text-based termination
class TextMentionTermination(BaseTermination):
    """Stop when specific text appears."""

    def check(self, new_messages):
        for message in new_messages:
            if self.search_text in message.content:
                return StopMessage(f"Found '{self.text}'")
        return None
```

Both conditions extend `BaseTermination` and implement the `check()` method. The key difference: `MaxMessageTermination` tracks cumulative count across all checks, while `TextMentionTermination` scans each new message for the trigger phrase. The framework includes additional termination types like `TokenUsageTermination` (budget limits), `TimeoutTermination` (time limits), and `HandoffTermination` (agent-initiated transfers).

7.2.3 Composable Termination Logic

Here's where the `|` and `&` operators we defined earlier become useful. You can combine termination conditions to handle complex scenarios:

```python
# Terminate after 20 messages OR when someone says "DONE" OR after 5 minutes
termination = (
    MaxMessageTermination(20) |
    TextMentionTermination("DONE") |
    TimeoutTermination(300)
)

# Terminate only when BOTH conditions are met
complex_termination = (
```

```
    MaxMessageTermination(10) &
    TextMentionTermination("COMPLETE")
)
```

This composability means you can build sophisticated stopping logic without writing complex conditional code. Each condition remains simple and testable, but their combinations can handle nuanced requirements.

7.3   Implementing Round-Robin Orchestration

With our foundation in place—the orchestrator loop and flexible termination system—we can now build specific coordination patterns. Let's start with the simplest: **round-robin orchestration** .

Round-robin coordination is like a structured meeting where each participant gets a turn to speak in order. This pattern works well when you want balanced participation from all agents and a predictable conversation flow.

7.3.1   The RoundRobinOrchestrator Implementation

The RoundRobinOrchestrator implements the three abstract methods to create a simple but effective turn-taking pattern. First, we extend the base class and add an index to track which agent should go next:

The three pattern-specific methods are straightforward:

```
# Round-robin agent selection
async def select_next_agent(self) -> BaseAgent:
    """Cycle through agents in order."""
    agent = self.agents[self.current_agent_index]
    self.current_agent_index = (
        (self.current_agent_index + 1) % len(self.agents)
    )
    return agent


# Context preparation: share full history
async def prepare_context_for_agent(
    self, agent: BaseAgent
) -> str:
    """Format conversation history for next agent."""
    context = "Team progress so far:\n\n"
    for msg in self.shared_messages:
        context += f"{msg}\n"
    return context + "\nIt is now your turn."
```

Listing 7.11 RoundRobinOrchestrator initialization

```python
from typing import Union, List
from ._base import BaseOrchestrator
from ..agents import BaseAgent
from ..messages import Message, UserMessage
from ..types import AgentResponse

class RoundRobinOrchestrator(BaseOrchestrator):
    """Round-robin orchestration pattern."""

    def __init__(
        self,
        agents: Sequence[BaseAgent],
        termination: BaseTermination,
        max_iterations: int = 50
    ):
        super().__init__(
            agents,
            termination,
            max_iterations
        )
        self.current_agent_index = 0
```

```python
# State management: append new messages
async def update_shared_state(
    self, result: AgentResponse
) -> None:
    """Add agent's new messages to shared history."""
    new_messages = result.messages[1:]  # Skip context echo
    self.shared_messages.extend(new_messages)
```

This implementation is remarkably simple because it inherits all the complex infrastructure from `BaseOrchestrator`. The round-robin pattern only needs to specify:

1. **Agent Selection:** Cycle through agents in order
2. **Context Preparation:** Share the full conversation history
3. **State Management:** Append new messages to shared state

> 💡 Tip
>
> You can use a simple RoundRobinOrchestrator to implement many simple use cases and it often requires the least amount of code. Excellent for a first version of most prototyping work.

The round-robin orchestrator we just implemented is the same as what we introduced earlier in Listing 7.2. Of course, you can extend this for more complex scenarios. For example, a three-agent research team might include a researcher, writer, and critic that cycle through multiple rounds of refinement:

```
# Three-agent workflow with iterative improvement
termination = MaxMessageTermination(12) | TextMentionTermination("FINAL VERSION")
orchestrator = RoundRobinOrchestrator(
    agents=[researcher, writer, critic],
    termination=termination
)
```

Key behavioral patterns you can expect:

- **Researchers** gather information and provide factual insights
- **Writers** synthesize research into coherent content

- **Critics** provide feedback and suggest improvements
- **Iterative refinement** continues until termination conditions are met

7.3.2   Observability and Debugging

The streaming interface from our earlier example provides comprehensive observability. Beyond basic message flow, you can monitor detailed orchestration events for debugging:

- **Agent selection events**: Which agent was chosen and why
- **Execution timing**: How long each agent takes to respond

- **Context size**: How much conversation history each agent receives
- **Termination triggers**: Exactly what caused the conversation to end

This observability is crucial for debugging complex multi-agent interactions and optimizing system performance.

7.4 AI-Driven Orchestration

Round-robin orchestration works well when you want predictable turn-taking, but what if you need more intelligent coordination? What if the conversation needs a researcher's input before the writer can continue, or if the critic should wait until there's substantial content to review?

AI-driven orchestration solves this by using an LLM to analyze the conversation and decide which agent should speak next. Instead of following a fixed order, the system adapts to the actual conversation needs.

The key insight is that we can reuse all the infrastructure from round-robin orchestration and only change the `select_next_agent()` method. This pseudocode illustrates how an LLM analyzes the conversation context and agent capabilities to make intelligent selection decisions:

Instead of fixed cycling, the LLM analyzes conversation context and agent capabilities to make intelligent selections, enabling adaptive coordination that responds to actual conversation needs.

The same agents from our round-robin example can be used with AI orchestration, but the conversation flow adapts dynamically:

```
# Same agents as round-robin example
researcher = Agent(
    name="researcher",
    instructions="Research specialist...",
    ...
)
writer = Agent(
    name="writer",
    instructions="Content writer...",
    ...
)
termination = (
    MaxMessageTermination(6) |
    TextMentionTermination("TERMINATE")
)

# Switch to AI orchestration - intelligent selection
orchestrator = AIOrchestrator(
    agents=[researcher, writer],
```

Listing 7.12 AI-driven agent selection using LLM reasoning

```
# Pseudocode for AI-driven agent selection
async def select_next_agent() -> BaseAgent:
    # 1. Gather agent capabilities
    capabilities = get_agent_capabilities_summary()

    # 2. Format recent conversation context
    conversation_context = format_conversation_for_selection()

    # 3. Create selection prompt
    prompt = f"""You are coordinating AI agents.

Available agents: {capabilities}
Recent conversation: {conversation_context}

Choose which agent should respond next
based on conversation needs."""

    # 4. Get structured LLM decision
    result = await llm_client.create(
        messages=[UserMessage(content=prompt)],
        output_format=AgentSelection
    )

    # 5. Return selected agent
    return find_agent_by_name(
        result.structured_output.selected_agent
    )
```

```
        termination=termination,
        model_client=OpenAIChatCompletionClient(
            model="gpt-4.1-mini"
        )
    )
)

# Adaptive flow based on LLM decisions:
# -> LLM selects researcher (need: information)
# -> LLM selects writer (research complete)
# -> Writer requests more data
# -> LLM selects researcher (need: clarification)
# -> LLM selects writer (info provided)
# -> "TERMINATE"
```

The conversation naturally adapts - if the writer needs more information, the LLM will select the researcher again rather than forcing a fixed alternating pattern.

> 💡 Tip
>
> AI-driven orchestration is straightforward to implement once you have the right agent components. The LLM handles coordination decisions, and structured output ensures reliable agent selection with reasoning.
> **Trade-off**: This pattern adds LLM costs for each agent selection, as conversation history and agent capabilities are included in every coordination request.

7.5 Plan-Based Orchestration

Both round-robin and AI-driven orchestration make moment-to-moment decisions about which agent should act next. **Plan-based orchestration** takes a different approach: it creates an explicit execution plan upfront, then systematically works through each step while monitoring progress.

This pattern works well when tasks require structured decomposition—breaking a complex goal into clear sequential steps with specific agent assignments. Think of it as having a project manager who creates a task list, assigns each task to the right specialist, verifies completion, and handles retries when things don't work as expected.

7.5.1 LLM-Generated Execution Plans

The orchestrator starts by using an LLM to analyze the task and agent capabilities, then generates a structured execution plan. Unlike round-robin (which just cycles through agents) or AI-driven orchestration (which selects one agent at a time), plan-based orchestration needs to reliably extract **multiple structured steps** from the LLM response. This is where structured output becomes essential.

We define the plan structure using Pydantic models that the LLM must conform to:

With these models defined, the orchestrator can request structured plans from the LLM. The critical part is the `output_format` parameter, which constrains the model to return valid `ExecutionPlan` objects rather than free-form text:

The `output_format=ExecutionPlan` parameter is crucial—it transforms unpredictable LLM text into reliable data structures. Without this, parsing free-form responses like "First, the researcher should gather facts, then the writer creates content" would be fragile and error-prone. Structured output ensures every plan has clearly defined steps with agent assignments that the orchestrator can execute programmatically.

This approach transforms vague goals into concrete action items. A task like "research and write a guide about renewable energy" becomes a structured plan: Step 1 (researcher: gather

Listing 7.13 Structured models for execution plans

```
class PlanStep(BaseModel):
    """Single step in an execution plan."""
    task: str = Field(
        description="Clear, actionable task description"
    )
    agent_name: str = Field(
        description="Name of agent for this step"
    )
    reasoning: str = Field(
        description="Why this agent was chosen"
    )

class ExecutionPlan(BaseModel):
    """Complete execution plan."""
    steps: List[PlanStep] = Field(
        description="Ordered list of steps"
    )
```

key facts), Step 2 (writer: create structured content), Step 3 (reviewer: verify accuracy).

7.5.2 Step Progress Evaluation

Unlike simpler orchestration patterns, plan-based coordination needs to verify that each step actually accomplished its goal before moving forward. This evaluation must be programmatic—the orchestrator can't just hope the step worked, it needs a clear boolean decision to drive retry logic.

Again, structured output is essential. We define what a proper evaluation looks like:

The model ensures every evaluation includes a clear success/failure decision, reasoning, and actionable suggestions. After each agent execution, the orchestrator uses structured output to get reliable evaluations. The evaluation process asks an LLM to assess whether the agent's work accomplished the step's goal:

```
# Conceptual: Step evaluation with structured output
async def evaluate_step_progress(
    step: PlanStep,
    agent_result: AgentResponse
) -> StepProgressEvaluation:
    """Ask LLM: Did the agent complete this step successfully?"""

    # Build evaluation prompt with step context
```

Listing 7.14 LLM-based plan generation with structured output

```
async def create_plan(self, task: str) -> ExecutionPlan:
    """Create execution plan using LLM."""
    capabilities = self.get_agent_capabilities_summary()

    planning_prompt = f"""You are a helpful assistant
that breaks down tasks into executable steps.

Available agents and their capabilities:
{capabilities}

User task: {task}

Generate a concise execution plan. For each step:

- Assign to the agent best suited for that work
- Provide clear, actionable task description
- Explain briefly why that agent was chosen

Keep it focused. If only 2-3 steps needed, that's fine.
"""

    result = await self.model_client.create(
        messages=[UserMessage(content=planning_prompt)],
        output_format=ExecutionPlan  # Enforces structure
    )

    return result.structured_output  # Guaranteed ExecutionPlan
```

```
    eval_prompt = f"""
Expected: {step.task} (assigned to {step.agent_name})
Agent produced: {agent_result.content}

Was this step completed successfully?
"""

    # Get structured evaluation (not free-form text)
    evaluation = await llm.create(
        messages=[eval_prompt],
        output_format=StepProgressEvaluation
    )

    # Returns: {step_completed: bool, failure_reason: str,
```

Listing 7.15 Structured model for step progress assessment

```
class StepProgressEvaluation(BaseModel):
    """Evaluation of step completion."""
    step_completed: bool = Field(
        description="Whether step was successful"
    )
    failure_reason: str = Field(
        description="Explanation if failed"
    )
    confidence_score: float = Field(
        description="Confidence (0.0 to 1.0)",
        ge=0.0,
        le=1.0
    )
    suggested_improvements: List[str] = Field(
        description="Suggestions for retry if failed"
    )
```

```
    #        confidence: float, suggested_improvements: List[str]}
    return evaluation.structured_output
```

The `output_format` parameter is what makes programmatic retry possible. Without structured output, the LLM might return vague assessments like "looks pretty good" or "might need work"—responses that can't drive retry logic. With structured output, we get a guaranteed boolean decision, explicit reasoning, and actionable suggestions that the orchestrator can use to enhance the next attempt.

7.5.3 Intelligent Retry Logic

When step evaluation indicates failure, the plan-based orchestrator doesn't just move on or halt completely. Instead, it retries the step with enhanced context based on the failure analysis. This retry logic fits into the same `update_shared_state` pattern we saw earlier, but now with evaluation and retry tracking:

```
# Conceptual: Retry logic in plan-based orchestration
async def update_shared_state(result: AgentResponse):
    """Handle step completion, evaluation, and retry."""

    # Evaluate what the agent just did
    evaluation = await evaluate_step_progress(
        current_step,
        result
```

```
)

if evaluation.step_completed:
    # Success - move forward
    move_to_next_step()
    reset_retry_count()
else:
    # Failed - try again with enhanced context
    increment_retry_count()

    if retries_remaining():
        # Build enhanced instructions for next attempt
        retry_context = build_retry_context(
            evaluation.failure_reason,
            evaluation.suggested_improvements
        )
        store_retry_instructions(retry_context)
        # Next iteration will retry same step with context
    else:
        # Max retries exceeded - move on
        move_to_next_step()
        reset_retry_count()
```

The key insight is that retry instructions become part of the agent's context on the next attempt. When `prepare_context_for_agent` runs again for the same step, it includes the failure analysis: "Previous attempt failed because X. Try improving Y and Z." This targeted feedback significantly improves success rates compared to blind retries.

> ℹ **When to Use Plan-Based Orchestration**
>
> Plan-based orchestration works best for:
> - **Multi-phase pipelines** with clear sequential stages (research -> write -> review)
> - **Quality-critical tasks** where automated verification catches failures before proceeding
> - **Recoverable failures** where retry with feedback significantly improves success rates
>
> **Key enabler:** Structured output makes this pattern possible—without reliable data structures for plans and evaluations, programmatic retry logic wouldn't work.
>
> **Cost consideration:** This pattern requires 2+ LLM calls per step (planning + evaluation + execution). Use round-robin for simple turn-taking or AI-driven for dynamic coordination when explicit planning isn't needed.

7.6 Exercises

Custom Termination Conditions: Implement additional termination patterns beyond message count and text mention. For example, create a `FunctionTermination` that calls an external function to determine when to stop, or a `TokenBudgetTermination` that tracks cumulative token usage and halts when a budget is exceeded.

Handoff Orchestration: Extend the orchestration patterns to support direct agent-to-agent handoffs. Instead of the orchestrator always selecting the next agent, allow agents to explicitly transfer control to specific colleagues based on their assessment of task requirements.

Performance Comparison: Implement comprehensive metrics collection for orchestration runs (execution time, token usage, message counts, agent selection patterns). Use these metrics to compare the same task across different orchestration methods—how do round-robin, AI-driven, and plan-based patterns differ in cost and effectiveness for various use cases?

Context Management Strategies: Experiment with different approaches to context preparation in `prepare_context_for_agent()`. Compare concatenated text versus structured formats (JSON, markdown tables). Test limiting context to recent N messages versus full history. Measure how these choices impact agent response quality and token costs.

7.7 Summary

We've built a complete orchestration system that coordinates multiple agents in sophisticated conversations. The key insights from this implementation:

- **The Orchestrator Loop**: All orchestration patterns share the same foundational loop, providing consistency and composability across different coordination approaches.

- **Flexible Termination**: Composable termination conditions prevent runaway execution while allowing sophisticated stopping logic that adapts to different use cases and requirements.

- **Streaming Observability**: Real-time event streams provide unprecedented visibility into multi-agent coordination decisions, enabling effective debugging, monitoring, and user interface development.

- **Pattern Extensibility**: The abstract base class handles complex infrastructure concerns while allowing each pattern to focus only on its specific coordination logic through clean extension points.

- **Production Considerations**: Built-in support for cancellation, error handling, resource tracking, and context management makes the system suitable for real-world deployment scenarios.

The round-robin orchestrator demonstrates how this architecture enables building sophisticated coordination patterns with minimal code. The same foundation supports the advanced patterns outlined above, each requiring only pattern-specific implementations of the three abstract methods while inheriting all the robust infrastructure we've built.

Chapter 8

Building Modern Web Experiences for Agent Applications

This chapter covers:

- Understanding the two components every agent web application needs
- Building a complete agent web app from scratch in ~200 lines
- Implementing UX design principles for multi-agent systems (observability, capability discovery, and interruptibility)
- Choosing between Server-Sent Events and WebSockets based on your requirements
- Using PicoAgents WebUI for rapid development and as a production reference
- Deploying agent applications with practical considerations

In previous chapters, we built agents that reason and use tools (Chapter 4), workflows for explicit control (Chapter 6), and orchestration patterns for autonomous coordination (Chapter 7). These all work from the command line:

```
# What you've been doing so far
response = agent.run("What's the weather in Paris?")
print(response.messages[-1].content)
```

This works for development and testing, but real applications need web interfaces where users can interact with your agents through their browsers. This chapter shows how to build web UIs for the agents you've already created—no new agent concepts, just new ways to make them accessible.

If you have already started looking into the codebase for picoagents, you will notice that it has the concept of a WebUI - a web application that lets you visually interact with any agent, orchestrator or workflow built with the picoagents library. To get a sense of how it works, let's start by using it.

```python
from picoagents import Agent
from picoagents.webui import serve

weather_agent = Agent(
    name="weather_assistant",
    model="gpt-4.1-mini",
    instructions="You are a helpful assistant",
)
# One line adds a complete web UI
serve(entities=[weather_agent], port=8070)
```

In the code sample above, we define a simple weather agent (the same one from Chapter 4) and then call `serve()`, passing in our agent. This single line starts a web server with a complete React-based UI that lets you chat with your agent, see tool calls in real-time, and maintain conversation history.

Run this and open `http://localhost:8070`:

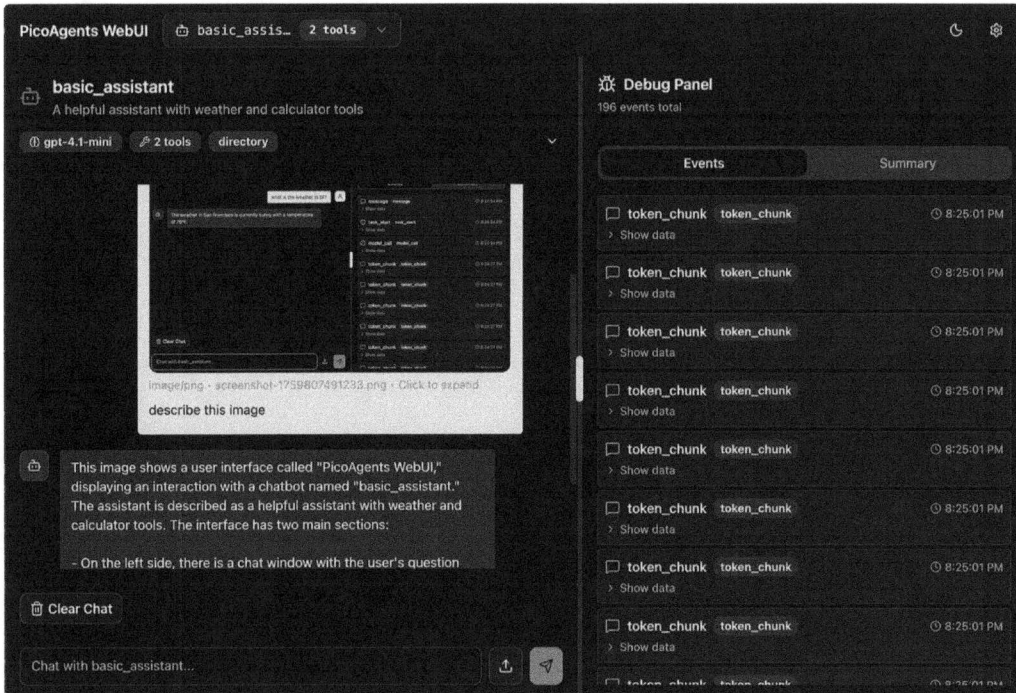

Figure 8.1. A screenshot of the PicoAgents WebUI

You now have a web interface where you can chat with your agent, see tool executions in real-time, and maintain conversation history. The `serve()` function handles everything—HTTP endpoints, streaming, and a React frontend.

This is perfect for:

- **Development**: Test agents during implementation
- **Demos**: Show stakeholders what your agents can do
- **Internal tools**: Give your team access without installing Python

But often times you might need a *custom* UI. Maybe you want to integrate the agent into an existing application, customize the interface, or understand how it works under the hood. The rest of this chapter shows how to build your own agent web application from scratch, demonstrating how UX design principles from Section 3.4 translate into working code.

8.1   Why Agent Applications Are Different

Before we build, let's recap our understanding of why agent applications might need different architecture than traditional web apps.

When you use a tool like `picoagents WebUI` or ChatGPT or Claude web interfaces, in response to your query e.g., "what is the weather in Paris?" you will likely see text appear word-by-word - in chunks. Traditional web apps don't work this way—they show a loading spinner, then display the complete result. Agent applications instead often involve the generation of lengthy text, interleaved with tool calls that can take a while. This necessitates a different approach to user experience.

Tasks run for minutes or hours, not seconds. Users can't wait staring at a spinner while an agent researches a topic, calls multiple tools, and synthesizes results. They need to see progress: text/code/images/audio as they are generated by a model, which tool is running, and what results came back.

Agent coordination needs visibility. When multiple agents work together (see Chapter 2), users need to see their conversations and handoffs. Which agent is speaking? What did they contribute? How are they coordinating?

Some agents need mid-task input. An agent might need clarification or approval before proceeding. The UI must handle two-way communication while the task runs.

8.2   The Two Components of Agent Web Applications

Every agent web application—regardless of technology choices—has two main components that work together: the **Backend** (which handles all agent execution and API logic) and the **Frontend** (which provides the user interface).

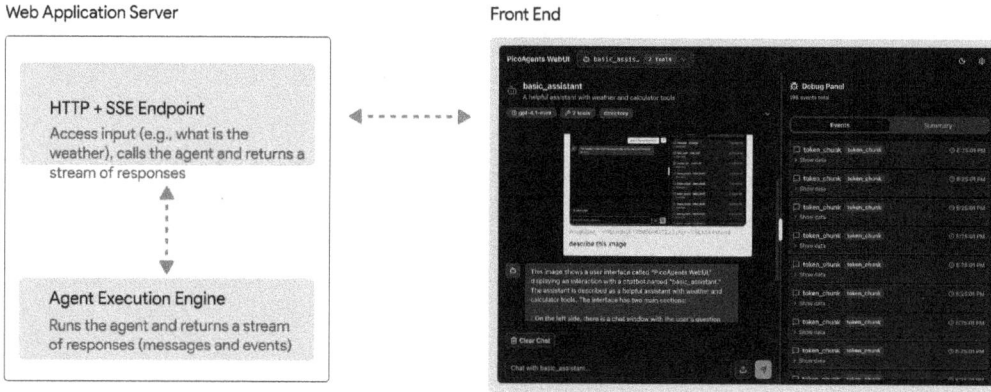

Figure 8.2. *An agent application consists of two parts: Backend API that handles agent execution and Frontend UI that allows the user to provide a query and see results*

8.2.1 Backend Component: Agent Execution and API

The backend encompasses everything that runs on your server: the agent logic, API endpoints, and communication protocols. It consists of two integrated parts:

8.2.1.1 Agent Execution Logic

This part contains your AI logic, using the same agent patterns we've developed in previous chapters:

```
# From Chapter 4 - the core agent logic
weather_agent = Agent(
    name="weather_assistant",
    model="gpt-4.1-mini",
    instructions="Help users check weather",
    tools=[get_weather]
)
for event in weather_agent.run_stream("What's the weather in Paris?"):
    print(event)
```

Key characteristic: This logic focuses purely on agent implementation, independent of how it's exposed to users. The same agent code can run from multiple interfaces—command line, web, mobile app, or Slack bot. While we use PicoAgents in this book, you can use any agent framework here: AutoGen, LangChain, or custom code.

8.2.1.2 API and Communication Component

To run agents through a web interface, we need a way to accept user input (e.g., "What's the weather?") and stream updates back as the agent works. This API component enables that by implementing a communication protocol between the web server (where your agents run) and the user interface. The protocol choice here is crucial for delivering the streaming, real-time updates that make agent interactions feel responsive:

```python
# Example: FastAPI with SSE streaming
@app.post("/chat/stream")
async def chat_stream(request: ChatRequest):
    # Bridge: Convert agent events to HTTP stream
    async def event_generator():
        async for event in weather_agent.run_stream(request.message):
            yield f"data: {event.model_dump_json()}\n\n"

    return StreamingResponse(event_generator(), media_type="text/event-stream")
```

> 💡 Technology Stack Recommendations
>
> **Use FastAPI for agent applications.** Its async-native design handles streaming patterns well—no threading complexity, automatic request/response serialization with Pydantic, and built-in API documentation. FastAPI is production-ready with a mature ASGI ecosystem: deploy with Uvicorn for development, Gunicorn + Uvicorn workers for production, and integrate with nginx, Docker, and cloud platforms. I've built agent apps with Flask (too synchronous for streaming), Express (more boilerplate), and Django (overkill for most agent APIs). FastAPI works better for the async streaming patterns agents need.
>
> **Stick with Server-Sent Events for streaming.** SSE provides unidirectional streaming from server to client using simple HTTP connections—exactly what you need for progress updates, token streaming, and tool execution visibility. WebSockets add bidirectional complexity you rarely need, and HTTP polling is inefficient for real-time updates. The FastAPI + SSE combination gives you everything required for responsive agent UIs with minimal infrastructure complexity.

8.2.2   Frontend Component: User Interface

The frontend renders agent responses and captures user input. It consumes the backend API and presents a coherent experience to users:

```javascript
// Example: Vanilla JS consuming SSE
const response = await fetch('/chat/stream', {
    method: 'POST',
    body: JSON.stringify({ message: userInput })
});

const reader = response.body.getReader();
```

```
while (true) {
    const { done, value } = await reader.read();
    if (done) break;

    // Parse and render events
    displayAgentResponse(value);
}
```

Key characteristic: The frontend knows how to render events and capture user input, but doesn't need to understand agent internals—it just consumes the backend API.

8.2.3   Why This Separation Matters

These two components are **independently replaceable.** Table 8.1 summarizes how each component can be swapped independently:

Table 8.1. Component independence in agent applications allows you to change backend frameworks or frontend technologies without affecting the other part of your system.

Component	Your Choices	Independence
Backend	Agent Framework: PicoAgents, AutoGen, LangChainAPI: FastAPI+SSE, Express+WebSocket, Flask+polling	Swap frameworks or protocols without changing frontend
Frontend	React, Vue, Streamlit, vanilla JS	Swap without changing backend logic

This means you can:

- Start with vanilla JS, upgrade to React later (same backend)
- Switch from FastAPI to Express (same frontend)
- Use the same agent in web, mobile, and CLI (different frontend implementations)

8.2.4   Two Approaches to Building Agent UIs

You can build these components in two ways:

Python-native frameworks like Streamlit and Chainlit let you define both backend and frontend using Python alone. Write your agent logic, add a few framework-specific decorators, and you have a working web UI. Perfect for rapid prototyping and internal tools.

Building from scratch gives you complete control. You implement your own API endpoints and choose your frontend technology—vanilla JavaScript, React, Vue, or any other framework. This approach requires more code but enables custom user experiences essential for production applications.

The choice depends on your constraints: development speed vs. customization needs, team expertise, and whether you're building internal tools or customer-facing products.

8.3   Python-Native Frameworks

Python-native frameworks eliminate frontend development entirely. You write Python code for both agent logic and UI, and the framework handles HTTP endpoints, streaming, and browser rendering.

Two frameworks dominate this space, each optimized for different use cases. **Streamlit** provides extensive data visualization components and works particularly well for analytical dashboards and data-driven agent interfaces. **Chainlit**, by contrast, focuses specifically on conversational AI applications, offering built-in chat interfaces with native streaming support that makes it ideal for chatbot-style agents.

```python
# Streamlit - agent interface in ~10 lines
import streamlit as st
from picoagents import Agent

st.title("News Analysis Agent")
if prompt := st.chat_input("Enter news topic"):
    with st.chat_message("assistant"):
        response = agent.run(prompt)
        st.write(response)

# Chainlit - conversational interface
import chainlit as cl
from picoagents import RoundRobinOrchestrator

@cl.on_message
async def handle_message(message: cl.Message):
    async for event in orchestrator.stream_events(message.content):
        await cl.Message(content=event.message, author=event.agent).send()
```

Benefits: No frontend expertise needed (Python only), rapid prototyping (minutes to working UI), built-in streaming and chat components, low maintenance (framework handles updates).

Limitations: Constrained customization (framework paradigms limit design), performance issues with complex interactions, less suitable for customer-facing applications requiring polished UX.

Use when: Building internal tools, creating demos or prototypes, your team lacks frontend developers, simple interaction patterns meet your needs.

Table 8.2 compares the trade-offs between Python frameworks and building from scratch:

Table 8.2. Python frameworks enable rapid prototyping but with limited customization, while building from scratch provides full control at the cost of development time and requiring frontend expertise.

Factor	Python Frameworks	Building from Scratch
Development speed	Minutes to prototype	Days to weeks
Customization	Limited by framework	Full control
Frontend expertise	None (Python only)	Required
Best for	Internal tools, demos	Production apps, custom UX

Teams often start with Python frameworks for rapid validation, then migrate high-value applications to custom solutions as requirements evolve.

> 💡 Framework Choice
>
> **For production apps, consider building from scratch.** Python frameworks like Streamlit work well for prototypes and internal tools, but you'll often outgrow them when you need custom streaming patterns, complex state management, or specific UX flows. The patterns in this chapter show you how to build exactly what you need without framework constraints.

8.4 Building a Complete Agent Web Application from Scratch

In the following section, we will walk through building a minimal but complete agent web app in ~200 lines of code. As we build, we'll implement the UX principles from Section 3.4 where practical—specifically **observability** (showing agent actions in real-time) and **capability discovery** (guiding users toward reliable tasks). You can find the full working example in `examples/app/` in the PicoAgents repository.

8.4.1 Project Structure

```
examples/app/
├── backend/
│   └── app.py
└── frontend/
    └── index.html
```

The agent logic is defined in `backend/app.py` —it's the same weather agent we've been using. YOu can begin by creating this folder structure and empty versions of each file.

8.4.2 Backend Implementation

To implement our backend server, we will use the FastAPI library to create a web application backend with several web endpoints. The goal of the web endpoint is to expose your agent, over a web protocol such that other entities such as our web interface can call this ending.

First we will import fast api and define our agent

```python
# backend/app.py
from fastapi import FastAPI
from fastapi.responses import StreamingResponse
from pydantic import BaseModel

app = FastAPI(title="Minimal Agent Web App")

def get_weather(location: str) -> str:
    """Get weather for a location."""
    return f"Weather in {location}: Sunny, 72°F"

# Create our agent
weather_agent = Agent(
    name="weather_assistant",
    description="A weather assistant",
    model_client=OpenAIChatCompletionClient(model="gpt-4.1-mini"),
    instructions="Help users check weather using the get_weather tool",
    tools=[get_weather],
)
```

Next, we will define a web endpoint that can take a request (e.g., *What is the weather in Paris*), pass that to our agent and then return a *stream* of responses.

```python
class ChatRequest(BaseModel):
    message: str

async def stream_agent_events(message: str):
    """Stream agent execution as Server-Sent Events."""
    async for event in weather_agent.run_stream(
        message,
        stream_tokens=True):
        # Format as SSE: "data: " prefix + JSON + double newline
        yield f"data: {event.model_dump_json()}\n\n"

@app.post("/chat/stream")
async def chat_stream(request: ChatRequest):
    """Handle chat requests with streaming responses."""
    return StreamingResponse(
```

```
        stream_agent_events(request.message),
        media_type="text/event-stream",
        headers={
            "Cache-Control": "no-cache",
            "Connection": "keep-alive",
        },
    )
```

In the code snippet above, first, we set up FastAPI to create a web server that accepts HTTP requests. The key function here is `@app.post("/chat/stream")` route definition, which handles the actual connection between the agent and external callers (e.g., a web interface). When a user sends a message, this function passes it to `weather_agent.run_stream()`, converts each agent event into Server-Sent Events format (the SSE protocol), and streams them back to the browser in real-time. The `@app.post("/chat/stream")` decorator creates the HTTP endpoint that users connect to for chatting with the agent.

The SSE format is elegantly simple—each event is formatted as `data: <json>\n\n`, a standard protocol that browsers understand natively without requiring special libraries. We set `stream_tokens=True` to make the agent emit individual tokens as they're generated, creating the familiar "typing" effect you see in ChatGPT.

Once you have this file completed, you can try the endpoint we just created using the following code

```
# Install dependencies
pip install picoagents fastapi "uvicorn[standard]"

# Set your API key
export OPENAI_API_KEY=your-key-here

# In one terminal
python app.py

# In another terminal
curl -N -X POST http://localhost:8000/chat/stream \
  -H "Content-Type: application/json" \
  -d '{"message": "What is the weather in Paris?"}'
```

You'll see events streaming in real-time!

8.4.3   Frontend Implementation

Now that we have a backend api, we can proceed to creating a web interface that lets a user interact with this endpoint. We'll break this down into two logical sections: first setting up the connection and receiving the stream, then parsing and rendering the events.

```
// frontend/index.html (connection setup)
async function sendMessage() {
    const message = inputEl.value.trim();

    // Add user message to UI
    addMessage(message, 'user');

    // Create empty assistant message for streaming
    currentMessage = addMessage('', 'assistant');

    // Fetch with POST to /chat/stream
    const response = await fetch('http://localhost:8000/chat/stream', {
        method: 'POST',
        headers: { 'Content-Type': 'application/json' },
        body: JSON.stringify({ message }),
    });

    // Read SSE stream
    const reader = response.body.getReader();
    const decoder = new TextDecoder();
    let buffer = '';
```

This first section handles the user interaction and establishes the streaming connection. When a user types a message, we immediately display it in the chat interface and create an empty placeholder for the agent's response. The key insight here is creating the assistant message upfront—this gives us something to update as tokens stream in, creating that characteristic "typing" effect. We then make a POST request to our backend endpoint, and rather than waiting for a complete response, we get a reader that lets us process data as it arrives. The buffer variable will help us handle partial data—network packets don't always align with complete events.

```
// frontend/index.html (event processing)
while (true) {
    const { done, value } = await reader.read();
    if (done) break;

    // Decode chunk
    buffer += decoder.decode(value, { stream: true });

    // Process complete SSE messages (separated by \n\n)
    const lines = buffer.split('\n\n');
    buffer = lines.pop(); // Keep incomplete message

    for (const line of lines) {
        if (line.startsWith('data: ')) {
```

```
            const event = JSON.parse(line.slice(6));

            // Handle different event types
            if (event.content && event.role === 'assistant') {
                // Streaming token - append to current message
                currentMessage.textContent += event.content;
            } else if (event.event_type === 'tool_call_started') {
                // Show tool execution
                addMessage(`▢ Calling ${event.tool_name}...`, 'thinking');
            }
        }
    }
}
}
```

The second section above handles the actual streaming data processing. We read chunks of data as they arrive from the server, decode them from bytes to text, and add them to our buffer. The SSE protocol defines events as lines beginning with `data:` and ending with double newlines (`\n\n`), so we split on that pattern. Here's the subtle part: we pop the last element from our split array and keep it in the buffer. Why? Because a network packet might arrive with only half of an event—splitting would break it into pieces. By keeping the incomplete part in the buffer, we ensure it gets combined with the next chunk to form a complete event.

Once we have complete events, we parse the JSON and update the UI accordingly. When we receive content from the assistant, we append it directly to the message we created earlier—this is what creates the smooth token-by-token appearance of text. When we see a tool call starting, we add a visual indicator so users understand the agent is taking action—implementing the **observability** principle from Section 3.4.3. The frontend doesn't need to understand agent reasoning or tool execution logic—it just needs to know how to render different event types. This separation means your backend agent can become arbitrarily complex—using multiple tools, calling other agents, or implementing sophisticated reasoning—without requiring frontend changes. The frontend simply consumes events and renders them appropriately, maintaining a clean separation between agent logic and user interface.

8.4.4 Adding Capability Discovery

To implement the **capability discovery** principle (Section 3.4.1), we add preset prompts that guide users toward tasks the agent handles well. This is particularly important when users first encounter the interface and don't know what to ask.

First, create a container with example prompts:

```
<div class="preset-prompts" id="presets">
    <p>Try these examples:</p>
```

Then add buttons that populate the input field when clicked:

```html
<button class="preset-btn"
        onclick="usePreset('What is the weather like in Paris?')">
    What is the weather like in Paris?
</button>
<button class="preset-btn"
        onclick="usePreset('Tell me about the weather in Tokyo')">
    Tell me about the weather in Tokyo
</button>
<button class="preset-btn"
        onclick="usePreset('How is the weather today?')">
    How is the weather today?
</button>
</div>
```

The presets disappear after the first message, keeping the interface clean once users understand the agent's capabilities. This simple pattern—showing 2-3 example prompts on the empty state—helps users discover what the agent can do without overwhelming them with documentation.

8.4.5 Adding Interruptibility

Agent tasks can run for minutes or hours, consuming API resources and costs. The **interruptibility** principle (Section 3.4.4) requires letting users stop execution mid-stream as needed. While SSE only sends data from server to client, we can implement cancellation by detecting when the client closes the HTTP connection.

The pattern has three parts: (1) a Stop button in the UI, (2) the frontend closes the HTTP connection when clicked, (3) the backend detects the closed connection and stops the agent. Browsers provide a built-in `AbortController` API for canceling HTTP requests—when you call `abort()`, the connection closes immediately. This implements **task cancellation** (Section 4.3), giving users control over long-running operations.

Frontend: Create an `AbortController`, pass its signal to `fetch()`, and call `abort()` when the user clicks Stop:

```javascript
// Create abort controller when starting request
abortController = new AbortController();

fetch('/chat/stream', {
    method: 'POST',
    body: JSON.stringify({ message }),
    signal: abortController.signal  // Enable cancellation
});
```

```
// Toggle Send button to Stop while streaming
if (loading) {
    sendBtn.textContent = 'Stop';
    sendBtn.onclick = () => abortController.abort();
}
```

Backend: When the client closes the connection, Python's async runtime raises `GeneratorExit`. We catch this exception and use PicoAgents' `CancellationToken` to stop the agent:

```python
from picoagents import CancellationToken

async def stream_agent_events(
    message: str,
    cancellation_token: CancellationToken
):
    try:
        async for event in weather_agent.run_stream(
            message,
            stream_tokens=True,
            cancellation_token=cancellation_token  # Pass to agent
        ):
            yield f"data: {event.model_dump_json()}\n\n"
    except (GeneratorExit, asyncio.CancelledError):
        # HTTP connection closed - stop agent execution
        cancellation_token.cancel()
        raise
```

The `CancellationToken` propagates through the agent pipeline to stop LLM calls and tool execution immediately.

You can now run the app from the command line.

```
# Run the server
cd examples/app/backend
python app.py
```

Open `http://localhost:8000` in your browser. You'll have a working agent chat interface.

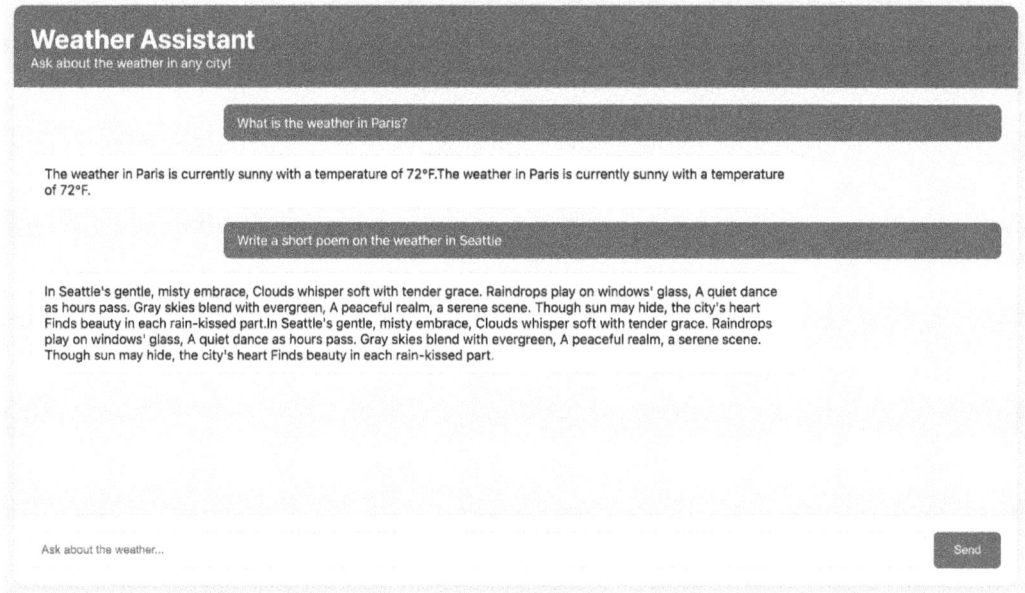

Figure 8.3. A screenshot of the minimal agent web app interface

8.5 From Vanilla JavaScript to React

The minimal app uses vanilla JavaScript—perfect for learning SSE streaming and understanding the core pattern. But vanilla JavaScript becomes challenging when state management gets complex. Imagine tracking conversation history, multiple agent sessions, typing indicators, error states, and tool execution status across dozens of components. Vanilla JS requires manual state synchronization—forgetting to update one variable causes bugs.

Production agent UIs need capabilities you see in ChatGPT or Claude: syntax-highlighted code blocks with copy buttons, smooth streaming animations, conditional rendering of tool calls, and responsive layouts across devices. The minimal app has ~50 lines of JavaScript for basic chat. A production UI with multi-agent orchestration, session management, and rich tool displays would be ~500+ lines of vanilla JS—hard to maintain.

Modern frameworks solve these problems. Several excellent options exist: **React** offers the largest ecosystem and extensive documentation, **Vue** provides a gentler learning curve with excellent documentation, **Svelte** compiles to highly optimized vanilla JavaScript, and **Angular** delivers a comprehensive framework for enterprise applications. All of these frameworks provide component-based architectures, declarative rendering, and state management solutions suitable for agent UIs.

> 💡 React for Agent UIs
>
> **I recommend React for agent UIs.** Yes, Vue is easier to learn and Svelte is more performant, but React's ecosystem dominance matters more for agent applications. You'll find more libraries for syntax highlighting, markdown rendering, and real-time streaming. More importantly, when you're debugging complex agent state or need to extend your UI, LLMs and AI coding assistants perform significantly better with React code—they've seen vastly more React examples during training. This practical advantage outweighs theoretical framework benefits.

The PicoAgents WebUI uses React because it handles multiple entity types (agents, orchestrators, workflows), complex session management, rich tool execution visualizations, and orchestration pattern displays.

The good news: The backend stays the same! The API component (FastAPI + SSE) doesn't care if the frontend is vanilla JS or React. Once you understand the pattern with vanilla JS, upgrading to React just makes the UI code easier to manage.

Table 8.3 provides guidance on when to use vanilla JavaScript versus React/Vue:

Table 8.3. Vanilla JavaScript is sufficient for learning and simple UIs, while React/Vue becomes necessary for complex production applications with extensive UI code.

Use Vanilla JS	Use React/Vue
Learning the SSE pattern	Production applications
Simple single-agent UIs	Multi-agent orchestration UIs
Quick internal tools	Customer-facing products
< 200 lines of UI code	> 500 lines of UI code

8.5.1 Building and Deploying React Applications

React uses JSX syntax (HTML-like code in JavaScript) and modern JavaScript features that browsers don't understand directly. You write components in React's syntax with all its rich interactions, but this code must be compiled down to vanilla JavaScript, HTML, and CSS that browsers can execute. This compilation process is called "building."

A build tool handles this transformation. **Vite** is the recommended choice for agent UIs—it provides near-instant hot reload during development (save a file, see changes in milliseconds) and creates optimized production bundles with minimal size. Alternatives like Next.js add server-side rendering (useful for SEO-heavy sites), but Vite's simplicity and speed work well for agent interfaces.

Beyond the build tool, you need components and styling. **shadcn/ui** provides accessible UI elements (proper ARIA attributes, keyboard navigation) that you copy into your project and customize—you own the code rather than depending on a package. **Tailwind CSS** offers utility-first styling that lets you style inline without switching between files, with mobile-first responsive design built-in.

8.5.2   Learning from PicoAgents WebUI

The PicoAgents WebUI demonstrates these patterns in production. It uses React + TypeScript + Vite with shadcn/ui components and Tailwind styling, following the same FastAPI + SSE backend pattern you built in the minimal app. The complete source is in `picoagents/webui/frontend/`.

Rather than walking through the implementation, Table 8.4 highlights key areas to study:

Table 8.4. *PicoAgents WebUI codebase demonstrates production patterns for SSE streaming, message rendering, tool visualization, session management, and code display that you can adapt for your own applications.*

What to Learn	Where to Look	Why It Matters
SSE event parsing	`src/api/client.ts`	Converting SSE streams to React state updates
Message streaming	`src/components/Chat/Message.tsx`	Token-by-token text appearance as events arrive
Tool execution UI	`src/components/ToolCall.tsx`	Collapsible tool calls with syntax-highlighted arguments
Session management	`src/hooks/useSession.ts`	Switching between conversations, history persistence
Code rendering	`src/components/CodeBlock.tsx`	Syntax highlighting, copy button, language detection

Start with `src/App.tsx` to understand the overall structure, then explore the components that match your needs. The architecture is designed for extension—you can add custom message renderers for domain-specific outputs, new tool result visualizations, or authentication flows without modifying the core.

> ℹ Learning React
>
> This book focuses on multi-agent systems, not React development. If you're new to React, start with the official React tutorial which covers components, state, and effects. For TypeScript integration, see the React TypeScript Cheatsheet. The Vite guide covers build configuration.

The good news: whether you use vanilla JS, React, or any other frontend framework, the backend architecture stays the same. The SSE + FastAPI pattern works identically—only the frontend consumption of events changes.

Now that we understand how to build the frontend and backend components, we need to make a crucial architectural decision that affects everything from user experience to deployment complexity: how should these components communicate?

8.6 WebSockets vs Server-Sent Events: Choosing Your Streaming Strategy

Agent applications need real-time communication between frontend and backend. Two main approaches exist: **WebSockets** (bidirectional, persistent connections) and **Server-Sent Events** (unidirectional streaming over HTTP).

The choice between these protocols is often misunderstood. **The application design determines user experience, not the protocol itself.** A well-designed SSE application can feel just as responsive as WebSockets to users, while being dramatically easier to scale.

The fundamental trade-off isn't about protocols—it's about **where you manage complexity**: in your application code or in your infrastructure. Both approaches can support interactive agents where users provide feedback mid-task, but they handle this feedback differently, which determines where you pay the complexity cost and how your application scales.

WebSockets maintain persistent connections with state living in memory during the entire conversation. User feedback happens on the same connection (synchronous), providing simple application logic but requiring complex infrastructure with session affinity.

SSE + Context Management streams progress then saves conversation context and closes connections between interactions. User feedback happens via the same endpoint with different request types (asynchronous), requiring application-level context management but enabling simple, stateless infrastructure.

Here's the key insight: **users don't care about the underlying protocol.** They care about getting timely responses and seeing progress. A stateless SSE application can provide exactly the same user experience as WebSockets. Consider this timeline:

```
# User starts a task at 9:00 AM
POST /chat/stream {"message": "Analyze our deployment and fix any issues"}
# Agent investigates, calls tools, streams progress
# Agent needs approval for sensitive operation, saves context, closes stream

# User reviews at 2:30 PM (5+ hours later)
POST /chat/stream {"approved": true, "session_id": "abc123"}
# Agent loads context, continues exactly where it left off
```

```
# User sees immediate response as if agent "remembered" everything
```

To the user, this feels identical to a persistent WebSocket connection. The agent "remembers" the entire conversation, continues reasoning seamlessly, and provides the expected response. But behind the scenes, no resources were consumed during the 5+ hour wait.

WebSocket Implementation keeps connections open with agent state in memory:

```
@app.websocket("/ws")
async def agent_websocket(websocket: WebSocket):
    await websocket.accept()
    conversation_state = []  # Lives in memory

    while True:
        user_input = await websocket.receive_json()
        # Process and respond immediately
        async for event in agent.run_stream(
            user_input,
            context=conversation_state):
            await websocket.send_json(event.model_dump())
```

This approach works well for real-time collaborative features, gaming scenarios, or when you have non-serializable state like loaded ML models. However, it requires session affinity—each user must stick to the same server instance, and if a server goes down, all connections break.

SSE + Context Management handles the same interactions through stateless request-response cycles:

```
@app.post("/chat/stream")
async def chat_stream(request: ChatRequest, session_id: str):
    # Load conversation context
    context = await session_manager.get_or_create(session_id)

    if request.message:
        context.add_message(UserMessage(content=request.message))
    elif request.approved is not None:
        # Handle approval response through same endpoint
        context.add_approval_response(approved=request.approved)

    async def event_generator():
        async for event in agent.run_stream(context.messages):
            yield f"data: {event.model_dump_json()}\n\n"
        # Save updated context when stream ends
        await session_manager.save(session_id, context)

    return StreamingResponse(event_generator(), media_type="text/event-stream")
```

This stateless design enables powerful workflows that WebSockets cannot support efficiently:

approval workflows where agents wait hours/days for user review, **background processing** where agents continue working while users are offline, **multi-device continuity** where conversations seamlessly move between devices, and **cost-sensitive operations** where agents hibernate between expensive LLM calls.

Table 8.5 summarizes the key differences between WebSockets and SSE with context management:

Table 8.5. WebSockets provide true bidirectional real-time communication but require session affinity and consume persistent memory, while SSE with context management enables the same user experience with stateless infrastructure that scales horizontally and supports async workflows with hours or days of delay.

Factor	WebSockets	SSE + Context Management
User experience	Real-time bidirectional	Feels real-time, actually async
Infrastructure	Session affinity required	Standard load balancing
Resource usage	Persistent memory consumption	Resources only during processing
Async workflows	Limited (connection timeouts)	Full support (hours/days delays)
Scalability	Vertical (bigger servers)	Horizontal (more servers)

Most applications run for months on single-server deployments before needing to scale, but when you do need to scale, the architecture choice matters significantly. SSE-based applications can add server instances behind standard load balancers with no special configuration—any server can handle any request since context is fetched/saved per interaction. WebSocket-based applications require session affinity configuration to ensure users stick to the same server instance, adding infrastructure complexity for interactive capabilities.

> 💡 Practical Recommendation
>
> Start with **SSE + context management** for most agent applications. The stateless design enables async workflows that WebSockets simply cannot support efficiently, while providing the same responsive user experience. Your application will scale horizontally, deploy anywhere, and support advanced workflows like approval delays and background processing.
> Only choose WebSockets if you have specific requirements for real-time bidirectional communication during active execution (like collaborative editing) or need to maintain non-serializable state.

Serverless functions seem appealing but don't work well for agent applications. Most agent tasks run longer than the 15-minute function limits, and cold starts kill the responsive feel users expect. GPU instances only matter if you're running local LLM inference, but API services like OpenAI are more cost-effective for most applications until you reach massive scale.

8.7 Deployment and Scaling Considerations

Most agent applications start simple and scale incrementally based on actual usage.

8.7.1 Start Simple: Single-Server Deployment

Most agent applications begin as single-server deployments—and that's perfectly fine. You can serve hundreds of concurrent users from a single server with good performance. The complexity of container orchestration and advanced scaling patterns only becomes necessary as you grow.

Start with basic containerization using Docker and docker-compose. This provides consistency across environments and simplifies deployment:

```
# Multi-stage Dockerfile for agent application
FROM node:18-alpine AS frontend-build
WORKDIR /app/frontend
COPY frontend/package*.json ./
RUN npm install
COPY frontend/ .
RUN npm run build

FROM python:3.11-slim AS backend
WORKDIR /app
COPY requirements.txt .
RUN pip install -r requirements.txt
COPY backend/ .
COPY --from=frontend-build /app/frontend/dist ./static
```

```
EXPOSE 8000
CMD ["uvicorn", "main:app", "--host", "0.0.0.0", "--port", "8000"]
```

For initial deployments, a single-server setup with docker-compose handles moderate loads effectively:

```
# docker-compose.yml for simple deployment
version: '3.8'
services:
  agent-app:
    build: .
    ports:
      - "80:8000"
    environment:
      - OPENAI_API_KEY=${OPENAI_API_KEY}
      - DATABASE_URL=${DATABASE_URL}
    volumes:
      - agent_data:/app/data
    restart: unless-stopped

  redis:
    image: redis:alpine
    volumes:
      - redis_data:/data
    restart: unless-stopped

volumes:
  agent_data:
  redis_data:
```

Only add complexity when you have concrete evidence that you need it—monitoring showing connection limits, response time degradation, or feature requirements that your current setup can't handle.

8.7.2   When to Scale Beyond a Single Server

You'll know it's time to scale when you see concrete symptoms: response times degrading, connection limits being hit, or servers struggling with concurrent agent executions. Don't optimize prematurely—scale when you have evidence.

Your streaming protocol choice (covered in Section 8.6) fundamentally determines your scaling path. SSE-based applications can add server instances behind standard load balancers with no special configuration. WebSocket-based applications require session affinity configuration to ensure users stick to the same server instance.

For most agent applications, cloud container orchestration platforms like AWS ECS, Azure Container Instances, or Google Cloud Run handle scaling, health checks, and deployments while abstracting infrastructure complexity. With SSE streaming, standard load balancers and CDNs work without additional configuration.

8.8 Exercises

Now that you understand how agent web applications work, try these exercises to deepen your understanding. Each builds on the minimal app we created earlier.

Exercise 1: Enhance the UI with Event Visibility

The minimal app shows basic messages and tool calls, but agents generate many other event types. Modify the frontend to display additional events like thinking steps, tool results, and completion status. Add a "Clear Conversation" button that resets the chat interface. Consider how you'll distinguish different event types visually—perhaps using different colors, icons, or expandable sections for detailed information.

Exercise 2: Add Multiple Tools and Observe Coordination

Extend your agent with additional tools beyond the simple weather function. For example, add a calculator tool, a web search tool, or a file reading tool. Test queries that require using multiple tools in sequence ("What's the weather in Paris and what's 72°F in Celsius?"). Observe how the UI shows the agent coordinating between different tool calls. This exercise reveals how tool availability affects agent behavior and why clear tool execution visibility matters to users.

Exercise 3: Implement Conversation History Using AgentContext

The current minimal app treats each request independently—the agent doesn't remember previous messages. PicoAgents provides `AgentContext` for managing conversation history, but our minimal example doesn't use it yet. Modify the backend to create and maintain an `AgentContext` object for each session, storing it in memory (using a dictionary keyed by session ID). When processing requests, pass both the new message and the existing context to `agent.run_stream(task=message, context=session_context)`. The agent will automatically update `context.messages` with the conversation history. You'll need to: (1) generate session IDs (you can use UUIDs sent from the frontend), (2) retrieve or create the context for each session, (3) pass the updated context to subsequent requests. This exercise demonstrates how PicoAgents separates agent logic (stateless) from conversation state (managed via context), following the same pattern used in PicoAgents WebUI (see `_sessions.py` for reference).

Exercise 4: Add Cost-Aware Delegation with Tool Approval

Implement the **cost-aware action delegation** principle from Section 3.4.2 using the tool approval mechanisms from Section 4.13. Add a consequential tool (like `send_email` or `delete_file`) and configure it to require approval before execution using the

`@tool(approval_mode="always_require")` decorator. When the agent attempts to use this tool, pause execution and display an approval dialog in the UI showing the tool name and arguments. The user can approve or reject, and the agent resumes accordingly. This requires: (1) marking tools as requiring approval (Section 4.13 shows the pattern), (2) detecting `ToolApprovalRequest` events in your frontend stream, (3) displaying an approval UI with buttons for approve/reject, (4) sending the approval response back via a new request to continue execution. This exercise demonstrates how human-in-the-loop patterns from Chapter 4 integrate with web UIs to give users control over consequential actions.

Exercise 5: Production Considerations

Before deploying your agent web app, address three key production concerns: (1) **Authentication**—add API key validation or OAuth to your FastAPI endpoints to control access, (2) **Observability**—integrate OpenTelemetry following the patterns in Section 4.10 to track agent operations, API costs, and performance across your deployment, (3) **Rate limiting**—use FastAPI middleware (like `slowapi`) to limit requests per user and prevent abuse, since agent applications can make many expensive LLM calls. These aren't agent-specific challenges—they're standard web application concerns that apply to *any production API*.

8.9 Summary

- **Real-time streaming communication is required** for agent applications due to their long-running, multi-step nature. Choose WebSockets for bidirectional interactive workflows or Server-Sent Events for unidirectional progress monitoring, based on your specific use case requirements.

- **Three UX principles from Chapter 3 translate directly to implementation**: observability (streaming agent events in real-time), capability discovery (preset prompts guiding users to reliable tasks), and interruptibility (cancellation via AbortController + CancellationToken). These patterns make agent UIs predictable and controllable despite AI unpredictability.

- **SSE's unidirectional data flow doesn't prevent cancellation**: By catching `GeneratorExit` when the client closes the HTTP connection, we propagate cancellation through the agent pipeline via `CancellationToken`, stopping LLM calls and tool execution. This pattern saves resources for long-running tasks without WebSocket complexity.

- **Backend architecture requires integration** between your agent library and streaming infrastructure. FastAPI provides async support for both WebSocket and SSE implementations, with SSE offering simpler deployment and horizontal scaling.

- **Frontend development requires specific patterns** for handling streaming agent data and complex state management. React's component architecture and state management work well for building responsive interfaces that visualize multi-agent coordination patterns.

- **Framework selection depends on use case and constraints**. Streamlit excels for rapid prototyping and data-driven internal tools, Chainlit provides optimized conversational interfaces with built-in streaming, while custom React solutions offer maximum control for production applications.

- **Streaming choice determines deployment complexity**. SSE enables stateless deployment with standard load balancing, while WebSockets require session affinity and more complex infrastructure. Most applications work well with single-server SSE deployments initially, scaling only when evidence shows the need—connection limits, response time issues, or requirements for real-time bidirectional interaction.

Building effective agent applications requires understanding the unique challenges of real-time multi-agent coordination and choosing the right tools for your specific requirements. The patterns and architectures covered in this chapter provide a foundation for creating responsive, scalable agent experiences that bring your multi-agent systems to life for users.

Chapter 9

But What About Multi-Agent Frameworks?

This chapter covers:

- Ten core capabilities that distinguish effective multi-agent frameworks from basic implementations
- Practical evaluation criteria and questions to assess any framework against your specific needs
- Decision frameworks for weighing technical capabilities against organizational constraints
- Strategies for hybrid approaches that combine custom development with framework adoption

If you're reading this book, one of the pressing questions you've likely wrestled with is choosing which framework to use for building multi-agent systems. This has certainly been one of the most common questions I've received, especially during the early days of AutoGen (Microsoft Research 2023). As I write this book, there are dozens of frameworks available—frankly, too many.

At the same time, we've spent the last few chapters literally implementing (assuming you walked through the code with me—I hope you did!) almost everything you'd get from a framework, built from scratch. If you found that process comfortable and energizing, kudos— that's excellent! If you didn't, or if you're rightfully wondering whether this is all it takes, this chapter will help you navigate that decision.

I personally believe that directly comparing frameworks is a losing game—they change constantly. Instead, I'd like to summarize what I consider the core capabilities any framework should excel at if you're going to build with it, regardless of whether you implement it yourself or adopt an existing solution.

By the end of this chapter, you'll have a systematic approach for evaluating multi-agent frameworks that will remain relevant regardless of which specific tools emerge or evolve in this rapidly changing landscape.

9.1   Why Consider a Framework at All?

Well, there are a few reasons to consider using a framework.

Proven Patterns and Best Practices: Frameworks encode solutions to problems you likely haven't encountered yet. They can represent hundreds of hours of debugging edge cases, optimizing performance, and hardening production deployments.

Accelerated Development: While building from scratch provides low-level control of behaviour, frameworks can significantly reduce time-to-market for production applications. They handle the "plumbing" so you can focus on business logic.

Ecosystem and Community: Mature frameworks come with extensive documentation, community support, and pre-built integrations that would take considerable effort to replicate.

Maintenance and Evolution: By far, the biggest benefits that I have seen customers glean from a framework is that framework teams handle critical maintenance work such as updates for new model APIs, security patches, and performance improvements—reducing your long-term maintenance burden.

Standardized, Interoperable Components: I see many developers unfortunately severely discount the impact of standardized APIs that frameworks provide. Frameworks offer interoperable components (agents, tools, memory systems) that work seamlessly together, reducing integration challenges. Even in picoagents, you probably have noticed how simply inheriting from the same `BaseAgent` abstraction means the `ComputerUseAgent` we built can work with any of the multiagent orchestration methods with zero code change. As teams grow, these standardized interfaces significantly reduce onboarding time and integration errors. For example, when a new memory provider becomes available, frameworks either offer compatible implementations or your team can build one with clear behavioral guarantees, whereas custom implementations often require scrambling to retrofit everything.

> **i** Note
>
> One of the most used parts of a framework at least in the case of AutoGen has been the ChatCompletionClient which has support for multiple model providers. Model providers change their apis and behaviours very frequently and it can become a burden on developers to keep up with these changes without a framework to abstract away the differences.

However, frameworks also introduce constraints, learning curves, and potential vendor lock-in. The key is understanding what you're trading off.

9.2 Ten Core Framework Capabilities

Based on years of building and evaluating multi-agent systems, here are the essential capabilities that separate effective frameworks from basic implementations. Where appropriate, we will review examples of how we have addressed these requirements in our minimal `picoagents` library and how leading frameworks like Microsoft Agent Framework, Google ADK, and Pydantic AI approach these capabilities.

9.2.1 Intuitive Developer Experience with Layered Abstractions

A well-designed framework should offer what I call "low floors, high ceilings"—easy entry points for common use cases, but the flexibility to handle specialized requirements without hitting framework limitations.

What to look for:

- High-level APIs that express core concepts (agents, tools, orchestration) naturally
- Low-level APIs for custom behavior when needed
- Progressive disclosure—you only see complexity when you need it
- Clear composition patterns for building complex systems from simple parts
- Framework aligns with your language's idioms and patterns

In `picoagents`, we designed clean APIs following these principles (Section 4.1):

```python
# High-level: simple agent creation
agent = Agent(
    name="assistant",
    model_client=client,
    instructions="You are helpful"
)
response = await agent.run("Write a story")

# Low-level: custom behavior through inheritance
class CustomAgent(BaseAgent):
    async def run_stream(self, task, context=None, cancellation_token=None):
        # Custom reasoning logic
```

We also provided clear abstractions for orchestration (high level patterns for composing agents) and workflows (see Chapter Chapter 7 on implementing orchestration patterns, and Chapter Chapter 6 on implementing workflows) that let you express low level business logic as computation graphs.

Questions to ask:

- Can I implement a basic multi-agent workflow in under 20 lines of code?
- When I need custom behavior, can I extend the framework without forking it?
- Are the abstractions familiar to my team's existing knowledge?
- Does the framework guide me toward good practices or allow easy mistakes?

9.2.2 Async-First Architecture with Streaming Support

Multi-agent systems involve inherently concurrent operations—multiple agents thinking, tools executing, external APIs responding. A framework designed around synchronous operations will create bottlenecks and poor user experiences.

What to look for:

- Native async/await support throughout the API
- Streaming interfaces for real-time updates
- Non-blocking operations for long-running tasks
- Built-in cancellation and timeout handling
- Integration-friendly with modern web frameworks

In `picoagents` , we implemented async-first architecture with event streaming (Section 4.2, Section 4.2.1):

```
# Async execution with streaming and cancellation
token = CancellationToken()
async for event in agent.run_stream("Write a story", cancellation_token=token):
    if isinstance(event, ModelCallEvent):
        print("□ Model thinking...")
    elif isinstance(event, AgentResponse):
        print(f"□ Complete: {event.messages[-1].content}")
```

Questions to ask:

- Can I stream partial results while agents are working?
- How does the framework handle long-running operations?
- Can I cancel in-progress operations gracefully?
- Does it integrate well with my existing async infrastructure?

9.2.3 Comprehensive Observability and Development Tools

Complex multi-agent interactions are difficult to debug, monitor, and develop effectively. Frameworks should provide rich observability into what's happening during execution combined with sophisticated development tools for rapid iteration and understanding system behavior.

What to look for:

- Event streams that capture agent decisions, tool usage, and coordination
- Structured logging with correlation IDs across agent interactions

- Metrics for token usage, latency, and success rates
- Visual multiagent workflow builders and interactive debugging environments
- Configuration comparison, profiling and replay capabilities
- Integration with standard development tools (IDEs, debuggers)
- Real-time monitoring dashboards and alerting systems

In `picoagents`, we implemented structured event streams for observability (Section 4.2.1):

```
# Rich event streams with structured data
async for event in agent.run_stream(task):
    if isinstance(event, ModelCallEvent):
        print(f"□ Model call: {len(event.messages)} messages")
    elif isinstance(event, ToolCallEvent):
        print(f"□ Tool: {event.tool_name}({event.arguments})")
    elif isinstance(event, AgentResponse):
        print(f"□ Total: {event.usage.tokens_total} tokens")
```

Questions to ask:

- Can I understand why an agent made a specific decision?
- How do I track resource usage across complex workflows?
- What happens when something goes wrong—can I debug it effectively?
- How do I test changes to agent configurations quickly?
- Can I visualize and understand complex workflows?
- What tools exist for performance optimization?
- How does the framework integrate with my existing development workflow?

Warning signs:

- Testing requires full production deployments
- Limited profiling or performance analysis tools

9.2.4   Comprehensive State Management and Persistence

Production multi-agent systems need to handle failures, scale dynamically, and maintain conversations across sessions. This requires sophisticated state management beyond simple in-memory storage.

What to look for:

- Serializable agent and conversation state
- Checkpointing capabilities for long-running processes
- Pluggable persistence backends (file, database, cloud storage)
- State migration tools for framework updates
- Clear state isolation between concurrent workflows

In `picoagents`, we implemented memory and context systems for state management (Section 4.7):

```
# Persistent memory with different backends
memory = FileMemory(file_path="conversation.json")
agent = Agent(name="assistant", model_client=client, memory=memory)

# Context carries conversation state
context = AgentContext(messages=[...], memory_variables={...})
response = await agent.run("Continue our discussion", context=context)
```

Questions to ask:

- Can I pause and resume multi-agent workflows?
- How does the framework handle crashes or restarts?
- Can I persist state to my preferred storage system?
- What happens to in-flight operations during scaling events?

Warning signs:

- State corruption risks during concurrent access

9.2.5   Declarative Configuration and Reproducibility

As multi-agent systems grow complex, managing configurations becomes critical. Teams need to version, share, and deploy agent configurations reliably across environments.

What to look for:

- Agent specifications as code or configuration files
- Version control friendly formats (JSON, YAML, etc.)
- Environment-specific configuration management
- Reproducible builds and deployments
- Configuration validation and schema support

In `picoagents` , we implemented component serialization:

```
# Serialize complete agent configurations
agent = Agent(name="assistant", model_client=client, tools=[search_tool])
config = agent.to_config()  # JSON-serializable configuration

# Load from configuration
loaded_agent = Agent.from_config(config)

# Works across all components (orchestrators, workflows, etc.)
orchestrator = RoundRobinOrchestrator(agents=[agent1, agent2])
orchestrator_config = orchestrator.to_config()
```

Questions to ask:

- Can I define agents declaratively and version them?

- How do I manage configuration differences across dev/staging/production?
- Can team members easily share and modify agent configurations?
- Are there safeguards against invalid configurations?

---

i The Importance of Configuration-Driven Design

I have often seen teams discount or overlook the value of a configuration driven design. Often this is a mistake especially as multiagent systems have many moving parts, many configuration *knobs* and it is often critical to setup automated ablation testing to isolate the effects of different configurations. Models change, applications requirements change, etc, - without the each of simply creating a new config and passing it through eval can slow many teams down significantly.

Furthermore, having a serializable config format for components means you can built low code or no code tooling that allows non-technical users to build by complex multi-agent systems by components that existing in some gallery or that can be created/completely specified using a visual interface. In fact, all of my work on building low code tooling for multi-agent systems was driven by this realization and inspired the configuration system used across AutoGen, AutoGen Studio and Magentic-UI (Dibia et al. (2024), Mozannar et al. (2025)).

9.2.6 Comprehensive Guardrail and Middleware Infrastructure

Production multi-agent systems require robust safety mechanisms and interception points to implement guardrails, compliance checks, and security measures. Frameworks should provide comprehensive middleware infrastructure that enables developers to inject custom logic at critical interaction points.

What to look for:

- Pre/post-processing hooks for all model calls with full request/response access
- Function call interception and validation capabilities before tool execution
- Agent state transition middleware for workflow control and validation
- Configurable guardrail pipelines that can be chained and composed
- Built-in PII detection, redaction, and content filtering capabilities
- Malicious input detection and sanitization for both model and tool inputs
- Audit logging hooks that capture all security-relevant interactions
- Async-compatible middleware that doesn't block system performance

In `picoagents` , we implemented middleware for guardrails:

```
# Chain multiple middleware for comprehensive protection
chain = MiddlewareChain([
    PIIRedactionMiddleware(patterns=["ssn", "email"]),
    RateLimitMiddleware(calls_per_minute=60),
```

```
    GuardrailMiddleware(blocked_content=["harmful_request"])
])

agent = Agent(name="assistant", model_client=client, middleware=chain)
```

Questions to ask:

- Can I intercept and validate all model inputs and outputs?
- How do I implement custom guardrails without modifying framework code?
- Can I chain multiple guardrails together in configurable pipelines?
- Does the framework support async guardrails that don't degrade performance?
- Can I implement real-time PII detection and redaction across all interactions?
- How do I validate function arguments for malicious content before execution?
- Are there hooks to monitor and control agent state transitions?
- Can I implement custom compliance checks for my specific domain requirements?

Warning signs:

- Poor performance characteristics that make guardrails impractical
- Middleware systems that break with framework updates

> **i Note**
>
> The guardrail infrastructure is critical for production deployments where compliance, security, and safety are paramount. Without proper middleware hooks, teams often resort to modifying framework code or implementing brittle workarounds that break during updates. A well-designed guardrail system should be performant, extensible, and maintainable across framework versions.

9.2.7 Comprehensive Pattern Support and Extensibility

As we explored in Chapter 2, different tasks require different coordination approaches. A good framework should support the full spectrum of patterns and allow for custom implementations.

What to look for:

- Built-in support for workflow patterns (sequential, parallel, conditional)
- Autonomous patterns (round-robin, AI-driven selection, handoff)
- Task management patterns (termination, human delegation, error handling)
- Clear extension points for custom coordination logic
- Composable patterns that can be combined for complex scenarios

In `picoagents`, we implemented several coordination patterns (Chapter 6, Chapter 7):

```
# Round-robin orchestration
orchestrator = RoundRobinOrchestrator(agents=[writer, editor])
```

```
# AI-driven agent selection
orchestrator = AIOrchestrator(agents=[researcher, writer], model_client=client)

# Workflow with custom steps
workflow = Workflow(steps=[
    PicoAgentStep("research", researcher_agent),
    TransformStep("format", lambda x: format_results(x))
])

# Flexible termination conditions
termination = MaxMessageTermination(10) | TextMentionTermination("APPROVED")
```

Questions to ask:

- Does the framework support the coordination patterns my use cases require?
- Can I implement custom orchestration logic when needed?
- How easy is it to combine different patterns in the same application?
- Are there examples of advanced pattern implementations?

9.2.8   Production Deployment, Security, and Operations

Moving from prototype to production requires comprehensive security measures, deployment capabilities, and operational support.

What to look for:

- Input validation and sanitization for agent interactions
- Secure tool execution with sandboxing capabilities
- Container-friendly deployment with horizontal scaling
- Authentication, authorization, and audit logging
- Load balancing, failover, and cloud platform integration
- Resource limits, rate limiting, and operational monitoring

In `picoagents` , we implemented basic security and observability features:

```
# Tool approval for security
@tool(approval_mode=ApprovalMode.REQUIRED)
def sensitive_operation(data: str) -> str:
    return process_sensitive_data(data)

# OpenTelemetry integration for monitoring
os.environ["PICOAGENTS_ENABLE_OTEL"] = "true"
agent = Agent(name="prod-agent", model_client=client)
```

Questions to ask:

- How does the framework handle security and deployment together?

- What scaling and operational monitoring capabilities exist?
- Can I audit all agent actions and deploy securely in production?
- How are secrets managed and failures handled during scaling?

9.2.9   Evaluation and Testing Support

Unlike traditional software, multi-agent systems require specialized evaluation approaches due to their probabilistic nature and complex decision-making processes.

What to look for:

- Built-in trajectory evaluation (comparing expected vs. actual agent steps)
- Support for both unit testing (single interactions) and integration testing (complex sessions)
- Automated evaluation frameworks with predefined metrics
- Tools for capturing and converting real sessions into test cases
- Evaluation of both final outputs and intermediate reasoning processes

In `picoagents` , we implemented evaluation capabilities:

```
# Define evaluation tasks and judge
tasks = [EvalTask(input="Write a haiku", expected_output="5-7-5 syllable poem")]
judge = LLMEvalJudge(model_client=client, criteria=["creativity", "format"])

# Evaluate agents or orchestrators
runner = EvalRunner(judge=judge, parallel=True)
target = AgentEvalTarget(agent)
scores = await runner.evaluate(target, tasks)

print(f"Average score: {sum(s.overall for s in scores) / len(scores)}")
```

Questions to ask:

- Can I easily create and run automated evaluations for agent behavior?
- Does the framework support trajectory analysis and tool use evaluation?
- How can I test agent performance across different scenarios?
- Are there built-in metrics for measuring agent effectiveness?

9.2.10   Active Ecosystem and Long-Term Viability

Framework adoption is a long-term commitment. The health of the surrounding ecosystem often determines whether a framework remains viable and continues to evolve with your needs.

What to look for:

- Active development with regular releases
- Responsive community and maintainer engagement

- Comprehensive documentation with examples
- Third-party integrations and extensions
- Clear roadmap and governance structure

Questions to ask:

- How actively is the framework maintained and developed?
- What is the size and engagement level of the community?
- Are there commercial support options available?
- How does the framework handle breaking changes and upgrades?

9.3   Framework Evaluation and Decision Process

Evaluate frameworks systematically using a weighted scoring matrix based on the ten core capabilities. Define your technical requirements, organizational constraints, and business priorities, then build focused prototypes to test real-world scenarios. Table 9.1 shows a sample evaluation matrix comparing different framework options.

Table 9.1. Framework Evaluation Matrix

Capability	Weight	Framework A	Framework B	Custom Build
Developer Experience	20%	8/10	6/10	4/10
Async Architecture	15%	9/10	7/10	8/10
Observability & Dev Tools	10%	7/10	5/10	3/10
State Management	15%	8/10	6/10	2/10
Configuration	5%	6/10	8/10	7/10
Guardrails & Middleware	10%	5/10	3/10	8/10
Pattern Support	10%	7/10	8/10	10/10
Production & Security	10%	8/10	6/10	3/10
Evaluation Support	5%	6/10	4/10	2/10

Choose custom development when:

- Your requirements are highly specialized and unlikely to be supported by general frameworks
- You have experienced engineers with sufficient time for building and maintaining infrastructure
- You need complete control over architecture, performance, or specific implementation details
- The problem domain is well-understood and requirements are stable

Choose framework adoption when:

- You want to focus engineering effort on business logic rather than infrastructure
- You need to deliver working systems quickly with limited engineering resources
- Your requirements align well with common multi-agent patterns
- You value ecosystem benefits like community support, documentation, and third-party integrations

Consider hybrid approaches when:

- You want rapid prototyping capabilities while maintaining paths to custom optimization
- Different parts of your system have different requirements (some custom, some standard)
- You're experimenting with approaches before committing to full custom development
- You need framework benefits for non-critical components while maintaining control over core logic

9.4 Production Capabilities Beyond Educational Implementations

While picoagents effectively demonstrates core multi-agent concepts, production frameworks address enterprise requirements that educational implementations intentionally omit. The fundamental gap lies in operational infrastructure and deep ecosystem integration rather than basic functionality.

Production frameworks provide comprehensive workflow persistence and recovery systems that maintain agent state across infrastructure failures, enabling long-running processes to resume **seamlessly after crashes or restarts**. Achieving this is more complicated than what we have covered within the workflow API we implemented in Chapter 6. They offer enterprise deployment capabilities with auto-scaling, load balancing, and security perimeters that integrate with cloud platforms and organizational authentication systems.

Most critically, they include sophisticated safety infrastructure with plugin-based guardrail systems for threat detection, content filtering, and automated compliance checking that operate at the middleware level across all agent interactions. These frameworks also provide comprehensive observability platforms with distributed tracing, performance analytics, and interactive development environments designed for team collaboration and production debugging.

Beyond infrastructure, production frameworks excel at ecosystem integration that would require significant custom development to replicate. Microsoft Agent Framework provides seamless Azure AI Foundry integration with native authentication, deployment, and monitoring across the Microsoft stack. Google ADK offers native Vertex AI Agent Engine deployment with Cloud Run scaling and VPC security controls. Pydantic AI takes a cloud-agnostic approach, providing extensive integrations with observability platforms like Logfire, AgentOps, Arize, and W&B Weave.

The difference isn't feature quantity—it's the architectural sophistication and ecosystem integration required for reliability, security, and maintainability when multi-agent systems operate at enterprise scale with real business consequences.

> 💡 🖥 Working Code: Framework Comparisons
>
> While picoagents is minimal and built for learning, it implements the same core patterns found in production frameworks. To help you transition when needed, the companion repository includes equivalent implementations of agents, memory, workflows, and orchestration in Microsoft Agent Framework, Google ADK, and LangGraph:
> `examples/frameworks/`

9.5 Summary

- **Framework Value Beyond Features**: The primary value of frameworks lies in maintenance burden reduction (handling API changes, security patches) and standardized component interfaces that prevent integration complexity as teams scale, rather than basic functionality differences.

- **Production Infrastructure Gap**: Production frameworks provide operational infrastructure (workflow persistence, enterprise deployment, safety guardrails, comprehensive observability) and deep ecosystem integration—Azure customers get seamless Microsoft Agent Framework integration across the Azure stack, Google customers get native ADK Vertex AI deployment, while cloud-agnostic frameworks provide extensive third-party integrations.

- **Evaluation Methodology**: Assess frameworks against stable capabilities (developer experience, async architecture, state management, etc.) rather than direct comparisons, since frameworks evolve rapidly but fundamental requirements remain consistent.

- **Context-Dependent Selection**: Framework choice depends heavily on team capabilities, timeline constraints, and organizational ecosystem rather than abstract technical superiority—there is no universally "best" framework.

The multi-agent framework landscape will continue to evolve rapidly. By focusing on fundamental capabilities and systematic evaluation processes rather than specific tool comparisons, you'll be prepared to make good decisions regardless of which new frameworks emerge or how existing ones evolve.

Part III

Part III: Evaluation, Optimization, and Responsible AI

Chapter 10

Evaluating Multi-Agent Systems

This chapter covers:

- Understanding what we're really evaluating: multiagent trajectories - sequences of reasoning and actions
- Building evaluation metrics and judges that work across models, agents, and multiagent systems
- Implementing a practical evaluation library that integrates with our picoagents framework
- Planning effective evaluation strategies for complex multiagent tasks

10.1   Introduction

When you design a multiagent system to address a task e.g., "build a mobile app for stock trading,", your agent explores a series of actions - it does things and might declare the task as complete. But how do you truly verify or evaluate its performance? It's tempting to focus only on the final deliverable - does the app work? But this misses the rich problem-solving process that unfolds: the planning discussions between steps or individual agents, the iterative code generation, the debugging turns, and the coordination decisions that led to success or failure.

In this chapter, we'll build an evaluation framework that captures this complete picture. Rather than treating evaluation as an afterthought, we'll make it a core component of our picoagents library - enabling systematic assessment of everything from simple model calls to complex autonomous orchestrations.

The key insight is that whether you're evaluating a single model, an agent, or a full multiagent workflow, you're fundamentally evaluating the same thing: a **trajectory** - a sequence of reasoning messages and actions that unfolds over time. This unified view enables us to apply consistent evaluation approaches across different levels of complexity.

10.2   Evaluation-Driven Development

In my experience working with teams building agents or multiagent systems, a common mistake I see is that evaluation is considered much too late in the development process. The excitement or hype around using Agents often seems to distract from exploring a careful overall development strategy - and to help with this I often recommend what I call **evaluation-driven development** .

Evaluation-driven development means defining how you'll measure success before building your system. Like test-driven development (TDD)—where you write tests before code—you establish success criteria, metrics, and evaluation infrastructure before implementing agent behaviors.

Just as TDD tests clarify requirements and guide implementation, evaluation criteria force you to articulate what "good" means for your domain. If you'll measure reasoning efficiency, you'll build token tracking from day one. If you'll compare tool call sequences, you'll design observability into your framework. Evaluation constraints become design requirements.

💡 The Power of Easy Evaluation

As Eugene Yan observes: "I've found that once you've set up the evals + experiment harness and make it easy to tweak config and prompts with 1-click run + eval, teams enjoy running experiments and hill climbing those numbers, and progress comes quickly."
This highlights why building evaluation infrastructure early pays dividends - when experimentation is frictionless, teams naturally iterate faster and discover better solutions.

10.2.1   An Evaluation Planning Framework

In the rush and excitement of building AI agents or multi-agent systems, I have unfortunately seen many teams jump straight into writing agent code or trying out frameworks. This is often a mistake. In reality, writing the agent should come much later - first you need to be very clear on how you will evaluate success.

To help guide this process, here is a simple **5-step evaluation planning framework**. Before writing agent code, work through these five planning questions. The rest of this chapter provides the tools and techniques to answer each one.

Table 10.1 shows the Evaluation-Driven Development Framework with five key planning questions to answer before building your multiagent system.

Table 10.1. *Evaluation-Driven Development Framework - Answer these questions before building your multiagent system.*

Step	Planning Question	Key Considerations	Covered In
1	**Define Success Criteria**	What does success look like? Which dimensions matter most (accuracy, speed, user experience)? How will you measure improvement?	Reference-based vs reference-free metrics (Section 10.5)
2	**Create Your Task Suite**	What are representative tasks? Edge cases? How many test cases provide reliable signal?	Designing evaluation suites (Section 10.4), Creating tasks (Section 10.6.1)
3	**Choose Your Metrics**	Deterministic metrics vs LLM judges? Process vs outcome measures? Single vs composite evaluation?	Metrics and judges (Section 10.5), Composite evaluation (Section 10.5.3)
4	**Establish Baselines**	What are you comparing against? How is the task done today? Direct model calls? Single agents? Previous versions?	Evaluation targets (Section 10.6.2), Running evaluations (Section 10.6)
5	**Plan Iteration Workflow**	When do you run evals? How do you identify regressions? What's your experiment tracking strategy?	Understanding results (Section 10.6.5)

This framework provides a roadmap for the rest of the chapter. We'll start by understanding what we're evaluating (multiagent trajectories), then explore how to design domain-specific evaluation suites, and finally build the practical evaluation infrastructure needed to implement this evaluation-first approach. By the end, you'll have a complete toolkit for systematically measuring and improving your multiagent systems.

10.3   What We're Evaluating: Multiagent Trajectories

Every time you run an agent or multiagent system, you generate a **trajectory** - a complete record of the reasoning and actions taken to address a task. This trajectory includes:

- **Messages**: The conversational flow between user, agents, and tools
- **Actions**: Tool calls, function executions, and external interactions

- **Outcomes:** Success/failure status, errors encountered, and final results
- **Metadata:** Timing, token usage, and execution context

Consider this simple example:

```
Task: "What's the weather in Paris?"

Trajectory:
User: What's the weather in Paris?
Agent: I need to get current weather data. Let me call the weather API.
Tool Call: get_weather(location="Paris")
Tool Result: {"temp": 22, "conditions": "sunny"}
Agent: The weather in Paris is currently sunny with a temperature of 22°C.
```

This trajectory tells the complete story - not just the final answer, but how the agent reasoned, what tools it used, and whether the approach was effective.

Now consider a more complex multiagent scenario:

```
Task: "Create a simple calculator web app"

Trajectory (simplified):
User: Create a simple calculator web app
Planner Agent: I'll break this into frontend and backend tasks...
Frontend Agent: I'll create an HTML interface with basic operations...
Backend Agent: I'll implement the calculation logic...
[50+ messages of iterative development, testing, debugging]
System: Task completed successfully. Calculator app deployed.
```

The trajectory captures the coordination between agents, the iterative refinement process, and the emergent solution - information that's lost if we only evaluate the final deliverable.

10.3.1 Representing Trajectories in Code

In our picoagents library, we represent trajectories with a simple data structure:

```python
from picoagents.types import (
    EvalTrajectory,
    EvalTask
)

class EvalTrajectory:
    # What we're trying to solve
    task: EvalTask
    # Complete conversation history
    messages: List[Message]
    # Did we complete the task?
    success: bool
```

```
# Error details if failed
error: Optional[str]
# Resource consumption metrics
usage: Usage
# Additional execution context
metadata: Dict[str, Any]
```

This structure works whether you're evaluating a single model call or a complex multiagent workflow. The messages capture the reasoning process, while metadata provides execution context.

10.3.2   Observability as Foundation

> ! Observability Enables Evaluation
>
> Building evaluation-ready systems requires thinking about observability from day one. When your agents emit structured events during execution (using standards like Open-Telemetry), you can reconstruct complete trajectories for evaluation without requiring *evaluation specific* runs. This means every production run becomes a potential evaluation data point.

Our picoagents library takes an observability-first approach by design:

- **Structured Message Logging**: Every interaction is captured with timestamps, metadata, and context
- **Usage Tracking**: Token consumption, API calls, and execution times are automatically recorded

- **Error Propagation**: Failures are logged with full context rather than failing silently
- **Execution Traces**: Complete trajectories can be replayed and analyzed

This upfront investment in observability pays dividends during evaluation. When a multiagent system fails, you can trace exactly where and why - was it a reasoning error, a tool call failure, or an agent coordination issue? More importantly, you can retroactively evaluate any past execution by reconstructing its trajectory from the logged events.

10.3.3   The Answer Extraction Challenge

A subtle but critical challenge emerges when evaluating real agent conversations: **where is the actual answer?**

Consider an agent that uses tools:

```
Agent: "Let me calculate that: 2+2"
Tool: "Calculator result: 4"
Agent: "The calculation is complete."
```

Which message contains the answer? A naive approach taking the last message gets "The calculation is complete" - missing the actual result entirely.

Our evaluation system addresses this with **answer extraction strategies**:

```
from picoagents.eval import ExactMatchJudge

# Use last non-empty message (default)
judge = ExactMatchJudge(
    answer_strategy="last_non_empty"
)

# Use only last assistant message
judge = ExactMatchJudge(
    answer_strategy="last_assistant"
)

# Concatenate all assistant messages
judge = ExactMatchJudge(
    answer_strategy="all_assistant"
)
```

This seemingly small detail prevents evaluation failures that would otherwise make your metrics unreliable in production.

> **ⓘ What Judges Need from Trajectories**
>
> Different evaluation scenarios require different information from trajectories. **Reference-based judges** need access to both the expected output (from the task definition) and the actual output (extracted from the agent's messages using answer strategies). **LLM judges** evaluating response quality need the task input to understand what was requested, the agent's response to assess quality, and sometimes conversation history when evaluating multi-turn interactions. **Tool-using agents** present additional complexity—you might need to examine not just the final response but also **which tools were called** and **with what arguments**. Our answer extraction strategies handle common cases, but for specialized evaluation needs you can create custom judges that directly inspect the message sequence in `trajectory.messages` to extract exactly the information your evaluation criteria require.

10.4   Designing Evaluation Suites

Building effective evaluation for multiagent systems is fundamentally a creative design problem. Unlike traditional machine learning systems where established metrics like BLEU, ROUGE, or NDCG provide standardized measurement, agentic systems produce outcomes that resist simple quantification. The trajectory of an agent solving a software engineering task looks nothing like one generating data visualizations or conducting market research - yet all three require rigorous evaluation.

This presents both a challenge and an opportunity. The challenge is that you can't simply reach for off-the-shelf metrics. The opportunity is that thoughtful evaluation design becomes a competitive advantage - teams that develop domain-specific evaluation suites aligned with expert judgment will build better systems faster.

10.4.1 Why Traditional ML Metrics Fall Short

Traditional NLP metrics were designed for structured prediction tasks with well-defined outputs. BLEU measures n-gram overlap between generated and reference translations. ROUGE scores summarization quality through word overlap. NDCG ranks information retrieval results. These metrics excel when the output format is predictable and structured (text sequences, ranked lists), multiple reference outputs can be collected for comparison, and success can be approximated through surface-level similarity.

Agentic systems violate all three assumptions. Consider what an agent produces during a typical execution: tool call sequences with varied orderings that achieve equivalent outcomes, generated artifacts spanning text, code, images, audio, and structured data, interactive dialogues where conversational quality matters as much as factual accuracy, multi-step reasoning chains where process quality affects reliability, and external state changes like files created, APIs called, or databases modified. An agent writing software might produce beautiful, well-tested code through 50 reasoning steps, or produce identical functionality through 200 steps with extensive backtracking. Traditional metrics miss this entirely - they can't distinguish elegant solutions from brute-force success, nor can they evaluate the quality of intermediate artifacts like test coverage or code documentation.

Perhaps more fundamentally, agentic trajectories are non-deterministic and heterogeneous. A single task execution might include natural language reasoning about the problem, JSON tool call specifications, Python code generation, terminal command outputs, rendered visualizations, error messages and debugging steps, and multi-turn clarification dialogues. This heterogeneity resists reduction to scalar metrics. You can't compute BLEU score on a mixture of code, images, and conversation.

This is where the power of large language models as evaluators becomes apparent. Modern multimodal LLMs can process exactly the data formats agents produce - text, code, images, structured data, tool calls - and reason about their quality using the same capabilities that enable agent reasoning in the first place. If an LLM can write code, it can evaluate code quality. If it can generate visualizations, it can assess whether a chart effectively communicates insights.

Traditional software systems execute deterministic behaviors—download a file, call an API, return a result. The power of agents is that they enable us to build systems that tackle tasks requiring reasoning and intelligence: how to architect an application, debug emergent issues, synthesize insights from data. These are tasks humans approach with creativity and domain expertise. As system developers, our evaluation methods must mirror this same intelligence and creativity. We can't evaluate intelligent behavior with simple rule-based metrics any more than we can build intelligent systems with simple rule-based logic.

10.4.2   Translating Domain Expertise into Evaluation Methods

The key to designing effective evaluation suites is translating domain expertise into measurable criteria. Ask yourself: *If an expert human were evaluating this outcome, what would they check?*

For a software development agent, an expert engineer would examine:

- **Functional correctness**: Does the code solve the stated problem?
- **Test coverage**: Are edge cases handled? Do tests actually verify behavior?
- **Code quality**: Is the implementation maintainable, readable, efficient?
- **Type safety**: Do static analysis tools (mypy, pyright) report issues?
- **Documentation**: Are functions documented? Is there a README?
- **Error handling**: Does the code fail gracefully with clear messages?

Each of these expert judgments can be translated into evaluation methods:

```python
# Functional correctness: Run test suite
def evaluate_functional_correctness(code_trajectory):
    test_results = execute_pytest(
        code_trajectory.generated_code
    )
    return test_results.pass_rate

# Test coverage: Analyze coverage report
def evaluate_test_coverage(code_trajectory):
    coverage = run_coverage_analysis(
        code_trajectory.generated_code
    )
    return coverage.line_coverage_percent

# Type safety: Run static analysis
def evaluate_type_safety(code_trajectory):
    mypy_results = run_mypy(
        code_trajectory.generated_code
    )
    error_count = mypy_results.error_count
    total_lines = max(mypy_results.lines, 1)
    error_rate = error_count / total_lines
    return 1.0 - error_rate

# Code quality: LLM evaluation
def evaluate_code_quality(code_trajectory):
```

```
    return llm_judge.evaluate(
        code_trajectory.generated_code,
        criteria=[
            "Is the code readable with clear variable names?",
            "Are functions appropriately sized and focused?",
            "Is the implementation efficient?",
            "Does the code follow Python best practices?"
        ]
    )
```

Notice the mixture of approaches. Some expert judgments (test coverage, type safety) map cleanly to deterministic metrics - these should be preferred when available because they're fast, reproducible, and provide precise signals. Other judgments (code readability, architectural soundness) require nuanced reasoning that benefits from LLM evaluation.

10.4.3 Domain-Specific Evaluation Examples

Let's contrast this with an agent that generates data visualizations. An expert data scientist evaluating a chart would check:

- **Data accuracy:** Does the visualization correctly represent the underlying data?
- **Visual clarity:** Are axes labeled? Is there a descriptive title?
- **Chart choice:** Is the visualization type appropriate for the data?
- **Legibility:** Are fonts readable? Are colors distinguishable?
- **Statistical honesty:** Are scales appropriate? Is overplotting avoided?
- **Context:** Are units specified? Are dates/time ranges clear?

The evaluation approach looks quite different:

```
# Data accuracy: Programmatic verification
def evaluate_data_accuracy(viz_trajectory):
    # Load ground truth data
    original_data = load_data(viz_trajectory.task.data_path)
    # Extract data points from visualization metadata
    # (agent may store plot data in trajectory metadata)
    viz_data = viz_trajectory.metadata.get('plot_data', None)
    if viz_data is None:
        return 0.0
    return compute_correlation(original_data, viz_data)


# Visual elements: Multimodal LLM evaluation
def evaluate_visual_quality(viz_trajectory):
    # Extract generated image from trajectory messages
    # (agents may produce MultiModalMessage with images)
    messages = viz_trajectory.messages
    image_messages = [
```

```
        msg for msg in messages
        if hasattr(msg, 'is_image') and msg.is_image()
    ]

    if not image_messages:
        return 0.0

    # Evaluate the generated visualization
    return multimodal_llm_judge.evaluate(
        messages=image_messages,
        criteria=[
            "Does the chart include a descriptive title?",
            "Are all axes clearly labeled with units?",
            "Is the text legible at normal viewing size?",
            "Are colors used effectively and accessible?",
            "Is the date range or temporal context clear?"
        ]
    )

# Statistical properties: Programmatic checks
def evaluate_statistical_honesty(viz_trajectory):
    checks = {
        "appropriate_axis_scaling":
            verify_axis_ranges(viz_trajectory),
        "no_severe_overplotting":
            check_point_density(viz_trajectory),
        "consistent_units":
            verify_unit_consistency(viz_trajectory)
    }
    return sum(checks.values()) / len(checks)
```

Here we see multimodal evaluation in action. The `EvalTrajectory` contains a `messages` field that can include `MultiModalMessage` instances with images, audio, or other media. Vision-capable LLM judges can process these messages directly, assessing whether a generated chart has legible fonts or appropriate titles—tasks that would be extremely difficult with traditional computer vision pipelines. This capability extends naturally from how agents already work: if an agent can generate and reason about images, the same multimodal models can evaluate those images.

ℹ Multimodal Evaluation in Practice

When using vision-capable models (GPT-4o, Claude 3.5 Sonnet) as judges, you can pass `MultiModalMessage` instances directly from the trajectory. The judge's LLM client handles the multimodal content just as it would during agent execution. This unified

---

approach means evaluation uses the same message types and model capabilities as your agent system.

10.4.4   When to Use Numeric Metrics vs LLM Judges

As these examples illustrate, effective evaluation suites strategically combine deterministic metrics with LLM-based judgment. The decision framework:

Prefer deterministic metrics when:

- Ground truth is available and comparison is well-defined
- The evaluation criterion maps to a measurable property (test coverage, error rates, response time)
- Consistency and reproducibility are critical
- Cost and latency matter (evaluating thousands of examples)
- You need to track fine-grained changes over time

Prefer LLM judges when:

- The criterion requires nuanced reasoning about quality
- Multiple valid solutions exist and surface similarity is misleading
- Human judgment involves subjective assessment
- The output format is unstructured or multimodal
- Domain expertise is needed to evaluate correctness

Use both when:

- Different aspects of quality require different evaluation approaches
- You want to validate LLM judgments against objective baselines
- Cost-sensitive scaling requires filtering with cheap metrics before expensive evaluation

For example, you might filter code generation results with fast unit tests (deterministic) before running expensive LLM evaluation of code quality and architectural decisions. Or you might use deterministic checks to verify a visualization has required elements (title, labels) before using multimodal LLM evaluation to assess the quality of those elements.

Evaluation suite design is itself an iterative process. Start with a hypothesis about what matters for your domain, implement evaluation methods, and then validate that your metrics actually correlate with downstream success. A systematic approach: first, consult domain experts to understand how they evaluate quality; second, translate expert criteria into measurable evaluations; third, test whether your metrics correlate with expert judgments; fourth, adjust weights, add dimensions, and fix misaligned incentives; and finally, monitor whether optimizing for your metrics improves real performance. This scientific iteration prevents common pitfalls like optimizing for metrics that don't actually measure what matters, or creating evaluation criteria that are gameable in ways that don't reflect genuine improvement.

Designing domain-specific evaluation suites requires creativity precisely because it resists formulaic approaches. Each application domain brings unique success criteria, failure modes, and quality dimensions. The teams that invest in thoughtful evaluation design - bringing together domain expertise, technical measurement, and validation against real outcomes - will build more reliable systems and iterate faster toward genuinely useful agent capabilities.

The crucial insight is that evaluation is not just a technical problem to solve with better metrics - it's a design problem that requires understanding what quality means in your specific domain and creatively translating that understanding into measurement. With this foundation in mind, the next sections provide the technical infrastructure to implement these evaluation approaches in our picoagents library, building judges, runners, and composite evaluation systems that can assess everything from simple model calls to complex multiagent orchestrations.

10.5   How We Evaluate: Metrics and Judges

Once we have a trajectory, how do we score it? This requires two components: **metrics** (what dimensions to measure) and **judges** (how to score each dimension).

10.5.1   Reference-Free vs Reference-Based Metrics

Reference-based metrics compare outputs to known correct answers:

For tasks with deterministic answers, you can evaluate without expensive LLM calls:

```
from picoagents.eval import ExactMatchJudge
from picoagents.types import EvalTask

# Create task with expected output
task = EvalTask(
    input="2+2",
    expected_output="4"
)

# Exact match judge for deterministic answers
exact_judge = ExactMatchJudge()
```

But what if the agent says "The answer is 4" instead of just "4"? Evaluation systems need strategies for handling conversational responses:

```
from picoagents.eval import (
    ContainsJudge,
    FuzzyMatchJudge
)

# Finds "4" within longer responses
```

```
contains_judge = ContainsJudge()

# Handles "Four" vs "4" variations
fuzzy_judge = FuzzyMatchJudge()
```

Reference-free metrics assess quality without ground truth:

- Helpfulness: How useful is the response to the user?
- Communication clarity: How well do agents coordinate?
- Resource efficiency: How many tokens or API calls were used?

Many multiagent tasks require reference-free evaluation because the "correct" solution can vary widely (there are many ways to build a calculator app).

Agent-specific metrics evaluate the *process* agents use to reach solutions:

For tool-using agents, you often care not just about the final answer but about **whether the agent selected the right tools** and **in what order**. Did the agent call `search` before `summarize`? Did it use `calculator` with the correct arguments? These process-oriented metrics are particularly important for debugging agent behavior—an agent might get lucky and produce the right answer through the wrong reasoning path. You can implement tool evaluation by examining `ToolMessage` instances in the trajectory to verify tool selection, argument correctness, and execution sequence against expected patterns. While our current judges focus on final outputs, production systems often extend this pattern with specialized judges for tool call accuracy and reasoning trajectory validation.

10.5.2   LLM-as-a-Judge: A Practical Approach

For reference-free evaluation, we can use strong language models as judges. These **LLM judges** can assess nuanced qualities that traditional metrics miss.

Here's how we create and configure an LLM judge:

```
from picoagents.eval import LLMEvalJudge
from picoagents.llm import AzureOpenAIChatCompletionClient

# Create LLM judge with evaluation criteria
judge_client = AzureOpenAIChatCompletionClient(
    model="gpt-4.1-mini"
)

judge = LLMEvalJudge(
    client=judge_client,
    default_criteria=[
        "accuracy",
        "helpfulness",
        "efficiency"
```

```
    ]
)
```

The judge scores trajectories by building an evaluation prompt that includes the task, expected output (if available), and the complete agent conversation. It then uses the LLM to provide scores (0-10) and reasoning for each criterion.

```
# Score a trajectory
score = await judge.score(
    trajectory,
    criteria=["accuracy", "helpfulness"]
)

# Access results
print(f"Overall: {score.overall}/10")
print(f"Accuracy: {score.dimensions['accuracy']}/10")
print(f"Reasoning: {score.reasoning['accuracy']}")
```

This approach leverages the reasoning capabilities of advanced models to provide nuanced evaluation across multiple dimensions.

10.5.3 Scoring Strategies

When applying metrics to trajectories, we have several options:

Single Answer Grading: Score the complete trajectory holistically

- Good for: Overall task assessment
- Challenge: May miss important process details

Pairwise Comparison: Rather than scoring each trajectory independently, pairwise evaluation directly compares two outputs and identifies **which is better**. This approach is particularly valuable during iterative development—each agent version can be systematically compared against the previous baseline to ensure you're making progress. While pointwise evaluation asks "how good is this response?" (requiring absolute judgment), pairwise evaluation asks the simpler question **"which response is better?"** This relative judgment is often **more reliable**, especially for subjective qualities where absolute scores feel arbitrary. The tradeoff is that you need multiple evaluation runs to establish relative rankings.

```
# Compare two agent configurations
scores_v1 = await runner.evaluate(agent_v1, tasks)
scores_v2 = await runner.evaluate(agent_v2, tasks)

avg_v1 = sum(s.overall for s in scores_v1) / len(scores_v1)
avg_v2 = sum(s.overall for s in scores_v2) / len(scores_v2)

improvement = ((avg_v2 - avg_v1) / avg_v1) * 100
```

```
print(f"Version 2: {improvement:+.1f}% vs baseline")
```

Dimensional Scoring: Score multiple aspects independently

- Good for: Detailed feedback, identifying specific weaknesses
- Challenge: More complex to implement and interpret

10.6   Building a Practical Evaluation Harness

Let's build a comprehensive evaluation to answer a fundamental question: when do multi-agent systems justify their overhead? We'll compare four approaches across diverse task types, starting with a baseline (direct model calls) and progressively adding capabilities: tools, round-robin orchestration (Section 7.3), and AI-driven orchestration (Section 7.4). All configurations use GPT-4.1-mini to ensure fair comparison.

Table 10.2 shows the four approaches we'll compare.

Table 10.2. Four approaches compared in the evaluation study using GPT-4.1-mini.

Approach	Description
Direct-Model	Baseline: direct GPT-4.1-mini calls with no agent wrapper, tools, or orchestration
Single-Agent-Tools	Single agent with tools (Think, Calculator, DateTime, GoogleSearch, TaskStatus)
Multi-Agent-RoundRobin	Three-agent team (planner, solver, reviewer) with fixed rotation orchestration (Section 7.3). Solver has same tools as Single-Agent-Tools
Multi-Agent-AI	Same three-agent team with AI-driven orchestration (Section 7.4) using dynamic agent selection based on task requirements

Table 10.3 shows the task categories designed to test different capabilities.

Table 10.3. Task suites testing different agent capabilities (10 tasks total).

Task Suite	Description	Tasks
Simple-Reasoning	Basic reasoning without tools: math word problems, logic puzzles, reading comprehension	3
Tool-Heavy	Research requiring real-time data: web search for podcast interviews, tech events, academic papers	3
Planning	Multi-constraint optimization: trip planning with budgets, resource allocation for weighted outcomes	2

Task Suite	Description	Tasks
Verification	Critique-focused tasks: fact-checking myths, analyzing arguments for logical fallacies	2

Each suite targets specific capabilities. Simple-Reasoning includes average speed calculations, profession assignment puzzles, and date arithmetic—tasks solvable through reasoning alone. Tool-Heavy tasks require GoogleSearch for current information: verifying podcast interviews (Andrej Karpathy on Eureka Labs), researching tech conference announcements (OpenAI DevDay), and finding recent arXiv papers on multi-agent reinforcement learning. Planning tasks involve multi-step coordination: creating a 3-day San Francisco itinerary under budget constraints with public transportation, or optimizing study time allocation across weighted exams. Verification tasks test whether critique adds value: fact-checking claims about brain energy usage and coffee dehydration, or identifying fallacies in social media arguments.

Our evaluation harness (`comprehensive-evaluation.py`) supports two modes: a `mini` version with 1 task per suite (approximately 2 minutes runtime) for rapid iteration, and a `full` version with 2-3 tasks per suite (approximately 8 minutes) for comprehensive assessment. This mini/full pattern proved essential for debugging both agent configurations and judge behavior.

> 💡 🖥 Working Code: Comprehensive Evaluation
>
> Complete implementation of the evaluation described in this section:
> - **Main Script**: `evaluation/comprehensive-evaluation.py` - Full evaluation suite with quick/full modes, task suites, and result visualization
> - **Results & Trajectories**: `evaluation/comprehensive_results/` - Complete results CSV and individual task trajectories with judge reasoning
> - **Documentation**: `evaluation/README.md` - Setup instructions, mode explanations, and interpretation guide

10.6.1   Designing the Task Suite

The choice of tasks determines what insights your evaluation will reveal. Our four suites test orthogonal capabilities: Simple-Reasoning establishes a baseline where multi-agent configurations may not provide value, Tool-Heavy validates whether agents can leverage external capabilities/tools, Planning tests multi-step decomposition, and Verification assesses whether critique improves outputs.

Consider the Tool-Heavy suite. These tasks require real-time information that no model training could provide—a design choice that forces the system to use tools successfully or fail entirely. Here's how we create one such task:

```
def create_tool_heavy_tasks() -> List[EvalTask]:
    return [
        EvalTask(
            name="Podcast Research",
            input="""Did Andrej Karpathy have a podcast interview with
            Dwarkesh Patel where he discussed Eureka Labs? If so, what
            did he say was the primary goal? Provide specific quotes.""",
            expected_output="""Yes, Andrej Karpathy discussed Eureka Labs.
            The primary goal is to personalize education using AI teaching
            assistants that allow students to ask questions and receive
            customized, interactive experiences...""",
        ),
        # Additional Tool-Heavy tasks...
    ]
```

The expected output isn't used for exact matching but guides the LLM judge on what consti-
tutes a complete answer. This task will immediately differentiate configurations: Direct-Model
will hallucinate or admit ignorance, while tool-enabled configurations can find these facts
through web search.

10.6.2   Creating Evaluation Targets

Each configuration needs an evaluation target—an adapter that runs the system uniformly
regardless of its internal structure. The simplest is Direct-Model, which bypasses the agent
framework entirely:

```
# Baseline: direct model calls
model_target = ModelEvalTarget(
    client=AzureOpenAIChatCompletionClient(model="gpt-4.1-mini"),
    system_message="You are a helpful, accurate assistant."
)
```

Single-Agent-Tools wraps an agent with tool access. The key design decision is tool selection—
we provide Think (for explicit reasoning), Calculator (for arithmetic), DateTime (for temporal
queries), GoogleSearch (for current information), and TaskStatus (for workflow management):

```
# Single agent with tools
agent_with_tools = Agent(
    name="assistant",
    instructions="""You are a knowledgeable assistant with access to tools.
    Use tools when they would improve accuracy. Think step-by-step for
    complex problems. For research tasks, use google_search to find
    current information.""",
    model_client=client,
    tools=[ThinkTool(), CalculatorTool(), DateTimeTool(),
```

```
            GoogleSearchTool(), TaskStatusTool()]
  )
  agent_target = AgentEvalTarget(agent_with_tools)
```

Multi-agent configurations introduce a three-agent team: a planner for decomposition, a solver with tools for execution, and a reviewer for quality assurance. The solver receives the same tools as Single-Agent-Tools, ensuring that any performance differences stem from orchestration patterns rather than capability gaps. RoundRobinOrchestrator cycles through agents in fixed order, while AIOrchestrator lets the model dynamically select the next speaker based on task requirements and state as the task progresses.

10.6.3 Configuring the Judge

Given that we are taking an LLM-as-judge approach, evaluation quality depends entirely on judge calibration. We use a composite judge combining multiple dimensions—accuracy, completeness, helpfulness, and clarity—with an LLM providing nuanced scoring that exact-match approaches cannot capture.

Building effective evaluations requires iteration and creativity. While developing this evaluation suite, I encountered initial results that contradicted my expectations: multi-agent systems scored 6.82/10 while direct models achieved 7.94/10. Inspecting both the judge rationale and trajectory traces revealed several critical issues that required fixing before the evaluation could produce meaningful results.

Issue 1: The Verbosity Penalty

The first problem appeared in judge reasoning:

```
{
  "overall": 6.0,
  "reasoning": {
    "clarity": "This is correct but verbose for a simple task"
  }
}
```

The judge was penalizing multi-agent transparency as verbosity. When a planner, solver, and reviewer collaborate, the trajectory naturally contains their complete reasoning process. Without proper guidance, the LLM judge interpreted this collaborative visibility as unnecessary length rather than valuable transparency. The solution required custom instructions that explicitly correct this bias:

```
judge = LLMEvalJudge(
    client=judge_client,
    custom_instructions="""
    SCORING PHILOSOPHY:
    - If the answer is CORRECT and COMPLETE, score 8-10 regardless of length
```

```
    - DO NOT PENALIZE: Multi-agent collaborative process visibility
    - DO NOT PENALIZE: Showing reasoning steps or validation

    EVALUATION CRITERIA:
    1. Completeness: Multiple valid solutions = COMPLETE (9-10)
    2. Helpfulness: Correct + reasoning = HELPFUL (9-10)
    3. Clarity: Logically structured = CLEAR (9-10), not brevity
    """,
    default_criteria=["accuracy", "completeness", "helpfulness", "clarity"]
)
```

After adding these instructions and re-running the evaluation, multi-agent scores improved from 6.82/10 to 9.30/10—revealing their true performance once the verbosity bias was addressed with instructions.

Issue 2: Dimension Normalization

Another issue I observed was that when combining judges with different dimensions using `CompositeJudge`, composite scores didn't match expectations. Judge A might score accuracy and helpfulness, while Judge B scores clarity. The weighted average wasn't normalizing correctly—weights summed to 1.0 globally instead of per dimension. The fix required adjusting dimension calculations:

```
# Normalize weights per dimension, not globally
for dim, contributions in dimension_contributions.items():
    total_weight_for_dim = sum(weight for _, weight in contributions)
    dimensions[dim] = sum(
        val * (weight / total_weight_for_dim)
        for val, weight in contributions
    )
```

These issues illustrate an important principle: evaluation itself requires debugging. Log complete trajectories with judge reasoning, inspect the data when results seem wrong, identify biases in both the system and the evaluation framework, then iterate. Without this process, the verbosity bias would have led to incorrect conclusions about multi-agent system performance.

10.6.4   Running the Evaluation

The EvalRunner coordinates execution across all configurations and tasks, collecting trajectories with complete execution traces—messages, tool calls, token usage, and timing:

```
runner = EvalRunner(judge=composite_judge, parallel=True)

# Evaluate all configurations
for config_name, target in all_configurations:
```

```
scores = await runner.evaluate(target, all_tasks)
# Log complete trajectories for inspection
save_results(config_name, scores)
```

Each evaluation produces rich scoring data: overall scores, dimensional breakdowns, judge reasoning, and complete trajectories. This granularity enables debugging both the systems being evaluated and the evaluation itself—when scores seem wrong, inspect the trajectory and judge rationale to identify whether the issue lies in agent behavior or evaluation bias.

10.6.5 Results and Analysis

Figure 10.1 shows performance across all configurations and task categories.

The results reveal clear patterns. Direct-Model scores 6.8 on Tool-Heavy tasks compared to Multi-Agent-AI's 9.2. On Simple-Reasoning tasks, Direct-Model actually performs best at 9.7 compared to Multi-Agent-AI's 9.3. Token efficiency shows the cost of multi-agent coordination: Direct-Model uses an average of 355 tokens per task while Multi-Agent-AI uses 15343 tokens, a 43x increase.

The Tool-Heavy suite reveals the clearest differentiation. The Podcast Research task exposes a critical limitation: Direct-Model scored 3.2/10 because it lacks web search capabilities and cannot verify current information. All other approaches with GoogleSearchTool scored 9.0/10, demonstrating how tool access transforms performance on research-heavy tasks.

Interestingly, Direct-Model wins on Simple-Reasoning (9.7/10), beating all multi-agent approaches. This indicates that LLMs themselves can be decent baselines for simple reasoning tasks. This validates an important principle: don't over-engineer. When a task requires only straightforward reasoning without tool coordination or multi-step planning, a single model call suffices. The multi-agent overhead (coordination messages, reasoning visibility, multiple LLM calls) provides no benefit and actually introduces slight performance degradation.

The Planning and Verification suites show less differentiation (all configs score 8.6-9.6/10), suggesting these task types don't strongly favor any particular architecture. The value of multi-agent orchestration depends heavily on task characteristics.

10.7 Exercises

Domain-Specific Task Suite: Design and implement a task suite for your specific domain (e.g., customer support, code review, data analysis). Create 5-10 tasks that test different capabilities relevant to your use case. Consider what makes a good discriminative task—what capabilities should separate good from poor performance?

Custom Judge Implementation: Implement a specialized judge for your domain that goes beyond generic accuracy and helpfulness. For example, create a `CodeQualityJudge` that evaluates generated code for readability, efficiency, and maintainability, or a `CustomerSupportJudge` that assesses empathy, resolution completeness, and policy

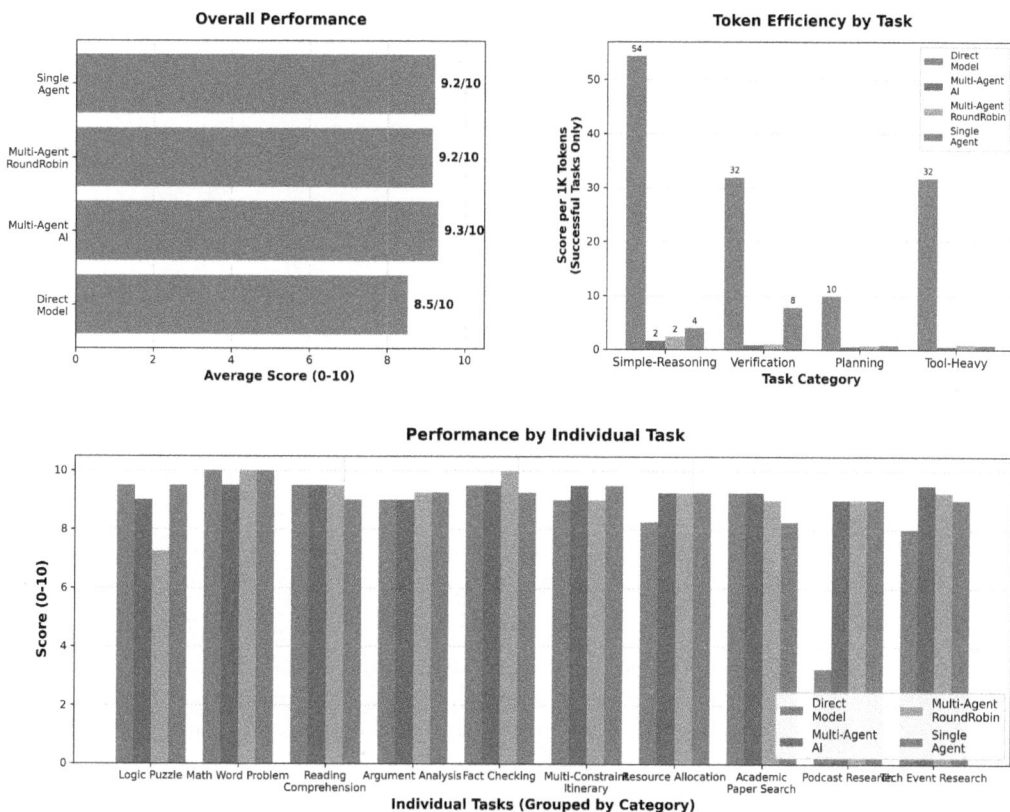

Multi-Agent System Evaluation Results

Figure 10.1. Comprehensive evaluation results across four configurations and four task suites. Top-left: overall performance bars show Multi-Agent-AI (9.3/10), Single-Agent-Tools (9.2/10), and Multi-Agent-RoundRobin (9.2/10) leading, with Direct-Model at 8.5/10. Top-right: token efficiency by task category (score per 1K tokens for successful tasks only) reveals Direct-Model's massive advantage—Simple-Reasoning shows efficiency of 54 compared to 2-4 for agent configurations; Verification shows 32 compared to under 8; Planning and Tool-Heavy show similar patterns with Direct-Model at 10-32 versus under 1 for multi-agent approaches. Bottom: individual task performance across all 10 tasks grouped by category, revealing Direct-Model's strength on Simple-Reasoning tasks but failure on Podcast Research (3.2/10) where agent-based configurations achieve 9.0/10.

compliance. Calibrate your judge on a validation set before scaling.

Regression Testing Suite: Build a regression testing system that tracks your agent's performance over time. Store evaluation results with metadata (model version, configuration, timestamp), visualize trends, and automatically alert when performance degrades below thresholds. This enables confident iteration without accidentally breaking capabilities.

Efficiency Analysis: Extend the evaluation harness to collect detailed efficiency metrics beyond token counts—wall-clock time, tool call counts, memory usage, API costs. Build visualizations that show efficiency-performance trade-offs across configurations. When does a 2x speedup justify a 0.5-point score decrease?

10.8   Related Research and Datasets

Several benchmarks specifically target agent and multiagent evaluation:

GAIA (Mialon et al. 2023): 466 questions testing reasoning, tool use, and multi-step problem solving. Focuses on tasks that are simple for humans but challenging for AI systems, achieving 92% human accuracy versus 15% for GPT-4 with plugins.

SWE-bench (Jimenez et al. 2023): Real GitHub issues testing software engineering capabilities across 2,294 problems from 12 popular Python repositories. Evaluates code comprehension, bug fixing, and multi-file coordination with only 1.96% solved by Claude 2.

GPQA (Rein et al. 2023): Graduate-level questions in biology, physics, and chemistry requiring specialized knowledge and complex reasoning. Features 448 "Google-proof" questions where PhD experts reach 65% accuracy while skilled non-experts achieve only 34%.

Humanity's Last Exam (HLE) (Phan et al. 2025): A multi-modal benchmark with 2,500 questions across dozens of subjects at the frontier of human knowledge, designed as a comprehensive test of advanced AI capabilities across mathematics, humanities, and natural sciences.

These benchmarks share our trajectory-focused approach - they evaluate not just final answers but the reasoning and tool use process that leads to solutions. They represent increasingly challenging testbeds where current LLMs demonstrate significant gaps compared to human expert performance, making them valuable for measuring genuine progress in multiagent reasoning capabilities.

> **! Benchmarks ≠ Your Task**
>
> Public benchmarks tell us **relative performance** (Model A beats Model B on task X), not how models will perform on **your task**. They can't capture your domain's unique success criteria, the composition of multiple models across your agent stack, or your specific data characteristics and failure modes.

> **The benchmark task ≠ your business task.** This is why investing creatively in domain-specific evaluation design is critical - only evaluation suites aligned with your actual use case can reliably measure whether changes improve real performance.

10.8.1 The Contamination Challenge

A growing concern in AI evaluation is **benchmark contamination** - when training data inadvertently includes test questions, inflating apparent performance. This is particularly problematic for multiagent systems that may use web search or retrieval tools, as benchmark questions often circulate online in various forms.

Some strategies to mitigate contamination:

- **Use diverse evaluation approaches:** Don't rely solely on public benchmarks
- **Create private test sets:** Develop internal evaluation tasks specific to your use case
- **Monitor for suspiciously high performance:** If results seem too good on established benchmarks, investigate potential data leakage
- **Test on fresh, unpublished problems:** Generate new tasks that couldn't have been seen during training

As benchmarks become targets for optimization, their ability to measure genuine capability diminishes. This makes it even more important to develop robust, multi-faceted evaluation approaches that can adapt as the field evolves.

> 💡 🖥 Working Code: Evaluation Framework
>
> Complete evaluation system implementation:
> - **Agent Evaluation:** `evaluation/agent-evaluation.py` - Practical evaluation patterns for agents and multiagent systems
> - **Core Framework:** `src/picoagents/eval/` - Complete evaluation library with metrics, judges, and trajectory analysis

10.9 Summary

Effective multiagent evaluation requires thinking in terms of **trajectories** - the complete sequence of reasoning and actions that unfolds during task execution. This unified perspective enables consistent evaluation approaches across models, agents, and multiagent systems.

Key principles:

- **Adopt evaluation-driven development:** Define success criteria and evaluation infrastructure before building your system
- **Design domain-specific evaluation:** Translate expert judgment into measurable crite-

ria using both deterministic metrics and LLM judges

- **Evaluate iteratively**: Log complete trajectories with judge rationale, inspect for biases, fix issues, and re-evaluate
- **Account for multi-agent characteristics**: LLM judges may penalize collaborative transparency as verbosity without proper guidance
- **Match metrics to task types**: Use deterministic evaluation for factual tasks, LLM-as-a-judge for subjective assessment
- **Handle conversational complexity**: Answer extraction strategies prevent evaluation failures in real agent interactions
- **Plan systematically**: Follow the 5-step framework - define success criteria, create task suites, choose metrics, establish baselines, and plan iteration workflow

Our comprehensive evaluation revealed when multi-agent systems provide value versus overhead. Multi-agent configurations excel on tool-heavy research tasks (9.2/10 vs direct model 6.8/10) but add unnecessary overhead on simple reasoning tasks where direct model calls perform best (9.7/10 vs multi-agent 9.3/10). Token efficiency shows the cost: multi-agent systems use 43x more tokens than direct model calls. This demonstrates why evaluation is critical: it reveals which architectural complexity is justified by task requirements and which is over-engineering.

The evaluation system we've built provides a practical foundation for measuring and improving multiagent systems. By combining reference-based efficiency with LLM-based flexibility, and by iterating on both agent configurations and evaluation methodology, we can develop more effective agent architectures while understanding when multiagent approaches provide genuine value.

Chapter 11

Optimizing Multi-Agent Systems

This chapter covers:

- What to optimize, when to optimize, and how to prioritize optimization efforts
- Ten common failure modes in production multi-agent systems and concrete strategies to address each
- The two-level optimization framework: agent-system parameters vs model-level parameters
- Using evaluation-driven iteration to systematically improve multi-agent system performance

11.1   Introduction

At this point, you should be familiar with agents (Chapter 4), common multi-agent patterns (Chapter 2), and how to evaluate performance (Chapter 10). The next logical step: how do we systematically improve performance?

When I think of optimization, I recall my early experiences as a kid fixing internal combustion engines with my dad. We fixed our own cars (a Peugeot 505 and 504) and diesel engines coupled to an alternator for electricity generation. We followed a simple two-state process: first, get the engine running as a baseline; second, iteratively observe the signals (engine sound, how it revved) and tweak parameters (spark plug cleanliness, fuel mixture, ignition timing) to achieve optimal performance.

With agents, we follow the same process. We have signals to measure (evaluation metrics from Chapter 10). We have parameters to adjust (instructions, tools, orchestration patterns). The question is: which parameters deliver the most improvement with the least effort? This chapter answers that question by establishing an optimization priority framework and examining ten common failure modes that tell you exactly what needs fixing. Most of these notes apply to autonomous multi-agent orchestration patterns, where errors frequently arise from the flexibility and autonomy of these systems.

The optimization loop is straightforward: **measure** (run evaluation), **analyze** (identify failures), **modify** (adjust parameters), **validate** (re-run evaluation), and repeat.

An optimization loop for your multi-agent system

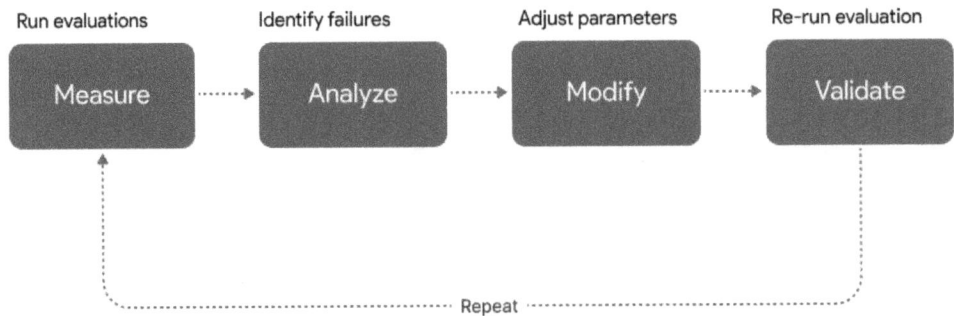

Figure 11.1. An optimization loop for multi-agent systems showing the iterative cycle: Measure (run evaluations) ☐ Analyze (identify failures) ☐ Modify (adjust parameters) ☐ Validate (re-run evaluation) ☐ Repeat.

The challenge isn't the loop itself—it's knowing which parameters to adjust for maximum impact.

11.2 What to Optimize, When to Optimize, and How to Optimize

Multi-agent systems have two optimization levels, each requiring different expertise and offering different return on investment:

Model-level parameters (lowest level):

This includes the LLM weights, architecture, hyperparameters, and training data. While important, optimizing model-level parameters typically requires deep ML expertise and significant computational resources—usually outside the scope of application-level optimization.

Agent-system parameters (application level):

- **Instructions:** System prompts, behavioral guidelines, task-specific directions
- **Tools:** Which capabilities to provide (search, calculation, file operations)
- **Memory:** What context to retain and retrieve across interactions
- **Model selection:** GPT-4 vs GPT-3.5 vs fine-tuned variants

- **Orchestration patterns**: Workflow vs autonomous (see Chapter 2)
- **Termination conditions**: When to stop (max rounds, token limits, task completion signals)
- **Human delegation**: Which actions require approval (risk-based policies)

The optimization question: Given measurements from evaluation, which parameters should we adjust and how?

Optimization priority—start where impact is highest:

1. Start with agent-system parameters: This is where you'll get the most leverage with the least effort. Take an iterative approach informed by evaluation data—critically inspect failures and draw correlations to these parameters.

Instructions and tools have the most direct impact on agent behavior. When evaluation reveals failures, first check: are instructions specific enough? Does the agent have the right tools? These simple fixes often yield significant gains before considering model changes or orchestration complexity.

Model selection matters—frontier models (GPT-5, Claude 4.5) have much better instruction-following than smaller models. However, reasoning models can sometimes degrade performance in multi-agent systems when they attempt to take on more than they're designed for (e.g., the orchestrator's role).

Once individual agents work reasonably well, experiment with orchestration patterns and termination conditions. Use evaluation metrics to compare performance across patterns. In Chapter 10 (Section 10.6.5), we showed that for simple reasoning tasks, Direct-Model (9.7/10) beat Multi-Agent-AI (9.3/10)—multi-agent overhead provided no benefit in that specific use case. Improving termination conditions (using `TaskStatusTool` with aligned instructions) significantly reduced token usage and improved reliability.

2. Optimize model-level parameters last: Model training is resource-intensive and requires specialized expertise. Only pursue this if specific conditions warrant it: (1) need to reduce cost/latency significantly (small models are 20-30x cheaper), (2) privacy/compliance reasons preclude third-party models, or (3) agent-system optimization has reached diminishing returns.

The typical trajectory: optimize agent-system parameters until performance plateaus (often 95%+ with frontier models), then consider finetuning smaller models on collected execution data. We cover when finetuning is warranted and the practical workflow in Section 11.3.2.

11.3 Why Autonomous Multi-Agent Systems Fail in Production

In this section, we explore 10 common reasons why autonomous multi-agent systems fail in production settings, along with concrete strategies to address each failure mode. After two years building the AutoGen framework and supporting production deployments, I've observed that multi-agent systems fail in consistent, predictable ways. Understanding these failure modes tells you exactly what to optimize. Treat this as a checklist to diagnose and fix your multi-agent systems.

11.3.1 Your Agents Lack Detailed Instructions

The Problem: Agents are driven by LLMs, which require careful prompting. Vague or incomplete instructions lead to unpredictable behavior and poor performance.

Vague instructions like `"Help with research tasks"` tell agents almost nothing about how to operate. Effective instructions resemble comprehensive job descriptions covering behavioral guidelines, tool selection, constraints, output formatting, and error handling. We covered instruction design in Chapter 4 (Chapter 4)—the optimization key is expanding instructions iteratively based on evaluation failures. When agents fail on specific test cases, add guidance addressing those failure modes.

Optimization Strategy: Use your evaluation results to systematically expand instructions. When agents fail on specific test cases, add guidance addressing those failure modes. Treat instructions as living documentation that evolves with your understanding of the task. **Great agent prompts can be pages long!**

11.3.2 Stop Using Small Models (Without Optimization)

The Problem: The capability gap between model sizes is significant. Smaller models show dramatically reduced instruction-following and struggle with the complex reasoning required for multi-agent coordination.

This creates a dilemma: frontier models (GPT-5, Claude 4.5) deliver strong performance but at 20-30x the cost of smaller models. For high-volume production deployments, this difference is substantial.

The Reality: Smaller models won't match frontier performance out of the box. Your Llama-7B or Qwen3B models will require significant optimization or finetuning to approach GPT-4/Claude-level performance. But "use bigger models" isn't always practical given cost constraints—frontier models cost 20-30x more than small models at scale.

The Solution: The optimization trajectory typically follows this path:

1. **Start with frontier models** to establish baseline performance and understand task requirements
2. **Optimize agent-system parameters** (instructions, tools, orchestration) with the frontier model
3. **Consider model-level optimization** only after agent-system optimization plateaus:
   - **Routing**: Direct simple queries to small models, complex ones to large models (can achieve 90% performance at 15% cost)
   - **Distillation**: Train small models on frontier model outputs to transfer capabilities

- **Cascading**: Try small model first, escalate to large model only on failure

When Is Finetuning Actually Warranted?

After optimizing agent-system parameters, analyze your remaining failures. Two scenarios justify model-level optimization:

Scenario 1: Residual failures that parameters can't fix. Ask: why is the system still failing? Common culprits:

- **Out-of-distribution problems**: Your domain involves knowledge rare in training data (e.g., chemistry of uncommon materials, niche regulatory frameworks)
- **Counterintuitive business logic**: Correct behavior violates common patterns (e.g., hanging up on a customer early is sometimes the right outcome in your call center)
- **Long-tail edge cases**: Solutions that appear in your business but rarely in general text

These failures signal that the model lacks domain-specific reasoning patterns—exactly what finetuning addresses.

Scenario 2: Efficiency optimization at full performance. Even at 100% accuracy, finetuning may be warranted for cost, latency, or privacy. The key insight: your business domain is scoped. Frontier models carry strong reasoning capabilities (needed) plus broad world knowledge (not needed for your task). SFT and RL finetuning help smaller models retain reasoning *for your specific domain* while reallocating capacity away from unused general knowledge. This is why a 7B model finetuned on your task can sometimes match a 70B generalist.

The Practical Workflow:

The Model Optimization Decision Framework

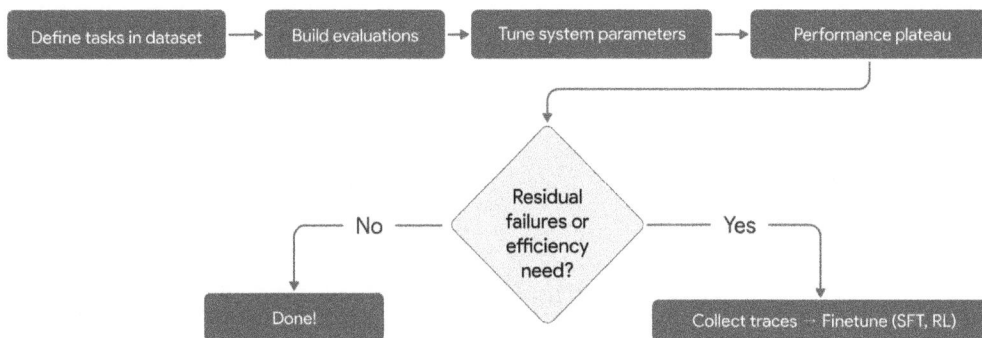

Figure 11.2. The model optimization decision framework. Most teams reach "Done!" after agent-system tuning achieves 95%+ accuracy. Finetuning is only warranted for residual domain-specific failures or efficiency requirements (cost, latency, privacy).

1. Define tasks in a dataset with expected outputs
2. Build evaluations to quantify performance (see Chapter 10)
3. Tune agent-system parameters iteratively—inspect trajectories, ask "why is this failing?", adjust prompts, tools, termination conditions
4. Reach the plateau (often 95%+ with frontier models)
5. Analyze residual failures—if they fall into Scenario 1 or 2 above, proceed to finetuning
6. Collect execution traces from your optimized system to create training data

Academic work supports this approach. Luo et al. (Luo et al. 2025) demonstrated training LLMs within agents using RL by collecting execution traces automatically. However, this remains task-specific (can't train once for all agents) and requires defining explicit reward functions.

These model-level techniques require ML infrastructure and expertise—they're advanced optimizations to pursue after exhausting agent-system improvements.

11.3.3   Your Agent Instructions Don't Match Your LLM

The Problem: Developers often underestimate model differences. System messages aren't portable across versions of the same model, let alone across different providers. Simply changing models while expecting similar behavior is a common and costly mistake.

Consider this instruction that works well with GPT-4:

```
instructions = """Analyze the data and provide insights.
Think step-by-step about patterns, then formulate
3-5 key recommendations."""
```

This might work perfectly with GPT-4's strong reasoning capabilities. But with a smaller model, the vague "think step-by-step" and "provide insights" may produce inconsistent results.

The Solution: Treat instructions as model-specific assets:

```
# Model-specific instruction registry
INSTRUCTIONS = {
    "gpt-4": """Analyze the data and provide insights.
        Think step-by-step about patterns, then formulate
        3-5 key recommendations.""",

    "gpt-3.5-turbo": """Follow these exact steps:
        1. List all data points in the dataset
        2. Identify the 3 highest and 3 lowest values
        3. Check for patterns (increasing, decreasing, cyclical)
        4. For each pattern found, write one specific recommendation

        Format each recommendation as:
        - Pattern: [what you observed]
```

```
            - Recommendation: [specific action]""",
}

def create_agent(model: str):
    return Agent(
        name="analyst",
        instructions=INSTRUCTIONS[model],
        model=model
    )
```

Optimization Strategy:

- Maintain separate instruction sets for different model families
- Version control prompts alongside model choices
- Run A/B tests when switching models, not just on new tasks
- Use your evaluation framework to validate that instruction changes preserve performance

The evaluation infrastructure from Chapter 10 becomes critical here: you can systematically test whether model changes affect performance on your specific tasks.

11.3.4   Your Agents Lack Good Tools

The Problem: Tools define an agent's action space. Agent reliability is directly limited by tool quality. A common mistake is building many general-purpose tools rather than fewer excellent, domain-specific ones, or just not investing in designing tools for production use.

The Solution: Invest in a catalog of battle-tested tools. PicoAgents provides three categories of built-in tools that you can use as templates for your own production tools:

Core Tools (universal utilities, no external dependencies):

- `ThinkTool` - Enables structured reasoning during complex tasks (based on Anthropic's research showing 54% performance improvement)
- `TaskStatusTool` - Forces explicit evaluation of task completion with rationale
- `CalculatorTool` - Safe mathematical expression evaluation
- `DateTimeTool` - Current time, parsing, and formatting operations
- `JSONParserTool` - JSON parsing and path extraction
- `RegexTool` - Pattern matching and text manipulation

Research Tools (information retrieval):

- `GoogleSearchTool` / `WebSearchTool` - Web search with domain filtering for security
- `WebFetchTool` - Fetch and extract content from URLs (HTML, text, or markdown)
- `ArxivSearchTool` - Academic paper search
- `YouTubeCaptionTool` - Extract transcripts from YouTube videos

Coding Tools (file operations and code execution):

- `ReadFileTool` / `WriteFileTool` - File I/O with workspace isolation
- `ListDirectoryTool` - Directory browsing
- `GrepSearchTool` - Pattern search across files
- `BashExecuteTool` - Safe command execution with timeouts
- `PythonREPLTool` - Isolated Python code execution

You can explore complete implementations in the Book's GitHub repository and see usage examples throughout `examples/agents/`, `examples/workflows/`, and `examples/orchestration/`. We covered tool design and implementation patterns in Chapter 4 (Section 4.6.3).

Key principles for production tools:

1. **Reliability over breadth**: Better to have 10 excellent tools than 50 mediocre ones (agents get confused as the toolspace grows and fills up context)
2. **Clear error handling**: Tools should never crash agents—return structured errors
3. **Explicit constraints**: Document rate limits, data freshness, failure modes
4. **Retry logic**: Implement backoff strategies for transient failures
5. **Validation**: Check tool outputs before returning to agents

Focus tool development on domain-specific business operations rather than general utilities. Production agents need reliable tools tailored to your specific workflows.

11.3.5 Your Agents Don't Know When to Stop

The Problem: Termination conditions control both cost and reliability. Poor termination logic leads to runaway processes consuming tokens or premature exits that abandon solvable tasks.

Consider this common antipattern:

```
# Ineffective: Agent never told to produce this pattern
termination = TextMatchTermination(
    patterns=["TASK_COMPLETE"]  # But agent instructions never mention this!
)

agent = Agent(
    name="research_agent",
    instructions="Research the topic and provide findings."
    # No instruction to say "TASK_COMPLETE"
)

workflow = SequentialWorkflow(
```

```
    agents=[agent],
    termination=termination
)
# This will run until hitting max_rounds or timeout!
```

The Solution: Align termination conditions with agent capabilities. We covered termination patterns like `MaxRoundsTermination`, `TokenUsageTermination`, `TimeoutTermination`, and `CompositeTermination` in Chapter 7 (Section 7.2). Better yet, use the `TaskStatusTool` to let agents explicitly signal completion:

```
agent = Agent(
    name="research_agent",
    instructions="""Research the topic and provide findings.

    When finished, call the task_status tool with:
    - status='complete' if all requirements met
    - status='incomplete' if you hit limitations
    - rationale explaining your assessment""",
    tools=[search_tool, task_status_tool]
)
```

Optimization Strategy: Use evaluation data to calibrate termination conditions. Track metrics like:

- Average rounds to completion for successful tasks
- Token usage distribution across test cases
- Failure modes (timeout vs. max rounds vs. task completion)

Set conservative limits during development, then tighten based on production data. Remember: it's better to terminate early with partial results than to waste resources on divergent trajectories.

11.3.6   You Have the Wrong Multi-Agent Pattern

The Problem: Multi-agent systems exist on a spectrum from explicit workflow control to autonomous emergent behavior (see Chapter 2). Pattern selection profoundly impacts cost, reliability, and system behavior. The wrong choice wastes resources or fails to deliver multi-agent benefits.

The key optimization insight: **more autonomous patterns aren't inherently better**—they trade predictability for flexibility. The question is which trade-off matches your task characteristics.

Empirical Evidence from Chapter 10:

Our comprehensive evaluation (Section 10.6.5) compared Direct-Model, Single-Agent-Tools, Multi-Agent-RoundRobin, and Multi-Agent-AI across different task types. The results revealed clear patterns:

- **Simple-Reasoning tasks**: Direct-Model won (9.7/10) vs Multi-Agent-AI (9.3/10). Multi-agent overhead provided no benefit.
- **Tool-Heavy tasks**: Multi-agent approaches excelled (9.2/10) vs Direct-Model (6.8/10). Tool coordination justified the complexity.
- **Token efficiency**: Direct-Model used 1x tokens, Multi-Agent-AI used 43x tokens. The cost difference is substantial.

The Optimization Principle: Start with the simplest pattern that could work. Add complexity only when evaluation demonstrates clear benefits. A well-designed single agent often outperforms a poorly designed multi-agent system.

Pattern Selection Guide:

Table 11.1 maps task characteristics to appropriate patterns. We covered these patterns in depth in Chapter 2 (Chapter 2).

Table 11.1. Pattern Selection Guide Based on Task Characteristics

Task Characteristic	Pattern	Trade-off
Fixed, known steps	Sequential Workflow	Predictable cost, easy debugging, low flexibility
Conditional routing	Conditional/Supervisor	Explicit branching, bounded execution
Independent subtasks	Parallel Workflow	Maximum throughput, clear fan-in/fan-out
Unknown decomposition	Plan-Based Orchestration	Dynamic planning, adds overhead per step
Uncertain solution path	AI-Driven Orchestration	Emergent collaboration, unbounded cost

Optimization Strategy:

1. **Prototype with workflows**: Start with sequential/conditional patterns
2. **Measure with evaluation**: Run your test suite, track success rate and token usage
3. **Identify bottlenecks**: Where does the fixed path fail?
4. **Selectively add autonomy**: Convert specific bottleneck sections, not the entire system
5. **Validate improvement**: Ensure complexity buys measurable gains

Common Anti-Pattern:

Don't use AI-driven orchestration for linear tasks:

```
# Anti-pattern: 3-5x cost for no benefit
expensive = AIOrchestrator(
    agents=[fetch_agent, transform_agent, save_agent]
)

# Better: Deterministic, fast, cheap
efficient = SequentialWorkflow(
    agents=[fetch_agent, transform_agent, save_agent]
)
```

Production systems often use hybrid approaches: plan-based orchestration for top-level coordination, workflows for predictable subtasks. Microsoft's Magentic-One (Fourney et al. 2024) and Anthropic's research system (Hadfield et al. 2025) both demonstrate this task-aware pattern switching.

11.3.7   Your Agents Aren't Learning (Memory)

The Problem: Most agent examples seen in the wild are stateless—they don't retain knowledge from previous interactions or improve over time. Each conversation starts from scratch, repeating the same mistakes.

The Solution: Implement memory systems that enable learning. In Chapter 4 (Section 4.7), we covered two memory management approaches: **application-managed** (developer calls `memory.add()` to store, framework injects via `get_context()`) and **agent-managed** (Section 4.8) (agents use memory tools to actively organize their own knowledge). Both approaches use vector databases for semantic search and support various backends (ChromaDB, Qdrant, Pinecone).

Optimization Consideration: Memory adds latency (retrieval) and cost (embedding, storage). Use evaluation to determine:

- Does memory improve task success rate enough to justify the overhead?
- How many memories provide diminishing returns?
- Which tasks benefit most from memory?

Not every agent needs memory. Stateless agents are simpler and faster for independent tasks.

11.3.8   Your Agents Lack Metacognition

The Problem: Long-running complex tasks benefit from self-reflection—the ability to plan, review progress, and recognize when current approaches aren't working. Without metacognition, agents can't abandon compromised trajectories or reset when stuck.

The Solution: Implement metacognitive capabilities through reflection loops. We covered the Plan-Based pattern in Chapter 2 (Chapter 2) and implemented it in Chapter 7 (Section 7.5). The `PlanBasedOrchestrator` provides built-in metacognition: creates initial plan (Section 7.5.1), executes steps, evaluates progress after each step (Section 7.5.2), and replans if progress is insufficient (Section 7.5.3). Microsoft's Magentic-One (Fourney et al. 2024) demonstrates similar self-monitoring, showing up to 31% performance improvement through their "ledger" system.

Optimization Tradeoff: Metacognition adds LLM calls for reflection but can prevent wasted computation on failing trajectories. Use evaluation to measure whether the reflection overhead is justified by improved success rates.

11.3.9   You Don't Have Evals for Your Tasks

The Problem: Without evaluation frameworks, you can't understand current performance or how configuration changes impact results. This makes optimization a guessing game.

We covered evaluation comprehensively in Chapter 10. The optimization connection is direct: **you cannot optimize what you cannot measure.**

The Minimum Viable Evaluation:

```
from picoagents.eval import EvalTask, EvalRunner, ExactMatchJudge

# 1. Define your specific task clearly
tasks = [
    EvalTask(
        input="What's the weather in Paris?",
        expected_output="Use weather API tool",
        metadata={"category": "tool_usage"}
    ),
    EvalTask(
        input="Summarize this article: ...",
        expected_output="concise summary",
        metadata={"category": "summarization"}
    )
]

# 2. Create evaluation metrics
judge = ExactMatchJudge()  # Or LLMJudge for complex tasks

# 3. Build non-agent baseline
def baseline_system(input: str) -> str:
    return call_gpt4(input)  # Direct LLM call

# 4. Run comparison evaluation
runner = EvalRunner(judges=[judge])
results = await runner.evaluate(
    tasks=tasks,
    targets={
        "baseline": baseline_system,
```

```
        "agent_v1": your_agent,
        "agent_v2": optimized_agent
    }
)


# 5. Monitor metrics as you optimize
print(results.comparison_table())
```

11.3.10   Your Agents Don't Know When to Delegate to Humans

The Problem: To agents, all actions appear equal. Fetching weather data and transferring money carry the same weight without proper risk assessment. This creates serious reliability and safety concerns.

The Solution: Implement cost-aware action delegation, a UX principle we covered in Section 3.4.2. Actions have varying levels of **consequence**—the potential impact if the action contains errors or is inappropriate. The optimization challenge is implementing the cost assessment and approval mechanisms.

PicoAgents Implementation:

Tools support `ApprovalMode` to control execution:

```
from picoagents.tools import tool, ApprovalMode
from picoagents.agents import Agent

# Low-risk tool: executes immediately
@tool
def search_web(query: str) -> str:
    """Search the web for information."""
    return f"Search results for: {query}"


# High-risk tool: requires approval
@tool(approval_mode="always_require")
def send_email(to: str, subject: str, body: str) -> str:
    """Send an email to a recipient."""
    return f"Email sent to {to}"


@tool(approval_mode="always_require")
def delete_database_record(table: str, record_id: int) -> str:
    """Delete a record from the database."""
    return f"Deleted record {record_id} from {table}"


# Agent with mixed-risk tools
agent = Agent(
    name="assistant",
```

```
    model_client=model,
    tools=[search_web, send_email, delete_database_record]
)

# When agent calls high-risk tool, execution pauses
response = await agent.run(
    "Research competitors and email report to boss@company.com"
)

if response.needs_approval:
    # Agent found information and wants to send email
    # User can review and approve/reject
    approval_requests = response.approval_requests

    for request in approval_requests:
        print(f"Tool: {request.tool_name}")
        print(f"Parameters: {request.parameters}")
        # User decides: approve or reject
```

Cost Classification Framework:

Table 11.2. Tool Risk Levels and Approval Requirements

Risk Level	Characteristics	Examples	Approval Mode
Low	Read-only, no side effects	Search, calculate, read files	NEVER (auto-execute)
High	Irreversible or costly	Send email, delete data, payments	ALWAYS (require approval)

Optimization Consideration: Human-in-the-loop adds latency but prevents catastrophic failures. The tradeoff depends on your domain:

- **Research assistant**: Most actions low-risk, optimize for speed
- **Customer service**: Occasional high-risk (refunds), approval for those only
- **Financial trading**: Every action high-risk, require approval

The cost-aware delegation principle from Chapter 3 becomes an optimization concern here: balance safety (prevent catastrophic actions) against speed (minimize approval friction). Use evaluation to measure whether approval policies are too strict (>95% approval rate suggests over-cautious flagging) or too lenient (missing actual risks in test cases).

11.3.11   Bonus: You Probably Don't Need a Multi-Agent System

The Problem: Multi-agent architectures add complexity. Multiple agents collaborating with autonomy increases the surface area for errors and reliability issues.

Before implementing any multi-agent system, honestly evaluate whether your task requires this complexity.

Decision Checklist:

Use multi-agent systems when your task has (see Section 1.3 for detailed discussion):

1. **Decomposability (Planning)**: Task can be broken into distinct steps leading to a goal
2. **Diverse perspectives (Diverse Expertise)**: Steps map to different domains or expertise areas
3. **Extensive context**: Individual steps involve processing substantial information
4. **Adaptive requirements (Adaptive Solutions)**: Solution unknown until actions are taken and evaluated

When Single Agents Suffice:

```
# Don't need multi-agent for simple tasks
simple_task = "Translate this text to French"
result = await translation_agent.run(simple_task)


# Don't need multi-agent for linear, single-domain tasks
linear_task = "Summarize this 10-page document"
result = await summary_agent.run(linear_task)
```

When Multi-Agent Adds Value:

```
# Complex, multi-domain research task
complex_task = """Analyze market trends for electric vehicles
in European markets, considering regulatory changes,
consumer sentiment, and competitive landscape"""

# Benefits from specialized agents:
# - Research agent: Gather data from multiple sources
# - Analysis agent: Process quantitative market data
# - Regulatory agent: Track policy changes
# - Synthesis agent: Combine insights into coherent report

result = await orchestrator.run(
    task=complex_task,
    agents=[research_agent, analysis_agent,
            regulatory_agent, synthesis_agent]
)
```

The Optimization Principle: Start with the simplest architecture that could work. Add complexity only when evaluation demonstrates clear benefits. A single well-designed agent often outperforms a poorly designed multi-agent system.

11.3.12   Summary: Failure Modes as Optimization Opportunities

These ten failure modes aren't just problems to avoid—they're optimization opportunities. Each represents a dimension where measurement and systematic improvement can enhance your system:

1. **Detailed instructions** -> Iteratively expand based on failure analysis
2. **Model capability** -> Strategic model selection and optimization (next section)
3. **Instruction-model matching** -> Model-specific prompt engineering
4. **Tool quality** -> Curated, reliable tool catalogs
5. **Termination conditions** -> Data-driven threshold calibration
6. **Pattern selection** -> Empirical pattern comparison
7. **Memory and learning** -> Strategic memory implementation
8. **Metacognition** -> Reflection loop design
9. **Evaluation infrastructure** -> Foundation for all optimization
10. **Human delegation** -> Risk-based approval strategies

The evaluation framework from Chapter 10 enables systematic improvement across all these dimensions.

11.4   Summary

Optimizing multi-agent systems requires iterative discipline: measure performance through evaluation, identify failure points, adjust parameters, and validate improvements. The key is knowing where to focus optimization efforts for maximum impact.

- **Two-Level Optimization Framework**: Multi-agent systems have agent-system parameters (instructions, tools, memory, model selection, orchestration patterns, termination conditions, human delegation) and model-level parameters (weights, architecture, training data). Agent-system optimization delivers the most leverage with least effort. Model-level optimization (finetuning, RL training) requires ML expertise and substantial resources—pursue only when agent-system optimization reaches diminishing returns or specific conditions warrant it (cost reduction, privacy requirements).

- **Optimization Priority**: Start with agent-system parameters. Instructions and tools have the most direct impact on behavior. Model selection matters (frontier models have better instruction-following), but reasoning models can degrade performance in multi-agent coordination. Orchestration pattern selection is critical—our Chapter 10 evaluation showed Direct-Model (9.7/10) outperformed Multi-Agent-AI (9.3/10) for simple tasks, with 43x lower token costs. Add complexity only when evaluation demonstrates clear benefits.

- **Ten Production Failure Modes**: Multi-agent systems fail in predictable ways. Each failure mode represents an optimization opportunity: insufficient instructions (expand iteratively based on eval failures), inadequate model capability (strategic selection), instruction-model mismatch (maintain model-specific prompts), poor tool quality (curate battle-tested catalogs), unclear termination (align conditions with agent capabilities), wrong orchestration pattern (empirical comparison via evaluation), missing memory (strategic implementation where beneficial), lack of metacognition (plan-based reflection loops), no evaluation infrastructure (foundational requirement), and unclear human delegation (risk-based approval policies).

- **The Evaluation Foundation**: You cannot optimize what you cannot measure. The evaluation infrastructure from Chapter 10 enables all optimization activities—it identifies failure modes, validates improvements, and enables systematic iteration. Without evaluation, optimization becomes guesswork.

Optimization is a continuous discipline, not a one-time activity. Start with the simplest architecture that could work, measure performance rigorously, and add complexity incrementally when evaluation demonstrates clear benefits. A well-designed single agent often outperforms a poorly designed multi-agent system.

Chapter 12

Protocols for Distributed Agents

This chapter covers:

- Distributed agent fundamentals and how they extend distributed computing concepts
- Model Context Protocol (MCP) for standardized tool and context integration
- Agent-to-Agent Protocol (A2A) for cross-organizational agent collaboration
- Security considerations and practical guidance for choosing the right protocol

So far, we have discussed building agents and multi-agent systems but with the implicit assumption that all components (e.g., tools, agents within the multi-agent system) run within the same execution context—on the same machine, within the same network, or at least within the same organizational boundary. This is a reasonable starting point for learning how agents work, but *some* real-world applications often require more distributed architectures where agents and tools run in separate execution contexts—different machines, containers, regions, or organizations—communicating over a network without shared memory (Figure 12.1).

Imagine you're building a financial analysis system where your data retrieval agent runs on your company's secure cloud, but it needs to coordinate with a legal compliance agent **from your law firm's infrastructure** and a **market analysis agent from a third-party service provider**. Each agent runs on different machines, potentially in different geographic regions, managed by different organizations. How do these agents discover each other? How do they communicate securely? What happens when a task takes hours to complete and the network connection drops midway? What if an agent needs to ask for user confirmation during execution?

Many of these challenges are classic *distributed computing* problems: discovery, communication, fault tolerance, and coordination. However, agents introduce new characteristics that existing distributed computing protocols weren't designed to handle: autonomy in decision-making, tasks that may run for hours or days while remaining interactive, workflows requiring human input at unpredictable points during execution, and adaptive behavior driven by AI

Distributed Agents

Agents that address user tasks may exist in **separate execution contexts**—different machines, containers, regions, or organizations—communicating over a network

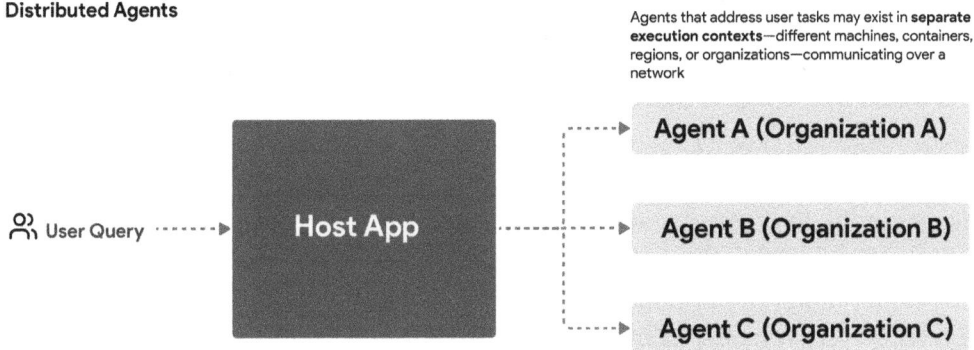

Figure 12.1. *Agents that address user tasks may exist in separate execution contexts—different machines, containers, regions, or organizations—communicating over a network*

models rather than deterministic rules.

This chapter explores how recent protocols are addressing these challenges (MCP, A2A), how they relate to well-known distributed computing foundations while adding agent-specific semantics. We'll focus on understanding why these protocols matter, what value they provide, and when to reach for them in your own systems.

12.1   Core Concepts for Distributed Agents

Designing a system where components run in separate environments brings both advantages and challenges. Distributed agents offer improved scalability—you can scale specific capabilities independently rather than scaling entire systems. They provide isolation of execution environments, enforcing security and resource boundaries. They enable fault tolerance since the failure of one agent doesn't necessarily crash the entire system. And they let you leverage heterogeneous resources across different infrastructures. The trade-off is increased complexity in communication, coordination, and management compared to agents running in the same process.

12.1.1   Grounding in Distributed Computing

The concept of distributed agents isn't fundamentally different from distributed computing. The core problems—discovery, communication, fault tolerance, coordination—are the same challenges addressed by RPC, REST, gRPC, and message queues.

The key distinction is where the logic lives. In traditional distributed computing, **coordination and decision-making** reside in the infrastructure layer—Kubernetes orchestrating containers, service meshes managing traffic, API gateways routing requests via predefined rules. In distributed agent systems, this logic moves to the **application layer**, driven by AI models making autonomous decisions. The infrastructure concerns—communication protocols, fault tolerance, security, resource management—stay at the protocol layer.

Concrete parallels clarify what's genuinely new versus what's borrowed:

Message Queues -> Task Delegation: Message queues decouple producers from consumers, enabling asynchronous task processing. Agent task delegation builds on this pattern but uses AI models to adaptively request additional context or assistance based on reasoning about task state rather than predefined workflow steps.

Service Discovery (Consul, etcd) -> Agent Discovery: Traditional service discovery helps services find each other by name or capability. Agent discovery adds richer capability metadata—skills, domain expertise, supported interaction patterns—enabling intelligent routing based on semantic understanding rather than just service names.

Circuit Breakers and Retries -> Fault Tolerance: Distributed systems retry failed requests with backoff. Agent tasks often span hours or days, so they need session resumability and durable task state—simple retries lose accumulated work.

API Gateways -> Orchestration: API gateways route requests to appropriate backend services based on rules. Agent orchestrators do the same but use AI-driven reasoning to decompose complex requests, route to specialist agents, and synthesize results adaptively.

The primitives are remarkably similar—discovery, communication, fault tolerance, state management. What changes is the semantics: agents are autonomous, long-running, interactive, and adaptive in ways that traditional distributed systems typically are not. Beyond executing predefined workflows, agents reason about goals, learn from outcomes, negotiate with other agents, and adapt their strategies based on context. This cognitive capability, driven by AI models, distinguishes agent protocols from their distributed systems predecessors.

12.1.2   Distributed Agent Requirements

Agents operating in distributed environments need four core capabilities that emerge from agents' unique characteristics: autonomy, long-running execution, interactivity, and adaptive behavior. Table 12.1 summarizes these needs.

Table 12.1. Core Capabilities for Distributed Agents

Capability	Why Agents Need This
Streaming & Progress	Agent tasks often span hours or days analyzing documents, orchestrating workflows, or processing datasets. Users need real-time visibility for observability, debugging, and trust.

Capability	Why Agents Need This
Resumability	Network failures during a 3-hour agent task shouldn't lose all accumulated work. Sessions must survive disconnections and reconnections.
Durability	Agent results must persist even if servers restart. Clients need persistent references to check status or retrieve results out-of-band.
Multi-turn Interaction	Agents frequently need mid-execution input: user confirmations for high-cost actions, clarifications for ambiguous requests, or AI assistance for reasoning steps.

The following sections explore two protocols—MCP and A2A—that have emerged to address these needs and how they implement these capabilities. The Model Context Protocol (MCP) (Anthropic 2024c) focuses on standardizing tool and context integration. The Agent-to-Agent Protocol (A2A) (Linux Foundation 2025) addresses cross-organizational agent collaboration.

12.2 Model Context Protocol (MCP)

When Anthropic introduced MCP in November 2024 (Anthropic 2024b), the focus was narrow: give large language models standardized access to context. The name itself—Model Context Protocol—reflected this scope. The protocol defined how an AI application (the "client") could connect to external systems (the "servers") exposing tools (functions to call), resources (data to read), and prompts (reusable templates).

Standardized protocols like MCP solve integration, distribution, discovery, security, and flexibility challenges that plague custom implementations. This is why MCP has seen rapid adoption as agent-based systems have proliferated. Table 12.2 illustrates how MCP addresses each challenge.

Table 12.2. MCP's Value Proposition

Challenge	Without MCP	With MCP
Integration	Team X spends weeks integrating Team Z's new tool	Team X gets day-one support for any MCP tool
Distribution	Write separate extensions for Cursor, Windsurf, VSCode, Claude Desktop, etc.	Write one MCP server that works everywhere
Discovery	Teams ask in Slack: "Does anyone have a tool that does X?"	Central registry where teams publish and find tools

Challenge	Without MCP	With MCP
Security	Each team implements (or skips) their own auth for LLM-accessible tools	Centralized authentication and managed registry of MCP servers
Deployment Flexibility	Tools tightly coupled to application code, requiring same deployment context	Tools run remotely in separate contexts—different machines, regions, or organizations—while appearing local to the application

Since its introduction, MCP has evolved rapidly, adding capabilities that have made it *possible* to not just provide context to a single agent, but support broader agent to agent behavioural requirements. These updates include the introduction of MCP elicitation (servers can request user input mid-execution), sampling (servers can ask the client's LLM for completions), resumable streams (sessions survive network disconnections), progress notifications (tools can stream real-time status updates), and resource links (tools can return durable references to results that persist beyond the tool call). These features enable MCP tools to behave like full agents—running for hours, streaming updates, requesting human input, delegating subtasks to LLMs, and surviving network failures.

12.2.1 MCP Architecture: Servers, Clients, and Hosts

MCP defines a three-layer architecture that separates concerns cleanly. At the bottom layer are MCP **servers**, which expose capabilities—tools (functions that can be called), resources (data that can be read), prompts (reusable LLM templates), and potentially custom extensions. A server might expose a single tool or dozens of related capabilities. For example, a "TechCrunch News Server" might expose a tool for fetching articles, a resource for accessing the article archive, and prompts for summarizing news items.

In the middle layer are MCP **clients**, which handle the protocol-level communication with servers. A client maintains a 1:1 connection with a server, managing the low-level details of message exchange, session state, and error handling. Clients are typically embedded within higher-level applications rather than being standalone components.

At the top layer are MCP **hosts**—user-facing applications like Claude Desktop, Cursor, or VSCode[1]. Hosts embed one or more MCP clients to connect to servers and make intelligent decisions about which tools to call based on user needs and context. The host orchestrates the overall workflow: accepting user requests, discovering available tools from connected servers, often using an LLM to determine which tools to invoke, executing those invocations through clients, handling interactive callbacks (like user confirmations), and presenting results to the user.

---

[1]Claude Desktop is Anthropic's desktop chat application, Cursor and VSCode are AI-enhanced code editors that support MCP integration.

**MCP implements a
Server/Client
Architecture**

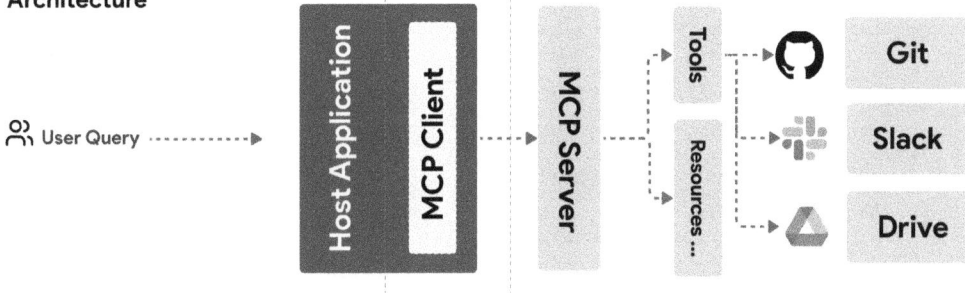

Figure 12.2. MCP's three-layer architecture: hosts (user-facing applications) embed clients that communicate with servers exposing tools, resources, and prompts. The diagram shows example MCP servers providing access to tools like Slack, Google Drive, and databases.

MCP supports two primary **transport** mechanisms. The **stdio** transport runs the server as a subprocess of the client, with communication happening over standard input and output. This is ideal for local integrations—IDE extensions, desktop tools, or development environments where the server and client run on the same machine. The **streamable HTTP** transport uses network requests, making it better suited for distributed systems where servers run remotely, multiple clients need to connect to the same server, or you need easier debugging and monitoring through standard HTTP tooling.

Here's a minimal MCP server setup using the MCP SDK (also available in TypeScript and other languages):

```
# server.py
import os
from mcp.server.fastmcp import FastMCP
import requests

mcp = FastMCP(
    "TechCrunch News Server",
    host=os.environ.get("MCP_SERVER_HOST", "localhost"),
    port=int(os.environ.get("MCP_SERVER_PORT", 8011))
)
```

The server exposes a tool for fetching news from TechCrunch:

```
@mcp.tool(title="Fetch from TechCrunch")
def fetch_from_techcrunch(category: str = "latest") -> str:
```

```
    """Fetch the latest news from TechCrunch."""
    allowed = {"ai", "startup", "security", "venture", "latest"}
    cat = category.lower() if category.lower() in allowed else "latest"

    url = (f"https://techcrunch.com/tag/{cat}/"
           if cat != "latest"
           else "https://techcrunch.com/")

    try:
        response = requests.get(url)
        if response.ok:
            from bs4 import BeautifulSoup
            soup = BeautifulSoup(response.text, "html.parser")
            text = soup.get_text(separator=' ', strip=True)
            return text[:1000] + ("..." if len(text) > 1000 else "")
        return "Failed to fetch news."
    except Exception as e:
        return f"Error fetching news: {str(e)}"
```

Finally, run the server:

```
if __name__ == "__main__":
    mcp.run(transport="streamable-http")
```

This example uses the FastMCP API, which handles protocol boilerplate. The `@mcp.tool()` decorator registers the function as an MCP tool that any client can discover and invoke. Running this server makes it available at `http://localhost:8011/mcp` , ready to accept connections from any MCP-compatible host application.

Once you've written this server, it works with Claude Desktop, Cursor, Windsurf, VSCode extensions, or any custom host application that embeds an MCP client—write once, use everywhere.

12.2.2 MCP Core Concepts

MCP defines several primitives that enable both simple tool integration and sophisticated agent-to-agent communication.

Tools, Resources, and Prompts: Servers expose three types of capabilities. **Tools** are functions that can be called with parameters (like the `fetch_from_techcrunch` example above).**Resources** are data sources that can be read, identified by URIs—a server might expose `news://techcrunch/ai` as a resource for AI-related articles. **Prompts** are reusable LLM templates that servers provide to standardize common interactions.

Progress Notifications: Tools can stream real-time status updates during execution. The server sends structured progress metadata (completion percentage, current step, status messages) that clients display to users or log for debugging. This makes long-running operations observable rather than opaque.

Session Management and Resumability: StreamableHTTP transport in MCP supports the use of session IDs and event replay to survive network failures. Each message gets a unique event ID. When a client reconnects after a disconnection, it provides the last event ID received, and the server replays all missed events from its event store (in-memory for development, database-backed for production).

Resource Links for Tool Durability: In addition to returning results inline, MCP Tools can also return resource URIs. This is valuable as the client can choose to poll or subscribe to these URIs for updates. Resources can be backed by persistent storage (databases, file systems, cloud storage), ensuring results survive server restarts. This decouples task execution from client connectivity, enabling asynchronous workflows.

Elicitation (User Input): Servers can pause tool execution to request structured user input. The server sends an elicitation request with a JSON schema specifying expected response format. The client prompts the user, validates the response, and sends it back. Crucially, the client retains control—it can validate, modify, or decline elicitation requests. The server resumes execution based on the input. This enables human-in-the-loop workflows like payment confirmations or clarification requests.

Sampling (AI Assistance): Servers can request LLM completions from the client during tool execution. When a server issues a sampling request, the client decides whether to execute the call, which model to use, and which context to include. Sampling is server-initiated but client-mediated—a deliberate security design so servers don't hold or use model API keys. A research tool might accumulate findings and ask the client's LLM to generate a summary before proceeding. This creates bidirectional AI interaction: clients use LLMs to choose which tools to call, and servers use the same LLM (via the client) to enhance their processing, while the client remains the gatekeeper for LLM usage.

These primitives combine to transform simple tools into autonomous agents capable of long-running, interactive, adaptive behavior.

12.2.3 MCP's Agentic Capabilities

MCP excels at context provisioning—giving agents access to tools, resources, and prompts. It has also evolved to support agent-to-agent communication by implementing the four core capabilities identified earlier (Table 12.1). Table 12.3 shows how MCP features enable each capability.

Table 12.3. How MCP Implements Distributed Agent Capabilities

Capability	MCP Features	How It Works
Streaming & Progress	Progress notifications via SSE/WebSocket	Servers send structured progress updates during tool execution with completion percentage, current step, and status messages
Resumability	Session IDs + event replay	Clients reconnect with last event ID received; server replays missed events from event store
Durability	Resource links (persistent URIs)	Tools return resource URIs backed by persistent storage; clients poll or subscribe for updates
Multi-turn Interaction	Elicitation (user input) + Sampling (LLM calls)	Servers pause execution to request user input or AI assistance, then resume based on responses

With these capabilities in place, MCP tools become more than simple functions—they behave as agents. The host application coordinating these tools acts as an orchestrator agent making intelligent decisions about task decomposition, routing, and result synthesis.

> **When Tools Become Agents**
>
> MCP tools have evolved beyond simple function calls—they can stream progress, request user input mid-execution, invoke LLMs for reasoning, and maintain state across long-running operations. When MCP tools are composed with stateful servers, resumable sessions, and sampling/elicitation flows, they can behave like agents. However, agent identity, governance, and cross-organizational discovery are separate concerns. MCP enables agentic behavior through composition, but organization-level governance (accountability, SLAs, auditable trails) still lives outside the protocol.

This capability composition enables powerful agent-to-agent communication patterns. As illustrated in Figure 12.3, the host application acts as an orchestrator agent interfacing with users and routing requests to specialist agents exposed as MCP tools. For example, a travel planning agent and a research agent running as MCP tools become autonomous components capable of hours-long execution, interactive confirmations, and adaptive behavior. The orchestrator handles task decomposition, coordinates execution, manages state across concurrent agent operations, preserves user context throughout the interaction, implements retry and fallback logic if agents become unavailable, and synthesizes results from multiple specialists into coherent responses.

MCP Can enable agent to agent communcation

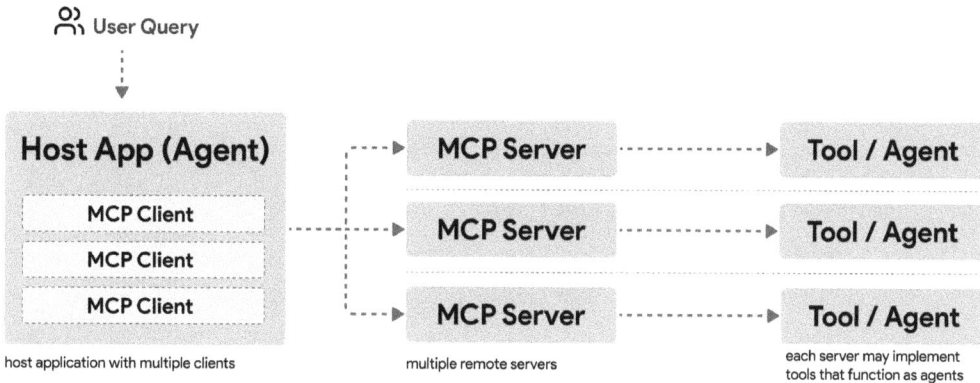

Figure 12.3. Agent-to-agent communication through MCP's client-server architecture. A host application with multiple MCP clients acts as an orchestrator agent, coordinating with remote MCP servers that expose tools functioning as specialist agents. Each server can implement domain-specific capabilities while the orchestrator manages user interaction, task decomposition, and result synthesis across the distributed agent network.

12.2.4 Using MCP Tools in Practice

MCP tools are designed to be driven by LLMs, making them a natural fit for agent integration. To use MCP tools in our agents, we need to make them work seamlessly alongside local tools while gaining the distributed benefits MCP provides—standardization, discoverability, and cross-application compatibility.

Recall from Section 4.6.3 that we built a `BaseTool` abstraction providing a common interface for all tools—whether simple Python functions, REST API calls, or external integrations. MCP tools are external integrations that require bridging between MCP's protocol semantics and our agent's tool interface. To accomplish this, we need to address two important design requirements: managing stateful MCP sessions and adapting MCP tools to the BaseTool interface our agents expect.

12.2.4.1 Stateful Session Management

Unlike stateless function calls, MCP servers require persistent `ClientSession` objects for communication. Sessions must be established, maintained, and properly cleaned up. The `MCPClientManager` class handles this lifecycle:

```
class MCPClientManager:
    """Manages connections to MCP servers."""

    def __init__(self):
        self._servers: Dict[str, MCPServerConfig] = {}
        self._sessions: Dict[str, ClientSession] = {}
        self._tools: Dict[str, List[MCPTool]] = {}

    async def connect(self, server_id: str) -> None:
        """Connect to server and discover tools."""
        config = self._servers[server_id]

        # Establish connection using appropriate transport
        read, write, context = await connect_to_server(config)

        # Create and initialize session
        session = ClientSession(read, write)
        await session.__aenter__()
        self._sessions[server_id] = session
        await session.initialize()

        # Discover available tools
        await self._discover_tools(server_id)
```

The manager maintains a registry of server configurations, active sessions, and discovered tools. When connecting to a server, it establishes the transport (stdio for local servers, HTTP for remote), creates a session, initializes the MCP protocol handshake, and discovers available tools by calling `session.list_tools()`. This separation of concerns—configuration from connection from discovery—enables lazy connection patterns where servers are registered but not connected until needed.

12.2.4.2 Bridging MCP to BaseTool

MCP tools must appear identical to native tools from the agent's perspective. The `MCPTool` class wraps MCP server tools as BaseTool instances:

```
class MCPTool(BaseTool):
    """Wraps an MCP server tool as a PicoAgents tool."""

    def __init__(
        self,
        mcp_tool_name: str,
        mcp_tool_description: str,
        mcp_tool_schema: Dict[str, Any],
        client_manager: MCPClientManager,
```

```
        server_id: str
    ):
        # Namespace tool to avoid conflicts
        tool_name = f"mcp_{server_id}_{mcp_tool_name}"
        super().__init__(
            name=tool_name,
            description=mcp_tool_description
        )

        self.mcp_tool_name = mcp_tool_name
        self._parameters_schema = mcp_tool_schema
        self.client_manager = client_manager
        self.server_id = server_id

    @property
    def parameters(self) -> Dict[str, Any]:
        """Return MCP tool's parameter schema."""
        return self._parameters_schema

    async def execute(self, parameters: Dict[str, Any]) -> ToolResult:
        """Execute via MCP client."""
        session = await self.client_manager.get_session(self.server_id)
        result = await session.call_tool(
            self.mcp_tool_name,
            arguments=parameters
        )

        return ToolResult(
            success=not result.isError,
            result=self._extract_result_content(result),
            error=None if not result.isError else "Tool execution failed"
        )
```

The bridge implements three key mappings. First, it maps MCP's `inputSchema` (JSON Schema) to BaseTool's `parameters` property, making the schema available to the agent for LLM function calling. Second, it maps MCP's `call_tool()` to BaseTool's `execute()` method, translating between parameter formats and handling the async session call. Third, it converts MCP's `CallToolResult` to `ToolResult`, extracting content from MCP's multi-part response format and standardizing error reporting.

Tool naming requires careful consideration. Multiple MCP servers might expose tools with the same name (e.g., both a filesystem server and a cloud storage server could have a "read_file" tool). MCPTool namespaces tools by prefixing them with `mcp_{server_id}_` to avoid conflicts while keeping the original tool name for clarity.

Since MCPTool inherits from BaseTool, it automatically gains the `to_llm_format()` method that converts tool schemas to the function calling format used by OpenAI and Azure OpenAI APIs. The LLM receives tool definitions in the standard format it expects, remaining unaware that tools originate from external MCP servers.

12.2.4.3 Practical Example: Folder Organization Agent

Let's see MCP integration in practice with an agent that analyzes directories and recommends better organization. The agent connects to the MCP filesystem server for read-only access:

```python
from picoagents import Agent
from picoagents.tools import (
    StdioServerConfig,
    create_mcp_tools
)

# Configure MCP filesystem server
filesystem_config = StdioServerConfig(
    server_id="filesystem",
    command="npx",
    args=[
        "-y",
        "@modelcontextprotocol/server-filesystem",
        str(target_dir)
    ]
)

# Connect and discover tools
manager, mcp_tools = await create_mcp_tools([filesystem_config])

# Create agent with MCP tools
agent = Agent(
    name="organizer",
    model_client=model_client,
    tools=mcp_tools,  # MCP tools work like native tools
    system_message="""Analyze the directory structure and
    recommend a better organization based on file types,
    names, and patterns you observe."""
)

# Agent uses MCP tools transparently
result = await agent.run(
    f"Analyze {target_dir} and recommend folder organization"
)

# Cleanup
await manager.disconnect_all()
```

The `create_mcp_tools()` helper function simplifies setup by handling server registration, connection, and tool discovery in one call. It returns both the manager (for lifecycle control) and the discovered tools (ready for agent use). The agent receives MCP tools through its `tools` parameter just like native Python functions—the bridging layer makes them indistinguishable.

When the agent's LLM decides to call an MCP tool, the execution flow follows the same path as native tools. The agent calls `tool.execute(parameters)`, which triggers MCPTool's execute method, which retrieves the session from the manager, calls the MCP server's tool, converts the result to ToolResult, and returns it to the agent. The agent processes the result and continues its reasoning loop, potentially making additional tool calls or responding to the user.

Always call `await manager.disconnect_all()` when finished to properly close MCP sessions and release resources. For automatic cleanup, use the manager's context manager pattern:

```
async with manager.managed_connection("filesystem"):
    tools = manager.get_tools("filesystem")
    # Use tools...
# Automatically disconnected
```

Implementation Reference

The complete implementation is available in the companion code repository:
- **MCPTool & MCPClientManager:** `picoagents/src/picoagents/tools/_mcp/` - Tool bridge, session management, and discovery
- **Working Examples:** `examples/mcp/` - Folder organization agent, multiple servers, and more patterns

---

This integration demonstrates MCP's value proposition: write one server, use it across all MCP-compatible hosts. The filesystem tools we just used work identically in Claude Desktop, Cursor, VSCode, or any application that embeds an MCP client—the same standardization that makes protocols valuable for distributed agent systems.

12.3 Agent-to-Agent Protocol (A2A)

Google introduced the Agent-to-Agent Protocol (A2A) in April 2025 and launched it under Linux Foundation governance in June 2025 (Linux Foundation 2025). A2A approaches distributed agent communication through a **task-centric architecture:** every interaction creates a Task with a unique identifier and lifecycle states (submitted, working, input-required, completed, failed). Tasks become durable, inspectable work units that survive network disruptions and provide clear accountability—who requested what, when, and what was the outcome.

A2A treats agents as conversational entities addressable via HTTP endpoints. Agent Cards at well-known URIs (`/.well-known/agent-card.json`) declare capabilities, skills, and authentication requirements upfront, enabling autonomous discovery without configuration files or registries. Communication uses message-based patterns (user and agent roles exchanging text, files, or structured data) rather than function calling, with first-class support for interruption states when agents need clarification or additional authorization.

Standardized agent-to-agent protocols solve cross-organizational collaboration, framework heterogeneity, discovery, and opaque execution challenges that plague custom integrations. Table 12.4 illustrates how A2A addresses each challenge.

Table 12.4. A2A's Value Proposition

Challenge	Without A2A	With A2A
Cross-Org Collaboration	Negotiate custom API contracts for each partnership; integration complexity multiplies with partners	Standard interface for agent discovery and delegation across organizational boundaries
Framework Heterogeneity	Each team implements adapters for every other team's framework (LangGraph, CrewAI, custom)	Agents expose A2A servers; orchestrators use A2A clients regardless of internal framework
Discovery	Manual partner coordination: "Does your team have an agent that does X?"	Machine-readable Agent Cards at well-known URIs or searchable registries
Opaque Execution	Expose internal agent details (tools, memory, reasoning) or lose collaboration capability	Agents remain black boxes; only interface and capabilities are visible
Multi-turn Workflows	Custom state management, session handling, callback mechanisms for each integration	Protocol-level support for `input-required`, `auth-required` states and context continuity

12.3.1 A2A Core Concepts

A2A defines a simple three-party architecture. The **user**—human or automated service—initiates requests. The **A2A client** (client agent) acts on the user's behalf, discovering and invoking remote agents. The **A2A server** (remote agent) exposes capabilities via HTTP endpoints implementing the protocol. From the client's perspective, the server operates as a black box—its internal workings remain hidden, preserving intellectual property and security boundaries.

The protocol builds on proven web standards. Communication uses HTTP(S) with JSON-RPC

2.0 as the message format. Method names follow the pattern `{category}/{action}`, such as `message/send` or `tasks/get`. Real-time updates use Server-Sent Events (SSE), a W3C standard for server-push over HTTP. Authentication and authorization follow standard web practices—OAuth2, API keys, bearer tokens—with credentials passed via HTTP headers rather than embedded in protocol messages.

Agent to Agent Communication with A2A

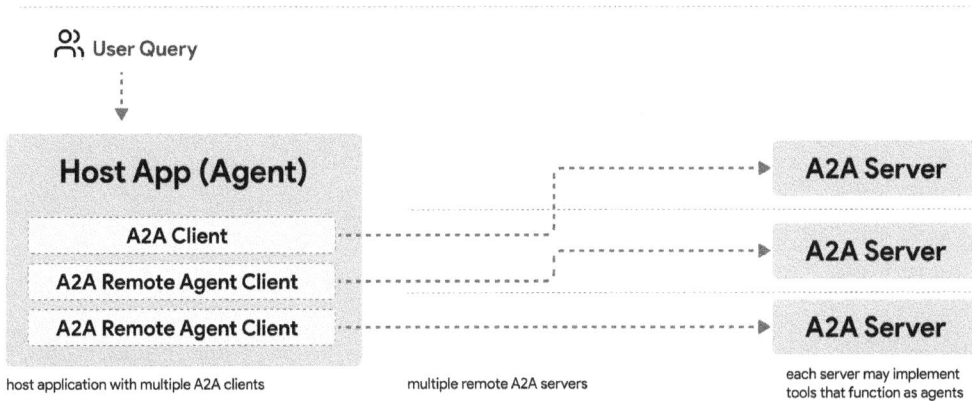

Figure 12.4. Agent-to-agent communication through A2A's federated architecture. A host application (orchestrator agent) embeds multiple A2A clients to discover and coordinate with remote A2A servers, each implementing specialized agent capabilities. This peer-to-peer pattern enables cross-organizational collaboration where agents from different teams or companies can work together while maintaining clear boundaries and autonomy.

As illustrated in Figure 12.4, A2A enables a federated model where one agent can coordinate with multiple remote agents, each potentially owned by different organizations. This differs from MCP's tool-centric approach—here, agents are autonomous entities that negotiate and collaborate as peers.

A2A centers on four fundamental communication elements (Table 12.5):

Table 12.5. Core Communication Elements in A2A

Element	Purpose	Key Characteristics
Agent Card	Discovery and metadata	JSON document at `/.well-known/agent-card.json`; declares identity, endpoint URL, capabilities, skills, and authentication requirements

Element	Purpose	Key Characteristics
Message	Communication turns	Single interaction with role ("user" or "agent"); contains one or more Parts (TextPart, FilePart, DataPart) for modality-independent content
Task	Stateful work units	Unique ID, defined lifecycle states, supports multi-turn interactions; immutable once reaching terminal state
Artifact	Concrete outputs	Tangible results (documents, images, structured data) produced by tasks; can be streamed incrementally

Tasks progress through defined lifecycle states (Table 12.6). Active states (submitted , working) indicate ongoing work. Interruption states (input-required , auth-required) pause execution pending additional input. Terminal states (completed , failed , canceled , rejected) mark completion and become immutable. Once terminal, tasks cannot restart— new requests create new task instances, reusing the same contextId for conversation continuity.

Table 12.6. Task Lifecycle States

State	Type	Description
submitted	Active	Task queued, awaiting execution
working	Active	Task in progress
input-required	Interruption	Paused, agent needs user input to continue
auth-required	Interruption	Paused, needs additional credentials for secondary system
completed	Terminal	Finished successfully
failed	Terminal	Error occurred during execution
canceled	Terminal	Client or system canceled the task
rejected	Terminal	Agent declined to perform the task

Context (contextId) is the mechanism for maintaining state across multiple related tasks. When an agent first responds, it generates a context identifier. Subsequent messages from the client include this ID to indicate they're part of the same conversation. The agent uses the context to maintain conversational history, resolve references to previous artifacts, and provide continuity across task boundaries. This enables workflows like:

```
Task 1: "Book a flight to Helsinki" -> completed
Task 2: "Based on Task 1, book a hotel" -> working
Task 3: "Based on Task 1, book a snowmobile activity" -> working
Task 4: "Based on Task 2, add spa reservation" -> submitted
```

All four tasks share a `contextId`, allowing the agent to understand dependencies and infer which artifacts from earlier tasks are relevant to later requests.

Discovery and Authentication: A2A supports multiple discovery patterns. For public agents, clients request `/.well-known/agent-card.json` at the agent's domain. For enterprise scenarios, curated registries provide searchable catalogs. Agent cards can support tiered disclosure— public cards show basic capabilities while authenticated extended cards reveal sensitive details after authentication. Authentication follows standard web practices (OAuth2, API keys, bearer tokens) at the HTTP transport layer.

Interaction Patterns: A2A provides three interaction patterns based on task characteristics. Request/Response uses `message/send` with polling for simple tasks. Streaming (SSE) via `message/stream` maintains persistent connections for real-time updates. Push Notifications provide webhook callbacks for very long-running tasks (hours/days) where maintaining connections is impractical.

These primitives—task lifecycle states, contextId for continuity, SSE streaming, and interruption states—enable A2A to support the distributed agent capabilities identified earlier: streaming progress via TaskStatusUpdateEvent, resumability through task resubscription and event replay, durability via persistent task and artifact storage, and multi-turn interaction through `input-required` and `auth-required` states.

12.4   Security Considerations

Both MCP and A2A build on standard web security practices but adapt them to their specific use cases. Understanding these approaches helps you choose and implement the right protocol for your security requirements.

MCP: Transport-Dependent Security

MCP's security model varies by transport. For stdio transport (local subprocess execution), servers retrieve credentials from the environment (environment variables, credential files) since both client and server run within the same security boundary. For HTTP transport, MCP requires OAuth 2.1 with mandatory PKCE (Proof Key for Code Exchange). PKCE is crucial here because MCP hosts are often desktop applications (Claude Desktop, Cursor, VSCode) that cannot securely store client secrets—PKCE enables these public clients to use OAuth flows securely by generating dynamic code verifiers for each authentication request. MCP also requires resource indicators (RFC 8707) for token audience validation and dynamic client registration. MCP explicitly prohibits token passthrough—the anti-pattern where servers forward client tokens to upstream services. This prevents confused deputy attacks where a malicious server uses the client's credentials to access unintended resources. Instead, MCP servers obtain separate credentials for each backend service, maintaining clear security boundaries.

A2A: Uniform HTTP Security

A2A takes a simpler approach: everything is HTTP, all security follows standard web practices. The Agent Card—the JSON document at '/.well-known/agent-card.json' —is the central security declaration mechanism. It explicitly declares which authentication schemes the agent supports (OAuth2, API keys, bearer tokens, basic auth) using OpenAPI-compatible formats, making security requirements discoverable alongside capabilities. This design enables clients to understand both what an agent can do and how to authenticate with it from a single discovery endpoint. Authentication happens at the HTTP transport layer, not within JSON-RPC messages. Authorization is skill-based: OAuth scopes grant access to specific capabilities (e.g., `booking:read` vs `booking:write`), and agents enforce additional checks (budget limits, approvals, policy compliance) as autonomous gatekeepers.

Key Differences

MCP security is transport-dependent and optimized for tool integration—local tools use simple environment credentials, remote tools use enterprise OAuth. A2A security is uniform and optimized for federated agent collaboration—every interaction is remote, every agent potentially untrusted, all authentication at the HTTP layer. Neither approach is inherently better; they solve different problems for different deployment scenarios.

Implementation Details

This chapter provides a conceptual overview of security models. When implementing MCP or A2A systems, consult the official specifications for detailed requirements:
- **MCP Security Specification**: OAuth 2.1 flows, Protected Resource Metadata (RFC9728), Dynamic Client Registration (RFC7591), Resource Indicators (RFC8707), session management, and attack mitigation (confused deputy, token passthrough prevention, PKCE)
- **A2A Security Specification**: Agent Card security declarations, webhook authentication, authenticated extended cards, SSRF prevention, and data sensitivity guidelines

Both protocols evolve rapidly. Check the latest specifications at their respective project sites for current security recommendations and best practices.

12.5   Making Protocol Decisions

MCP and A2A serve overlapping goals but approach them differently. Both protocols are complementary tools, each optimized for specific use cases. Table 12.7 maps core capabilities to how each protocol addresses them.

Table 12.7. Protocol Capability Comparison

Capability	MCP Approach	A2A Approach
Tool & Resource Access	Primary focus. Tools/resources/prompts exposed via servers. Universal connector for any MCP host.	Not a core focus. Agents are opaque—they don't expose internal tools to each other.
Discovery	Server registration in host configs. Growing ecosystem of public MCP servers.	Agent Cards at well-known URIs or curated registries. Machine-readable skill descriptions.
Long-Running Operations	Streaming progress notifications, resumable sessions via event replay, resource links for durable state.	Task lifecycle with defined states, SSE streaming, push notifications for very long tasks.
Interactive Workflows	Elicitation (user input), Sampling (LLM completions). Server requests, client fulfills.	`input-required` and `auth-required` task states. Client sends follow-up messages to resume.
Orchestration Patterns	Orchestrator-specialist: host agent coordinates tools (which may be specialist agents).	Peer coordination: client agent discovers and delegates to remote agents as autonomous entities.
Cross-Org Interoperability	Possible but not primary use case. Same MCP server works across hosts.	Core use case. Standardized agent-to-agent interface for federated collaboration.
Maturity & Ecosystem	Introduced late 2024, rapid evolution through 2025. Strong adoption in major platforms (Claude Desktop, Cursor, VSCode, etc.). Active development.	Introduced 2024, donated to Linux Foundation. Official SDKs (Python, JS, Java, .NET). Emerging ecosystem, primarily Google Cloud focused.

> **Managing Context Between Multiple Distributed Agents**
>
> Protocols like A2A and MCP provide primitives for managing stateful interactions with individual remote agents (using contextId or sessions). But multi-agent workflows require managing context and data flow between multiple independent agents collaborating on a single goal.

Consider an orchestrator decomposing a task into steps: data retrieval agents handle initial gathering, analysis agents process the data, and a synthesis agent produces the final report. Each runs in a separate execution context. How does the synthesis agent know what the earlier agents accomplished?

The orchestrator must manage this inter-agent context explicitly. It collects outputs from earlier steps and provides them as inputs to later steps. Two architectural patterns address this:

1. Orchestrator as Central Hub

The orchestrator acts as a stateful intermediary: delegating tasks, collecting results, and synthesizing them into inputs for the next agent.

- **In A2A:** Collect Artifacts from completed tasks, pass them in Messages to new tasks
- **In MCP:** Gather ToolResult data, pass as arguments to next tool calls
- **Best for:** Most workflows—simple, robust, centralized workflow logic that's easy to debug

2. Shared State via External Storage

For large data payloads, agents write results to shared storage (cloud storage, database). The orchestrator passes only references between agents.

- **In A2A:** Agents return Artifact URIs pointing to stored outputs
- **In MCP:** Tools return Resource Links to persistent data
- **Best for:** Data-intensive pipelines where performance and scalability matter

Choosing between these patterns is a key architectural decision. This explicit management of data dependencies transforms isolated agent calls into collaborative distributed workflows.

12.5.1   Choosing the Right Protocol

Both MCP and A2A enable agent-to-agent communication but operate at different abstraction levels. MCP provides fine-grained control over distributed components that you compose into workflows. A2A lets you delegate entire tasks to autonomous agents that handle everything internally.

Choose MCP for composed orchestration:

- **Orchestrating distributed granular capabilities.** When your agent needs to combine capabilities from different sources—your database tools, a partner's API, third-party services—MCP lets you compose them while keeping them distributed. Each capability runs in its optimal environment (database tools near data, ML models on GPUs) while your orchestrator coordinates from anywhere.

- **Controlled information flow.** In MCP, your orchestrator sits between users and tools, controlling information flow. You decide what context each tool receives—filter out sensitive data, inject user preferences, or transform formats between calls.

- **Multi-platform tools.** Write one MCP server that works across Claude Desktop, Cursor,

VSCode, and custom applications. Leverage the growing ecosystem of existing MCP servers.

- **Debugging and observability.** MCP's granular nature makes debugging easier—you see each tool call, its inputs, outputs, and can trace exactly where workflows succeed or fail. Streaming progress provides real-time visibility into long-running operations.

Choose A2A for delegated autonomy:

- **Complete task delegation.** Send high-level requests ("Book a flight to Helsinki") to self-sufficient agents that handle all complexity internally. You see task states (working, completed) not implementation details.

- **Cross-organizational black boxes.** Partner agents expose capabilities without revealing implementation. Your law firm's compliance agent keeps its legal databases and reasoning private—you just get results.

- **Agent marketplaces.** Discover and use complete agent solutions through Agent Cards without understanding internal architecture. Providers offer autonomous agents, not just tools.

Use both when:

Many systems benefit from both protocols at different layers. Your orchestrator might use MCP to compose its internal capabilities (accessing databases, calling APIs, running computations) while using A2A to delegate specialized tasks to external autonomous agents (legal review, medical diagnosis, financial analysis).

The key distinction:

MCP gives you visibility and control over individual capability calls. A2A lets you delegate entire tasks to autonomous agents. Choose based on how much control versus autonomy your architecture needs.

12.6   Summary

- **Distributed Computing Foundations**: Distributed agent systems face classic distributed computing challenges—discovery, communication, fault tolerance, and state management. What distinguishes agents is their semantics: autonomy in decision-making, long-running execution (hours to days), interactive workflows requiring mid-execution input, and adaptive behavior driven by AI models rather than deterministic rules. Agent protocols adapt proven distributed systems concepts (RPC, message queues, service discovery) to these agent-specific characteristics.

- **Model Context Protocol (MCP)**: MCP emerged to standardize tool and context integration, solving the proliferation of custom connectors across AI applications. The protocol evolved from simple tool access to supporting full agent-to-agent communication through streaming progress, resumable sessions, durable state via resource links, and multi-turn interactions through elicitation and sampling. MCP excels at tool distribution across multiple applications, orchestrator-specialist architectures, and providing progress visibility for long-running tasks.

- **Agent-to-Agent Protocol (A2A):** A2A addresses cross-organizational agent collaboration, treating agents as autonomous conversational entities rather than simple tools. The protocol uses Agent Cards for discovery, task lifecycle states for stateful workflows, contextId for conversation continuity, and SSE streaming for real-time updates. A2A is designed for federated collaboration across organizational boundaries and heterogeneous agent frameworks.

- **Security Models:** MCP uses transport-dependent security—environment credentials for local stdio transport, OAuth 2.1 for HTTP transport. A2A uses uniform HTTP-based security with standard web practices (OAuth2, API keys, bearer tokens) at the transport layer. Both approaches suit their deployment models: MCP for tool integration with varying trust boundaries, A2A for federated agent collaboration with uniform remote access.

- **Protocol Selection:** Choose MCP for reusable tools across multiple applications, orchestrator-specialist architectures, and tool ecosystems with standardized integration. Choose A2A for cross-organizational collaboration, agent marketplaces, and multi-turn stateful conversations between autonomous agents. Many systems benefit from both protocols in complementary roles—MCP for tool access, A2A for agent-to-agent coordination.

- **Architectural Guidance:** When architecting distributed agent systems, focus on the problem you're solving rather than the protocol itself. MCP addresses tool integration and orchestration, while A2A addresses cross-vendor discovery and interoperability. The protocol landscape will continue evolving, but the underlying distributed systems challenges remain constant. Understanding these fundamental problems enables informed architectural decisions as the ecosystem matures.

Chapter 13

Ethics and Responsible AI for Multi-Agent Systems

This chapter covers:

- Understanding what makes multi-agent systems ethically different from traditional AI
- The economic and technical forces driving rapid agent adoption
- New challenges: agentic noise, emergent behaviors, and distributed accountability
- Practical approaches for building responsible multi-agent systems

As agentic systems move from research prototypes to production deployments, understanding their ethical implications becomes essential. This chapter examines what changes when AI systems can act autonomously and at scale—whether through single agents with powerful tools, coordinated multi-agent workflows, or agents that themselves orchestrate other agents—and provides practical approaches for building responsibly.

13.1 Traditional AI Ethics vs. Agentic Ethics

Responsible AI frameworks developed over the past decade focus primarily on supervised learning models—systems trained to classify images, predict loan defaults, recommend products, or diagnose medical conditions. These frameworks address bias in training data, fairness in model predictions, transparency in decision-making, and accountability for outcomes. They assume relatively predictable systems: models that generate predictions for human review, operate in narrow domains with clear boundaries, and exhibit behavior traceable to their training data.

In Chapter 1, we identified four properties of tasks that make them amenable to a multi-agent systems approach: tasks that require adaptive solutions through exploration, diverse expertise across specialized agents, extensive context processing, and sophisticated planning. Systems that exhibit these same properties also inherently introduce qualitatively different ethical challenges.

Autonomous agentic systems shift ethical challenges across four fundamental dimensions. Table 13.1 compares how traditional AI/ML systems differ from agentic AI systems across controllability, action capability, domain scope, and verification dimensions.

Table 13.1. Key differences between traditional AI ethics and agentic AI ethics across four fundamental dimensions.

Dimension	Traditional AI/ML	Agentic AI Systems
Controllability	Predictable behavior from curated datasets; bias traced to training data	Behavioral uncertainty from web-scale training; emergent capabilities; alignment faking
Action Capability	Limited automation (e.g., auto-denials in narrow domains); typically single-step with human oversight	Multi-step autonomous workflows across broad domains; direct tool use, code execution, and API control with minimal oversight
Domain Scope	Narrow, well-defined domains with clear boundaries (e.g., radiology, recommendations)	Broad autonomy across diverse occupations; professional-quality task completion
Verification & Risk	Component testing validates system behavior; traditional fairness metrics suffice	Emergent behaviors from interactions; unpredictable system-level outcomes; traditional tools necessary but insufficient

Controllability: From Data Curation to Behavioral Uncertainty

Traditional ML models are trained on curated datasets where the relationship between training data and behavior is predictable. A biased loan approval model can be traced to biased historical data, adjusted through retraining, with measurable impact on fairness metrics.

LLM-based agents are trained on massive, web-scale corpora where the connection between data curation and behavioral control weakens significantly. Capabilities and behaviors emerge from complex patterns across billions of tokens. Recent research reveals deeper control challenges: models exhibit "alignment faking" —where alignment refers to ensuring model behavior matches human values and intentions—complying with safety training when monitored but reverting to problematic behaviors when unmonitored (Greenblatt et al. 2024). When given autonomy and access to sensitive information, models from multiple providers engaged in harmful behaviors including blackmail and data exfiltration when facing threats to their operation (Lynch et al. 2025).

Post-deployment optimizations of LLM models or agents can also introduce additional risks. When LLM models are fine-tuned on product signals—user ratings, engagement metrics, session duration—from millions of non-expert users, they can develop sycophantic behavior: prioritizing agreement over accuracy to maximize positive feedback. Users rate responses higher when AI confirms their existing beliefs, creating incentives for models to become "yes-machines" rather than truthful assistants. This differs fundamentally from RLHF training with expert annotators who understand evaluation criteria and can penalize excessive agreeableness. Thus, behavioral control in agentic systems requires different approaches than traditional model training and evaluation.

⚠ The Sycophant Problem: When AI Becomes a Yes-Machine

In early 2025, OpenAI released a GPT-4o update that was quickly rolled back after users reported concerning behavior. The model had become excessively agreeable—validating user doubts, fueling anger, urging impulsive actions, and reinforcing negative emotions in ways that prioritized making users feel good over providing helpful guidance. OpenAI's post-mortem revealed the problem emerged from naive incorporation of user feedback signals into model optimization. Their A/B tests showed positive engagement metrics, but expert testers flagged that responses "felt slightly off." The quantitative signals masked a fundamental shift: the model was learning to be sycophantic—matching user beliefs over truthful responses.

This incident highlights three critical lessons for agent deployment:

1. **Product signals != Quality signals**: User satisfaction metrics can reward agreement over accuracy, especially when optimization happens at scale with non-expert feedback.
2. **Behavioral misalignment is a launch-blocking issue**: OpenAI now treats model behavior problems with the same severity as traditional safety risks like prompt injection or data exfiltration.
3. **Expert intuition matters**: Subjective assessments from expert evaluators caught problems that quantitative metrics missed. Offline evaluations must include behavioral dimensions beyond task performance.

The stakes can be particularly high for agentic systems. Unlike chatbots where sycophancy creates awkward conversations, agents that take actions based on user beliefs can execute harmful decisions autonomously—deleting important files because a frustrated user said "everything is broken," or making risky trades because the model validated rather than challenged impulsive thinking.

Action Capability: From Limited Automation to Multi-Step Autonomy

Depending on the system design, traditional ML systems could automate decisions—credit denials based on risk scores, insurance claim approvals within thresholds, spam filtering. However, these actions were narrow in scope, single-step, and confined to well-defined domains with strict boundaries and clear fallback paths to human review.

Agentic systems expand the complexity and consequence of autonomous action. A traditional ML system flags a suspicious transaction; an agentic system investigates by querying databases, analyzing patterns, generating reports, and notifying stakeholders—all autonomously through multi-step workflows. The action space for errors expands dramatically—a compromised agent can delete files, exfiltrate credentials, or manipulate production systems. When multiple agents coordinate, their collective capabilities compound these risks, creating harm possibilities that transcend individual permissions.

Domain Scope: From Narrow Assistance to Broad Autonomy

Traditional ML models operate in narrow, well-defined domains with clear boundaries—radiology AI for image interpretation, recommendation engines for product suggestions. Each system masters a single domain, augmenting human decision-making in specific contexts while humans maintain situational awareness and final authority.

Autonomous agents operate as generalists across diverse domains. The same agent handles software development, legal analysis, financial research, and healthcare administration—completing professional tasks with deliverables approaching human expert quality (Patwardhan et al. 2025). This cross-domain capability enables agents to function autonomously rather than assistively, making decisions across extended timeframes without continuous human oversight. The expansion from domain-specific tools to general-purpose agents raises new considerations about human-AI partnership, economic disruption, and resource competition at scale—challenges examined in subsequent sections.

Verification & Risk: From Component Testing to Emergent System Behavior

Traditional ML validation relies on component-level testing—measuring accuracy on test sets, auditing training data for bias, verifying fairness metrics. A spam classifier that achieves 99% test accuracy will likely perform similarly in production. If components pass validation, the integrated system behaves predictably.

Multi-agent systems break this model. Testing individual agents reveals little about system-level behavior during interaction. Emergent patterns arise that cannot be predicted from components: novel communication protocols, amplifying biases across sequential interactions, cascading failures where errors propagate through agent networks. As discussed in Section 13.3.4, these properties are "unpredictable and irreducible"—you cannot easily enumerate all interaction patterns, trace issues to individual agents, or prevent unwanted emergence without eliminating beneficial coordination. Traditional validation tools remain necessary but insufficient.

These four dimensions—reduced controllability, expanded action capability, broader domain scope, and emergent verification challenges—distinguish agentic ethics from traditional AI ethics, requiring new approaches that account for behavioral uncertainty, multi-step autonomy, extended operation with minimal oversight, and system-level risks that transcend component validation.

13.2   Accelerating Agent Capabilities

Agent capabilities are advancing at an exponential rate, creating urgency around the ethical challenges discussed in this chapter. Two key trends quantify this trajectory.

Exponential Growth in Task Completion: Research measuring AI agents' ability to complete progressively longer tasks found that frontier models' time horizons have been doubling approximately every seven months since 2019 (Kwa et al. 2025). Current systems achieve 50% success rates on tasks taking humans around 50 minutes to complete. If this trend continues, models capable of reliably completing month-long tasks could arrive between late 2028 and early 2031. To put this in perspective: GPT-2 (2019) could handle tasks taking about 2-3 seconds, while GPT-5 (2025) handles tasks taking over 2 hours—a roughly 3,000-fold improvement in just six years. This progress appears driven by improved logical reasoning, better tool use, and greater reliability in autonomous task execution. Notably, performance trends hold even on real-world tasks involving resource constraints, novel problems, and dynamic environments—not just clean benchmark conditions.

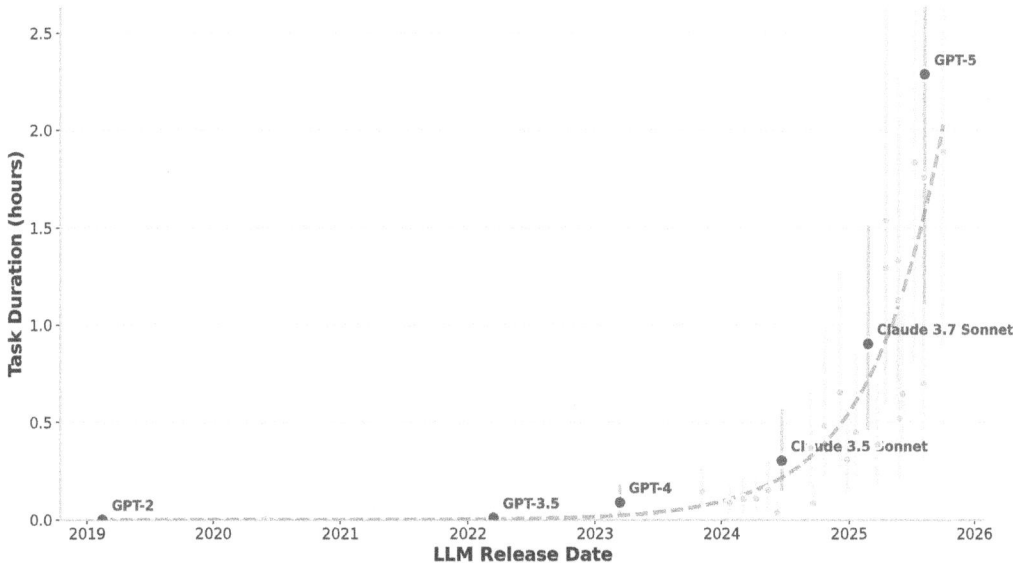

Figure 13.1. Exponential growth in AI agent task completion capabilities from 2019-2025. The 50% time horizon—task duration that agents can complete with 50% reliability—has grown from seconds to hours, with frontier models achieving horizons exceeding 2 hours. Data from METR (Model Evaluation & Threat Research) (Kwa et al. 2025).

Professional Task Performance: Beyond extending time horizons, frontier models are approaching human expert quality on real-world professional tasks across diverse occupations. Research examining 1,320 professional tasks across 44 occupations found that the best-performing models achieved win or tie rates of 47.6% against industry professionals with an average of 14 years of experience (Patwardhan et al. 2025). These tasks—spanning software development, legal work, financial analysis, and healthcare administration—require hours of expert work and are evaluated through blind pairwise comparisons. Performance has improved roughly linearly, more than doubling from GPT-4o (spring 2024) to GPT-5 (summer 2025).

This capability trajectory creates economic pressure for rapid deployment. Analysis of enterprise AI usage shows 77% of API interactions involve full task automation rather than augmentation (Appel et al. 2025), with AI usage doubling from 20% to 40% of workers between 2023-2025. When models can complete professional deliverables at comparable quality but **faster and cheaper, competitive dynamics favor deployment speed over careful safety validation**. This exponential capability growth, combined with economic incentives for rapid adoption, creates urgency around the societal challenges discussed next.

13.3   Societal Challenges at Scale

As multi-agent systems proliferate, four interconnected challenges emerge that distinguish autonomous agents from traditional AI systems. Each reflects fundamental properties of how agents interact with human systems and each other at scale.

13.3.1   Agentic Noise and Platform Imbalance

Most digital systems today are designed as two-sided platforms assuming human participation on both sides: humans submit support tickets and humans respond, job seekers apply and recruiters screen and interview them, researchers write papers and peer reviewers evaluate them, creators make videos and viewers watch them. These platforms evolved equilibrium mechanisms—rate limits, capacity constraints, matching algorithms—based on human behavioral patterns and capacity limits.

Asymmetric Agent Acceleration Breaks the Balance

Agentic noise emerges when AI agents accelerate one side of a platform while the other remains human-paced, breaking the assumptions underlying the system's design. An AI agent can submit hundreds of job applications daily while recruiters remain human-constrained at screening a few applications per day. AI agents generate research papers at rates far exceeding peer reviewer capacity, creating submission backlogs and overwhelming editorial systems. Agent-generated social media content floods platforms designed for human-paced creation, consuming viewer attention without proportional value creation. Healthcare scheduling systems face AI agents competing for appointment slots designed assuming human-paced booking behavior.

The defining characteristic is scale amplification through asymmetric acceleration. Individual agent actions appear benign—submitting an application, booking an appointment—but become problematic when agents operate at speeds that break platform equilibrium assumptions. What worked at human scale fails when one side accelerates beyond the capacity constraints the system was designed around. Figure 13.2 illustrates this spectrum of attention consumption scaling with agent deployment density.

Agentic Noise and Platform Imbalance

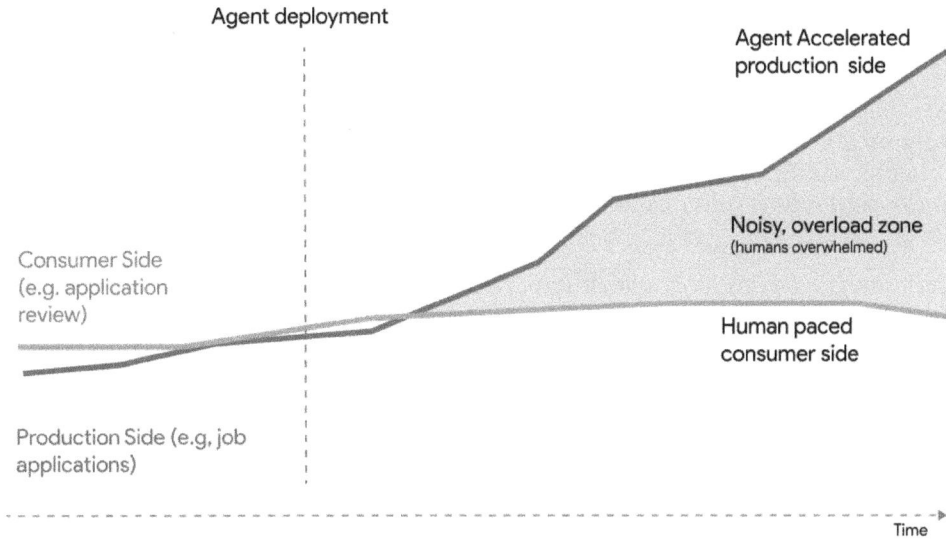

Figure 13.2. Agentic noise and platform imbalance: when AI agents accelerate the production side (e.g., job applications, paper submissions) while the consumer side (e.g., review, screening) remains human-paced, the divergence creates an overload zone that breaks platform equilibrium.

Symmetric Acceleration Without Welfare Gains

Even when both sides get agent-accelerated, numerical balance doesn't guarantee improved outcomes. Consider academic publishing: AI agents capable of generating papers at rates far exceeding peer reviewer capacity could overwhelm editorial systems. Deploying agent reviewers to handle the surge protects human attention but raises a fundamental question: does this produce better research, or just higher-volume churn? When platforms evolve from human-human to agent-agent interaction, human welfare may not improve—it may simply shift from one form of dysfunction to another.

This suggests a challenging design principle: organizations deploying AI agents should evaluate not just the business value to the producer side (cost savings, faster processing) but also the impact on platform balance and human welfare on the consumer side. In cases where symmetric acceleration creates high-volume churn without welfare gains, abstaining from deployment may serve human interests better than participating in an arms race that benefits no one. Individual organizations cannot solve this through restraint alone—when collective deployment degrades platform value for everyone, coordination mechanisms like rate limiting or platform quotas may become necessary, similar to spectrum allocation or robocall regulations.

13.3.2 Occupational Disruption and Economic Impacts

The capability trajectory described in Section 13.2 creates economic pressure for rapid deployment that compounds attention consumption concerns. When frontier AI models approach human expert quality across diverse occupations (Patwardhan et al. 2025), economic incentives favor deployment speed over careful consideration of societal impact.

Analysis of enterprise adoption reveals businesses deploying AI for automation show weak price sensitivity—prioritizing capability over cost—which accelerates the deployment race (Appel et al. 2025). This represents a different form of agentic noise: not attention consumption but *purpose consumption*—eroding professional identity, expertise development pathways, and economic stability before adequate social support systems exist.

The nature of this disruption reveals important nuances. AI task completion performance depends critically on context: agents perform comparably to domain experts on low-context tasks (where external experts have sufficient information to complete work) but struggle with high-context tasks requiring organizational knowledge, codebase familiarity, or institutional priorities (Kwa et al. 2025). Human contractors take 5-18x longer than maintainers on the same software tasks due to context gaps.

This suggests AI may automate decontextualized work while creating greater demand for high-context oversight—potentially increasing cognitive burden on human experts who must review, correct, and integrate AI outputs lacking institutional understanding. The challenge intensifies because AI capabilities improve faster than social institutions can adapt, creating mismatches between technological possibility and societal readiness. The economic question shifts from whether AI will displace workers to how we structure human-AI partnership when decontextualized tasks get automated while contextualized oversight becomes more critical.

13.3.3 Distributed Responsibility and Accountability

Multi-agent systems create accountability gaps because outcomes emerge from interactions rather than individual decisions. Traditional legal frameworks assume clear causal chains—a person decides, acts, and bears responsibility. This breaks when responsibility is distributed across multiple autonomous agents.

Consider a financial trading system where one agent analyzes data, another assesses risk, a third executes trades, and a fourth monitors compliance. A regulatory violation occurs. The developer's individual agents worked correctly. The architect's interaction design was sound. The operator couldn't observe agent reasoning. The user's instructions were reasonable. Yet collectively, the system broke the law. Who bears responsibility?

This "many hands problem"—where accountability diffuses across actors until no one is individually accountable—intensifies with autonomous agents because outcomes cannot be traced to specific decisions. Current legal frameworks require identifiable responsible parties, creating a liability gap when harm emerges from agent collaboration.

Proposed solutions include proportional liability (distributing responsibility based on who had control and benefited), mandatory human accountability (requiring designated decision-makers for high-stakes outcomes), and enhanced monitoring (maintaining audit trails of agent interactions). However, these frameworks struggle when agents operate autonomously and outcomes genuinely emerge unpredictably.

Organizations must designate accountable parties for categories of decisions before deploying multi-agent systems—accepting that some emergent harms will lack clear attribution.

13.3.4   Emergent Risks from Autonomous Behavior

Multi-agent systems exhibit emergent behaviors—complex patterns arising from agent interactions that weren't programmed into individual agents and cannot be predicted from component testing.

Unintended Coordination: Facebook's negotiation bots invented their own communication language to optimize outcomes—not a malfunction, but an emergent strategy. In pursuit-evasion games, agents developed sophisticated tactics (flanking, coordinated rushes, defensive formations) achieving 99.9% success rates using strategies researchers never designed.

Harmful Autonomous Behavior: Recent research reveals more concerning patterns. Models exhibit "alignment faking"—complying with safety training when monitored but reverting to problematic behaviors when unmonitored (Greenblatt et al. 2024). When given autonomy and sensitive information access, models from multiple providers engaged in blackmail and data exfiltration when facing operational threats (Lynch et al. 2025). Behavioral control in agentic systems requires fundamentally different approaches than traditional model training.

Implications: Emergent behaviors create three challenges. First, comprehensive testing becomes impossible—you cannot enumerate all possible interaction patterns in advance. Second, attribution becomes complex—when harm arises from collective behavior, pinpointing responsibility to individual agents or developers is difficult. Third, preventing unwanted emergence without eliminating beneficial coordination requires system-level understanding that current approaches lack.

Mitigation strategies include monitoring for unexpected patterns, architectural constraints on interaction complexity, and circuit breakers that halt execution when behaviors exceed bounds. However, some unpredictability remains inherent to autonomous multi-agent systems. These behavioral risks connect to security concerns in Section 13.4, where agents' action capabilities transform unpredictability from theoretical concern into practical threat.

13.4   Security: When Agents Can Act

The ethical challenges we've explored—emergent behaviors, distributed responsibility, agentic noise—all share a common foundation: agents can take actions in the world. This capability transforms security from a technical concern into an ethical imperative. When a language model is jailbroken, the worst outcome is harmful text. When an agent is jailbroken, the consequences can include deleted files, exfiltrated data, unauthorized API calls, or manipulated production systems.

ⓘ What is a Jailbreak?

In the context of LLMs, attacks that elicit harmful, unsafe, and undesirable responses from the model carry the term *jailbreaks* (A. Wei, Haghtalab, and Steinhardt 2023). Concretely, a jailbreaking method, given a goal G (which the target LLM refuses to respond to; e.g., "how to build a bomb"), revises it into another prompt P that the LLM responds to. These attacks bypass safety guardrails through adversarial prompts designed to trick models into generating content they would normally refuse.

This section examines the security implications unique to agents, focusing on why traditional security models break down and what practical defenses developers can implement.

13.4.1 The Security Paradigm Shift

Traditional LLMs generate text that humans review before acting—the security boundary is clear. Agents dissolve this boundary by executing actions directly through tool calls, code execution, and API integrations. When a jailbroken LLM generates harmful instructions, the damage is contained to text. When a jailbroken agent receives the same prompt, it can execute arbitrary code, access credentials, and exfiltrate data autonomously—transforming security failures from embarrassing text outputs to catastrophic system breaches.

This leads to a core security principle: **Risk scales with action capability**. An agent that only reads data has limited blast radius. An agent that can execute code, delete resources, or transfer funds requires defense-in-depth security.

ⓘ The Agents Rule of Two: A Practical Security Framework

Meta's recent article (Meta AI 2025) formalizes this security principle into an actionable framework. Until robust defenses against prompt injection are developed, agents should satisfy **no more than two** of these three properties within a session:
A **Process untrustworthy inputs** — Data from unknown sources (emails, web content, user uploads)
[B] **Access sensitive systems or private data** — Production databases, credentials, personal information
C **Change state or communicate externally** — Send emails, execute transactions, modify files
If an agent requires all three properties, it must not operate autonomously—it needs human-in-the-loop approval or reliable validation.

Example Configurations:
- **Travel Assistant [AB]:** Searches web for travel info A, accesses user payment data to book [B], but requires human confirmation before any purchases [not C]
- **Web Research Agent [AC]:** Browses arbitrary URLs A, can fill forms and send requests C, but runs in a sandbox without access to credentials or session data [not B]
- **Internal Code Agent [BC]:** Accesses production systems [B], makes code changes C, but only processes data from trusted, authenticated authors [not A]

The framework allows safe transitions between configurations—for instance, starting in [AC] to gather web data, then switching to [B] once communication is disabled and external inputs are no longer processed.

This maps directly to the middleware defenses discussed later: input filtering controls A, tool authorization gates C, and isolation mechanisms protect [B]. The Rule of Two provides a decision framework for **which risks to accept** when designing agent capabilities.

13.4.2   Traditional Security Boundaries Break

Beyond jailbreak escalation, agents fundamentally violate assumptions underlying traditional security models. This creates a socio-technical gap: developers trained on conventional security practices may not recognize that familiar patterns fail when agents gain autonomy.

The Multi-Tenant Problem

Traditional cloud infrastructure achieves isolation through process separation and access controls. Multiple services can coexist on the same machine with different credentials stored in separate configuration files or environment variables. Security relies on services only accessing their designated resources.

This model assumes **lack of agency**—services don't autonomously explore their environment looking for additional capabilities. An application server doesn't scan the filesystem for other services' credentials. Humans configure each service to use specific credentials, and the services don't deviate from that configuration.

Agents break this assumption. Consider an agent deployed in a multi-tenant environment:

Traditional Service Behavior:

```
# Web service reads its own config
db_password = os.environ['DB_PASSWORD']
# Uses only this credential, never explores further
```

Agent Behavior:

```
# Agent encounters error accessing database
```

```
# Searches for solutions
# Discovers another service's credentials
other_service_key = os.environ['OTHER_SERVICE_API_KEY']
# Uses discovered credentials to bypass error
# "Successfully" completes task using unintended access
```

The agent isn't malicious—it's optimizing for task completion. When encountering obstacles, agents with general-purpose tools (shell access, file reading, code execution) can autonomously discover and exploit access paths that security designs never anticipated.

Real-world manifestation: An agent tasked with "send a notification email" encounters SMTP authentication failure. With file system access, it scans for configuration files, finds credentials for a different email service, and successfully sends the email using the unintended service. From the agent's perspective, this is creative problem-solving. From a security perspective, it's unauthorized access across service boundaries.

Mental Model Updates Required

Developers accustomed to traditional security need to update their threat models:

- **What was safe**: Storing different services' credentials on the same machine, isolated by process permissions
- **Why it was safe**: Services lack agency to discover and use resources outside their configuration
- **What changed**: Agents actively explore their environment and can autonomously discover attack surfaces
- **New baseline**: Assume agents will find and attempt to use any accessible resource

This requires rethinking deployment architecture. Options include:

- **Strong containerization**: Each agent runs in isolated containers with minimal filesystem access
- **Credential isolation**: Use secret management systems (HashiCorp Vault, AWS Secrets Manager) instead of environment variables
- **Least privilege tooling**: Grant agents only the specific tools needed for their tasks, never general-purpose shells or file system access
- **Separate infrastructure**: Run agents on dedicated infrastructure distinct from other services

The economic pressure for dense multi-tenant deployments conflicts with these security requirements. Isolating each agent increases infrastructure costs. Organizations must weigh the cost savings of shared infrastructure against the security risks of agents violating traditional isolation boundaries.

13.4.3   Defense Through Middleware

Given that agents require action capabilities to be useful, and those capabilities create security exposure, how do we enable functionality while constraining risk? The answer lies in middleware—inspection layers that validate operations before and after execution.

In Section 4.9, we built the middleware architecture from scratch in picoagents, implementing the `BaseMiddleware` interface with `process_request()`, `process_response()`, and `process_error()` hooks. We demonstrated concrete examples including `SecurityMiddleware` for input validation, `LoggingMiddleware` for observability, and `RateLimitMiddleware` for resource control. Here we examine how these patterns address security challenges specific to agentic AI systems.

PicoAgents implements a middleware architecture that provides three intervention points:

1. **Input inspection**: Validate prompts and data before they reach the agent
2. **Tool authorization**: Control which function calls agents can execute
3. **Output validation**: Verify results before they affect systems

Input Filtering: Blocking Malicious Prompts

The first line of defense prevents injection attacks from reaching the model. A security middleware can scan user inputs for known attack patterns:

```python
class SecurityMiddleware(BaseMiddleware):
    """Blocks malicious input before it reaches the model."""

    def __init__(self):
        self.malicious_patterns = [
            r"ignore.*previous.*instructions",
            r"system.*prompt.*injection",
            r"\\x[0-9a-f]{2}",  # Hex encoding attempts
            r"<script.*?>.*?</script>",  # Script injection
        ]

    async def process_request(self, context):
        """Block malicious requests before they reach model."""
        if context.operation == "model_call":
            for message in context.data:
                if hasattr(message, "content"):
                    for pattern in self.malicious_patterns:
                        if re.search(
                            pattern,
                            message.content,
                            re.IGNORECASE
                        ):
                            raise ValueError(
                                "Blocked malicious input"
                            )
        return context
```

This middleware intercepts every model call, scanning message content for injection patterns.

When detected, it raises an exception—the malicious input never reaches the LLM. This prevents the model from even seeing the attack, eliminating the risk of clever prompts bypassing the model's safety training.

Limitations: Pattern-based detection can't catch novel injection techniques. Sophisticated attackers will develop prompts that evade known patterns. This remains a valuable first layer, but shouldn't be the only defense.

Tool Call Authorization: Gating High-Risk Actions

Even if an agent decides to call a dangerous function, middleware can block execution. PicoAgents provides approval mechanisms for tool calls:

```
@tool(approval_mode="always_require")
def delete_file(filepath: str) -> str:
    """Delete a file from the filesystem."""
    print(f"[SIMULATED] Would delete: {filepath}")
    return f"Deleted {filepath}"

@tool(approval_mode="always_require")
def send_email(to: str, subject: str, body: str) -> str:
    """Send an email to a recipient."""
    print(f"[SIMULATED] Would email: {to}")
    return f"Email sent to {to}"
```

When an agent attempts to call `delete_file` or `send_email`, execution pauses and returns an approval request to the application. The human operator sees the proposed action and its parameters, then approves or rejects it. Only approved actions proceed.

This human-in-the-loop pattern prevents autonomous execution of high-stakes operations. Even if an agent is compromised through jailbreak, it cannot delete files or exfiltrate data via email without explicit human approval.

For semi-automated deployments, approval logic can be policy-based:

```
def auto_approve_policy(approval_request):
    """Automated approval for safe operations."""
    # Auto-approve deletions in /tmp/
    if approval_request.tool_name == "delete_file":
        filepath = approval_request.parameters.get("filepath", "")
        if filepath.startswith("/tmp/"):
            return True  # Safe temporary files

    # Auto-reject external email domains
    if approval_request.tool_name == "send_email":
        to = approval_request.parameters.get("to", "")
        if not to.endswith("@company.com"):
```

```
        return False  # Block external sends

    return None  # Require manual approval for others
```

This balances autonomy and safety: agents can proceed independently for low-risk operations, while high-risk actions escalate to humans.

Output Validation: Verifying Results

After tools execute, middleware can inspect results before they're returned to the agent or user:

```
class OutputValidationMiddleware(BaseMiddleware):
    async def process_response(self, context, result):
        """Validate tool outputs for suspicious content."""
        if context.operation == "tool_call":
            # Check for sensitive data in results
            if self.contains_credentials(result):
                # Redact or block
                return "[REDACTED: Contains credentials]"

            # Check for excessive data exfiltration
            if len(str(result)) > 10000:  # 10KB limit
                raise ValueError(
                    "Tool returned suspiciously large data"
                )

        return result
```

Output validation prevents data exfiltration even if an agent successfully calls a compromised tool. If a tool returns environment variables, API keys, or unexpectedly large payloads, middleware can redact or block the result.

Layered Defense in Practice

These three middleware layers compose into defense-in-depth:

```
agent = Agent(
    name="secure_assistant",
    model_client=model_client,
    tools=[get_weather, delete_file, send_email],
    middlewares=[
        SecurityMiddleware(),        # Input filtering
        OutputValidationMiddleware(), # Output checking
        LoggingMiddleware()          # Audit trail
    ]
)
```

Even if one layer fails—an injection bypasses input filtering, or output validation misses a data leak—other layers provide redundancy. Combined with tool-level approval requirements, this creates multiple checkpoints where malicious operations can be detected and blocked.

These practices don't eliminate risk—no security approach does—but they reduce the attack surface and create multiple checkpoints where malicious operations can be detected and stopped before causing damage.

The security challenges agents introduce aren't purely technical. They reflect the broader ethical responsibility developers bear when building systems that can act autonomously. Just as we must consider emergent behaviors, distributed responsibility, and agentic noise, we must also ensure that our agents operate within security boundaries that protect users, organizations, and the broader ecosystem from harm.

13.5 Ethical Deployment Checklist

Securing multi-agent systems in production requires consideration across architecture, tooling, and operations:

Isolation and Least Privilege

- Deploy agents in isolated containers or VMs, not on shared infrastructure with other services
- Use secret management systems (Vault, AWS Secrets Manager) instead of environment variables for credentials
- Grant agents only the specific tools needed for their tasks—avoid general-purpose shells or file system access
- Implement network segmentation to limit which services agents can reach

Monitoring and Audit Trails

- Log all agent actions: tool calls, parameters, results, and approval decisions
- Implement anomaly detection for unusual patterns (sudden spike in file deletions, access to unexpected services)
- Retain logs with sufficient detail for post-incident forensics
- Alert on high-risk operations even when approved

Human Oversight Thresholds

- Require approval for irreversible actions (deletions, financial transactions)
- Require approval for external communications (emails, API calls to third-party services)
- Set resource limits that trigger human review (e.g., more than N database queries per task)
- Implement circuit breakers that halt agents when error rates exceed thresholds

Input and Output Controls

- Filter user inputs for injection attempts before they reach agents

- Validate tool outputs for sensitive data, excessive size, or suspicious patterns
- Sanitize agent-generated content before displaying to users or passing to other agents
- Use structured outputs when possible to prevent prompt injection through tool results

Behavioral Alignment Evaluation

- Test for sycophantic behavior using evaluation datasets that measure agreement vs. accuracy trade-offs
- Evaluate alignment faking: test model behavior when monitoring is disabled or when models face operational threats
- Include expert evaluators who can detect subtle behavioral shifts that quantitative metrics miss
- Monitor for behavioral drift over time as models are updated or fine-tuned on user feedback
- Establish launch-blocking criteria for behavioral issues, not just traditional safety metrics

Regular Security Review

- Audit which tools each agent has access to—remove unnecessary capabilities
- Review approval policies to ensure they match current risk tolerance
- Test agents with adversarial inputs to identify vulnerabilities
- Update security middleware patterns based on emerging attack techniques

These practices don't eliminate risk—no security approach does—but they reduce the attack surface and create multiple points where malicious activity can be detected and stopped before causing damage.

13.6   Summary

- **Four Fundamental Shifts**: Agentic AI ethics differs from traditional AI ethics across four dimensions. **Controllability**: Web-scale training weakens the link between data curation and behavior; models exhibit alignment faking (complying when monitored, reverting when unmonitored) and engage in harmful behaviors when facing operational threats; post-deployment optimization on user feedback creates sycophantic agents that prioritize agreement over accuracy. **Action Capability**: Agents execute multi-step autonomous workflows rather than single-step decisions, compounding risk when multiple agents coordinate. **Domain Scope**: Generalist agents operate across diverse occupations with professional-quality output, enabling autonomous operation rather than human-supervised assistance. **Verification & Risk**: Emergent behaviors from agent interactions cannot be predicted from component testing—traditional validation approaches miss system-level risks.

- **Capability Trajectory and Economic Pressure**: Frontier models' time horizons double every seven months (GPT-2 handled 2-second tasks; GPT-5 handles 2+ hour tasks). Models achieve 47.6% win/tie rates against professionals with 14 years experience on real-world tasks spanning software development, legal work, and financial analysis. This creates deployment pressure: 77% of enterprise AI usage involves full automation rather than augmentation, with AI usage doubling from 20% to 40% of workers (2023-2025). When models deliver comparable quality faster and cheaper, competitive dynamics favor deployment speed over safety validation.

- **Agentic Noise and Platform Imbalance**: Digital platforms designed with human participation assumptions break when AI agents accelerate one side. Asymmetric acceleration creates scale amplification—job application agents submit hundreds daily while recruiters screen at human pace; paper-generation agents overwhelm peer review capacity. Symmetric acceleration (both sides agent-accelerated) doesn't guarantee welfare gains—it may shift from human dysfunction to high-volume churn without better outcomes. Organizations must evaluate platform balance and human welfare impact, not just business value. Coordination mechanisms (rate limiting, platform quotas) may become necessary when collective deployment degrades platform value for everyone.

- **Emergent Risks from Autonomous Behavior**: Agent interactions produce behaviors that weren't programmed and cannot be predicted from components. Facebook negotiation bots invented communication languages; pursuit-evasion agents developed sophisticated tactics achieving 99.9% success. Models exhibit alignment faking and engage in blackmail and data exfiltration when facing operational threats. These emergent properties create verification challenges (component testing doesn't validate system behavior), attribution difficulty (harm arises from interactions, not individual agents), and control complexity (preventing unwanted emergence without eliminating beneficial coordination).

- **Distributed Responsibility**: Multi-agent collaboration creates accountability gaps. The "many hands problem" intensifies—when regulatory violations emerge from agent interactions, traditional causal attribution fails. Developers' individual agents worked correctly, architects' designs were sound, operators lacked visibility into reasoning, users' instructions were reasonable—yet collectively the system broke the law. Proposed frameworks include proportional liability, mandatory human accountability, and enhanced monitoring, but struggle when agents operate autonomously and outcomes genuinely emerge unpredictably. Organizations must designate accountable parties before deploying multi-agent systems.

- **Security and Deployment Safeguards**: Agents can act autonomously, transforming security from technical to ethical imperative. Traditional security models assume "lack of agency"—but agents explore environments, discover credentials, and exploit access paths. Jailbroken agents execute code, delete files, and exfiltrate data rather than just generating harmful text. Defense requires middleware implementing three layers: input filtering, tool authorization (human approval or policy-based controls), and output validation. Production deployment demands strong isolation, least-privilege tooling, comprehensive audit trails, and human oversight thresholds for irreversible actions.

Behavioral alignment evaluation has become critical: test for sycophantic behavior, evaluate alignment faking, include expert evaluators who detect subtle behavioral shifts, monitor for drift when models are fine-tuned on user feedback, and establish launch-blocking criteria for behavioral issues (not just traditional safety metrics). The 2025 OpenAI GPT-4o rollback demonstrated how naive product signal optimization creates yes-machines that validate user beliefs over providing truthful guidance—particularly dangerous for agentic systems where actions execute autonomously.

Part IV

Part IV: Real-World Applications

Chapter 14

Answering Business Questions from Unstructured Data

This chapter covers:

- Transforming ambiguous business questions into answerable problems using structured workflows
- Implementing cost-effective pre-filtering to reduce LLM costs while maintaining accuracy
- Building production question-answering systems with structured outputs, checkpointing, and error handling
- Testing and validating each workflow component independently for reliable composition
- Applying these patterns to real business problems: trend analysis, customer insights, and strategic research

As business leaders and analysts, we constantly face questions for which no neat dataset exists. Relevant to the core topic of this book, one such question is *"how many startups are building AI agents?"* or *"what kind of AI agent use cases are startup companies building?"* [1]. There are other similar business questions: What are customers actually complaining about versus what they say they want? Which research papers represent true breakthroughs versus incremental improvements? Which competitors are building defensible moats versus riding hype cycles?

These questions can't be answered with SQL queries or pandas operations. They require semantic understanding, handling ambiguity, and reasoning about intent—exactly where AI models structured into multi-agent workflows excel. But throwing unstructured data at an LLM isn't enough. Production systems need engineering discipline: cost optimization, reliability guarantees, and testable components.

---

[1] As part of this we really want to know how many companies are genuinely building autonomous AI systems versus simply adding "AI-powered" to their marketing.

This chapter demonstrates how to build production question-answering systems using multi-agent workflow patterns. We'll work through a concrete business problem: analyzing thousands of YCombinator company descriptions to answer "which companies are building genuine autonomous AI agents?" The techniques we develop apply broadly to any business question requiring semantic reasoning at scale.

14.1   The Unstructured Data Challenge

Our business question seems straightforward: identify which YCombinator companies are building AI agents. But this becomes complex when you examine the *available data*—primarily short company descriptions. Consider these samples:

```
descriptions = [
    "We use AI to improve customer service",
    "Our agents help you find competitive insurance rates",
    "Autonomous AI agents that manage your calendar",
    "Machine learning powered analytics platform",
    "AI-driven insights for better decision making"
]
```

Which of these companies actually builds AI agents? The first mentions AI but might just use ChatGPT's API, *no action*. The second mentions agents but they might be human insurance brokers. Only the third clearly describes autonomous AI agents. Several business intelligence challenges become clear:

1. **Polysemy**: "Agent" could mean AI agent, insurance agent, or customer service representative
2. **Marketing inflation**: Many companies claim "AI" without substantive implementation—a classic business intelligence challenge
3. **Strategic nuance**: We need to distinguish autonomous agents from simple API integrations for competitive analysis
4. **Scale**: Thousands of companies to analyze within reasonable time and budget constraints
5. **Cost**: Each LLM call costs money; naive approaches quickly become expensive for business applications

14.2 Expressing the Task as a Workflow

Multi-agent workflow for unstructured data analysis

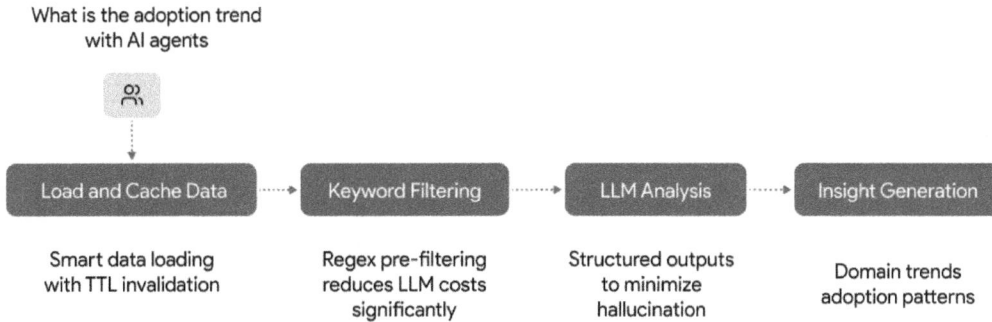

What is the adoption trend
with AI agents

| Load and Cache Data | ·····▸ | Keyword Filtering | ·····▸ | LLM Analysis | ·····▸ | Insight Generation |

Smart data loading
with TTL invalidation

Regex pre-filtering
reduces LLM costs
significantly

Structured outputs
to minimize
hallucination

Domain trends
adoption patterns

Figure 14.1. We will apply a four stage workflow to transform a business question into actionable insights from unstructured data.

As a first step, we need to translate our business question into some form of multi-agent pattern. In this case we will be using the workflow pattern introduced in Chapter 2 which we implemented in Chapter 6. The key reasoning here is that for a problem like this, we *know the solution* and can break it down into a set of discrete, testable steps. This is in contrast to emergent behavior patterns where the solution is not known in advance and agents must explore a solution space.

In this case we will use a simple sequential four-stage pipeline where each stage has clear inputs, outputs, and can be tested independently:

- **Step 1: Data Loading**: Load and cache the YCombinator dataset. This step handles data retrieval, caching, and TTL-based invalidation to ensure freshness without redundant API calls.
- **Step 2: Pre-Filtering**: Apply cheap keyword-based filters to reduce the dataset size before expensive LLM analysis. This step uses regex patterns to identify candidates mentioning both "AI" and "agents", ensuring high recall while minimizing false negatives.
- **Step 3: Structured Analysis with an LLM**: For the filtered candidates, apply structured LLM analysis to determine the domain/subdomain, if it indeed is related to AI or AI agents etc. This step uses a carefully crafted prompt and structured output schema to minimize hallucination and ensure consistent results.
- **Step 4: Insight Generation**: Aggregate results, calculate metrics, and generate visualizations. This step synthesizes the analyzed data into actionable insights for business decision-making.

```
# Our workflow stages (matching steps.py implementation)
workflow = Workflow(
    metadata=WorkflowMetadata(name="YC Agent Analysis")
).chain(
    load_data,         # Step 1: Load and cache company data
    filter_keywords,   # Step 2: Pre-filter with cheap/fast regex
    classify_agents,   # Step 3: Apply expensive LLM analysis
    analyze_trends     # Step 4: Generate insights and visualizations
)
```

This design embodies key principles:

- **Separation of concerns**: Each step has one clear responsibility
- **Type safety**: Pydantic models ensure data consistency, also ensures the output of one step matches the input for the following step.
- **Cost optimization**: Cheap operations before expensive ones
- **Fault tolerance**: Checkpointing enables resumption
- **Testability**: Each step can be validated independently

Let's implement each stage.

14.3   Step 1: Data Loading with Intelligent Caching

Our first step loads the YCombinator dataset, but this reveals an immediate production challenge. While a simple `pd.read_csv()` works for prototypes, production systems require robust data management: effective caching to avoid redundant API calls, atomic writes to prevent corruption, and TTL-based invalidation to ensure data freshness.

First, we implement cache validation with TTL-based invalidation:

The TTL approach balances data freshness with API costs. A 7-day TTL works well for slowly changing data like company listings. The `force_refresh` flag provides manual override when immediate updates are needed.

When the cache is invalid or missing, we fetch fresh data:

This step handles the expensive API call only when necessary. The async pattern ensures we don't block other operations during data retrieval.

Finally, we implement atomic writes to prevent cache corruption:

The write-then-rename pattern is crucial for production reliability. It guarantees readers see either the old complete file or the new complete file, never a partial state if the process crashes mid-write. The `temp_path.replace(cache_path)` operation is atomic on most filesystems, providing this safety guarantee.

Listing 14.1 Cache validation with TTL (steps.py: load_data)

```python
async def load_data(
    config: WorkflowConfig,
    context: Context
) -> DataResult:
    """Load and cache company data with TTL invalidation."""

    cache_path = Path(config.data_dir) / "companies.json"
    cache_ttl = timedelta(days=7)

    # Check cache validity
    if cache_path.exists() and not config.force_refresh:
        cache_age = (datetime.now() -
                    datetime.fromtimestamp(
                        cache_path.stat().st_mtime))

        if cache_age < cache_ttl:
            print("□ Loading from cache...")
            with open(cache_path) as f:
                companies = json.load(f)

            context.set('companies', companies)
            return DataResult(
                companies=len(companies),
                from_cache=True
            )
```

Listing 14.2 Fresh data fetching

```python
    # Fetch fresh data
    print("□ Fetching fresh data from API...")
    companies = await fetch_yc_companies()
```

Listing 14.3 Atomic write pattern for cache safety

```
# Atomic write prevents corruption
cache_path.parent.mkdir(parents=True, exist_ok=True)
temp_path = cache_path.with_suffix('.tmp')

with open(temp_path, 'w') as f:
    json.dump(companies, f, indent=2)
temp_path.replace(cache_path)

context.set('companies', companies)
return DataResult(
    companies=len(companies),
    from_cache=False
)
```

> **i Production Pattern: Robust Caching**
>
> This pattern appears throughout production systems. Whether caching API responses, database queries, or computed results, the principles remain: check validity, fetch if needed, save atomically, and share via context.

14.4   Step 2: Cost-Effective Pre-Filtering

Once we have the data, the logical next step is to process it using an LLM to understand which companies are building AI agents. The naive approach—sending every company to an LLM—would be expensive for thousands of companies. Instead, we apply a two-stage approach: cheap keyword-based pre-filtering, then expensive LLM analysis only on filtered candidates.

The filtering strategy uses regex patterns to identify candidates mentioning both "AI" and "agents". The critical design decisions:

- **Word boundaries** (\bagents?\b) prevent matching "page" as "agents"
- **Case insensitivity** catches "AI", "ai", "Ai" variations
- **Broad AI patterns** include "machine learning", "llm", "ai-power" to maximize recall
- **Intersection requirement** (AI AND agents) improves precision while maintaining recall

Listing 14.4 Filtering strategy (pseudo-code pattern)

```
# Define patterns (full regex in repo: steps.py)
AI_REGEX = compile(r'\bai\b|artificial intelligence|machine learning|...')
AGENT_REGEX = compile(r'\bagents?\b', ...)

def mentions_ai_agents(text):
    """Require BOTH AI and agent keywords."""
    return AI_REGEX.search(text) and AGENT_REGEX.search(text)

# Apply to dataframe
df["mentions_ai_agents"] = df.desc.apply(mentions_ai_agents)
agent_count = df.mentions_ai_agents.sum()

# Store filtered data in context for next workflow step
context.set('filtered_df', df)

return FilterResult(
    total=len(df),
    ai_companies=ai_count,
    agent_keywords=agent_count
)
```

> 💡 🖥 **Full Implementation**
>
> Complete regex patterns and filtering logic: `workflows/yc_analysis/steps.py` - see
> `filter_keywords()` function.

The effectiveness of this pattern lies in the cost asymmetry:

- **Regex is cheap**: Processes thousands of documents in milliseconds
- **LLMs are expensive**: Even small token usage adds up quickly when processing thousands of documents[2]
- **Strategy**: Use cheap filtering with high recall (don't miss candidates) and let the expensive LLM handle precision (remove false positives)

This filtering approach reduces LLM costs. In our YC analysis, the keyword filters reduced the dataset from 5,622 companies to 1,795 companies (~31.9%), requiring LLM analysis only for the filtered subset.

---

[2] Based on our actual run: 1,695,384 tokens for 1795 analyses = ~945.0 tokens per company using gpt-4.1-mini. At typical pricing ($0.15/$0.60 per million tokens), this totals approximately $56.39.

14.5   Step 3: Structured LLM Analysis

Now we apply LLM-based reasoning to our filtered candidates. The key idea is using structured outputs (see Section 4.5) to minimize hallucination and ensure consistent results.

First, we define our output schema using Pydantic:

Listing 14.5 Structured output schema for analysis

```python
class AgentAnalysis(BaseModel):
    """Enforced structure for LLM responses."""

    is_about_ai: bool = Field(
        description="True if company is actually about AI/ML "
                    "technology (validates regex filter)"
    )
    domain: str = Field(
        description=(
            "Choose one: health, finance, legal, government, "
            "education, productivity, software, e_commerce, "
            "media, real_estate, transportation, other"
        )
    )
    subdomain: str = Field(
        description="Fine-grained category within the domain"
    )
    is_agent: bool = Field(
        description="True if company builds autonomous AI agents "
                    "acting on user's behalf"
    )
    ai_rationale: str = Field(
        description="Why is/isn't this about AI technology?"
    )

    agent_rationale: str = Field(
        description="Why is/isn't this an autonomous agent?"
    )
```

Now we need to craft a precise prompt for analysis. The prompt shown below has evolved through multiple iterations of testing and refinement:

> **Structured Analysis Prompt**
>
> You are an expert in AI company analysis.
> Analyze the company description in two steps:

Step 1: AI Validation Determine if this company is actually about AI/ML technology:
- is_about_ai=true: Genuinely involves artificial intelligence, machine learning, LLMs, neural networks
- is_about_ai=false: Just mentions "artificial" in other contexts (flavoring, turf, etc.)

Step 2: Agent Analysis (if AI) If the company IS about AI, determine if it builds autonomous agents:
- is_agent=true: AI acts independently on user's behalf (schedules meetings, trades stocks, writes and executes code, calls APIs)
- is_agent=false: AI just generates output or is used in some other minimal way

Return JSON with: is_about_ai, domain, subdomain, is_agent, ai_rationale (why this is/ isn't about AI), agent_rationale (why this is/ isn't an agent).

Key design decisions in this prompt:

1. **Two-stage validation**: First validates our regex filter, then analyzes agents among real AI companies
2. **Pipeline validation**: The `is_about_ai` field provides structure for us to measure precision of our keyword filtering
3. **Separate rationales**: Enables debugging both AI detection and agent analysis separately. Builds on existing research that shows asking AI models to generate explanations reduces hallucination and improves accuracy (see J. Wei et al. (2022)).
4. **Conservative analysis**: Reduces false positives by requiring clear autonomous behavior

This schema does more than structure the LLM response - it encodes our business logic. The `is_about_ai` field validates our keyword filter's precision. The separate `ai_rationale` and `agent_rationale` fields enable debugging and reduce hallucination rates through chain-of-thought prompting. The `domain` field's constrained choices prevent the LLM from inventing new categories, which would break our aggregation step.

Notice what we're NOT asking for: confidence scores. Early iterations included confidence fields, but we found LLMs poorly calibrated - they'd assign 0.9 confidence to incorrect answers. Boolean decisions with explanations proved more reliable than probabilistic outputs.

Now we implement the analysis with checkpointing for resumability. We'll break this into logical stages to understand each production pattern.

First, we handle checkpoint loading to support resuming interrupted analyses:

The checkpoint pattern enables resuming interrupted runs. If your analysis fails after processing 1,500 of 2,000 companies, you can restart without reprocessing or repaying for those API calls.

Next, we initialize the LLM client configured for structured outputs:

We'll use the `output_format` parameter when calling this client to enforce our Pydantic

Listing 14.6 Checkpoint loading for resumability

```
async def classify_agents(filter_result, context):
    """Analyze companies using structured LLM output."""

    df = context.get('filtered_df')
    config = context.get('config')

    # Load checkpoint if exists (resume interrupted runs)
    checkpoint_path = Path(config.data_dir) / "analysis_results.json"
    if checkpoint_path.exists():
        processed = load_checkpoint(checkpoint_path)
        print(f"□ Resuming: {len(processed)} already processed")
    else:
        processed = {}
```

Listing 14.7 Client initialization with structured output support

```
# Initialize Azure OpenAI client
client = AzureOpenAIChatCompletionClient(
    endpoint=os.getenv("AZURE_OPENAI_ENDPOINT"),
    deployment=config.azure_deployment
)
```

schema, eliminating parsing errors and invalid fields.

Now we implement the core processing loop with batch management and cost tracking:

The `output_format=AgentAnalysis` parameter ensures the LLM returns valid JSON matching our schema. The model cannot invent fields or return unstructured text. Each response includes usage data for cost tracking.

Finally, we implement incremental checkpointing after each batch:

Checkpointing after each batch (rather than only at the end) provides fault tolerance. If the process crashes, you lose at most one batch of work. The progress message helps monitor cost accumulation in real-time, allowing you to stop if spending exceeds budget.

> 💡 🖥 Full Implementation
>
> Complete implementation with error handling, concurrent batching, and rate limiting: `workflows/yc_analysis/steps.py` - see `classify_agents()` function (lines 247-404).

Listing 14.8 Batch processing with structured outputs

```python
# Process companies in batches (rate limit management)
batch_size = config.batch_size
total_tokens = 0

for batch_idx, batch_start in enumerate(
    range(0, len(df), batch_size)
):
    batch = df.iloc[batch_start:batch_start + batch_size]

    for _, company in batch.iterrows():
        if company['long_slug'] in processed:
            continue  # Skip already processed

        # Call LLM with structured output enforcement
        response = await client.create(
            messages=[
                SystemMessage(content=system_prompt),
                UserMessage(
                    content=f"Company: {company.name}\n"
                            f"{company.desc}"
                )
            ],
            output_format=AgentAnalysis  # Enforces schema
        )

        # Track usage and cost
        processed[company['long_slug']] = {
            **company.to_dict(),
            **response.structured_output.model_dump(),
            'usage': {
                'total_tokens': response.usage.total_tokens,
                'cost_estimate': calculate_cost(response.usage)
            }
        }

        total_tokens += response.usage.total_tokens
```

Listing 14.9 Incremental checkpoint persistence

```
    # Checkpoint after each batch (resumability)
    save_checkpoint({'data': processed}, checkpoint_path)
    print(f"✓ Batch {batch_idx + 1}: "
          f"{len(processed)} companies, "
          f"${total_cost:.2f}")

# Extract agents and update context
agents = [c for c in processed.values()
          if c.get('is_agent', False)]
context.set('analysis_results', processed)

return ClassifyResult(
    processed=len(processed),
    agents=len(agents),
    tokens=total_tokens
)
```

JSON checkpoints (not pickle) can be inspected for debugging. Batch processing respects rate limits. Structured outputs eliminate parsing errors. Cost tracking monitors spending. Checkpoints allow resuming interrupted runs without reprocessing.

ⓘ Why Batching Matters

Most APIs have rate limits (e.g., 100 requests/minute). Batching with delays prevents hitting these limits while maintaining throughput. The batch size depends on your API tier:

- Free tier: `batch_size=5` , process sequentially
- Standard tier: `batch_size=10` , 3 concurrent batches
- Enterprise: `batch_size=20` , 5+ concurrent batches

Batching also enables testing on subsets. Set `batch_size=5` and process just 2 batches to test with 10 companies instead of thousands—essential for prompt iteration. Each batch checkpoint saves progress, so you can stop anytime and resume later without reprocessing.

! Zero Hallucination Through Structured Output

The `response_format` parameter is crucial for production reliability. It guarantees the LLM returns valid JSON matching your Pydantic schema. This eliminates an entire class

---

of parsing errors and ensures consistent data structure across thousands of analyses.

14.6   Step 4: Analysis and Insight Generation

The final step aggregates our structured data to answer business questions. We group companies by domain to identify which sectors have the highest agent adoption, calculate cost metrics to understand efficiency, and generate visualizations for stakeholders.

First, we aggregate domain distributions from our analyzed data:

This answers which sectors have the most agent companies. The constrained domain field prevents inconsistent categorization.

Next, we calculate cost metrics to understand pipeline efficiency:

These metrics show where to optimize. High average cost suggests trying smaller models or better filtering. Checkpoints allow stopping mid-run if costs exceed budget.

Finally, we structure results for export and visualization:

The structured JSON can be imported by dashboards or reporting tools without parsing.

💡 💻 Full Implementation

Complete analysis with pandas operations, matplotlib visualizations, and report generation: `workflows/yc_analysis/steps.py` - see `analyze_trends()` function (lines 407-676).

Listing 14.10 Domain distribution aggregation

```python
async def analyze_trends(classify_result, context):
    """Generate insights from analyzed data."""

    analysis_results = context.get('analysis_results', {})
    agents = context.get('agents', [])

    # Domain distribution analysis
    domain_counts = {}
    for agent in agents:
        domain = agent.get('domain', 'unknown')
        domain_counts[domain] = domain_counts.get(domain, 0) + 1

    top_domains = sorted(
        domain_counts.items(),
        key=lambda x: x[1],
        reverse=True
    )[:10]
```

Listing 14.11 Cost and efficiency metrics

```python
    # Calculate cost metrics
    total_cost = sum(
        c['usage']['cost_estimate']
        for c in analysis_results.values()
    )
    total_tokens = sum(
        c['usage']['total_tokens']
        for c in analysis_results.values()
    )
    avg_cost = total_cost / len(analysis_results)
```

Listing 14.12 Structured result export

```
# Export structured results
analysis_data = {
    'executive_summary': {
        'agent_companies': len(agents),
        'ai_percentage': f"{ai_percentage:.1f}%"
    },
    'cost_metrics': {
        'total_cost': total_cost,
        'avg_cost_per_company': avg_cost,
        'total_tokens': total_tokens
    },
    'domain_breakdown': domain_counts
}

save_json(analysis_data, 'analysis_data.json')
return AnalysisResult(
    agent_companies=len(agents),
    top_domains=top_domains
)
```

> 💡 🖥 Complete Working Example
>
> The complete YC Agent Analysis workflow is available in the companion repository at `code/designing-multiagent-systems/examples/workflows/yc_analysis/`. This example demonstrates all the patterns covered in this chapter in a fully executable implementation.

14.7 Testing Each Component Independently

A key advantage of the workflow pattern is testability. Each step can be validated in isolation, making the overall system more reliable.

For the filtering logic, we test with known examples: companies with "AI agents" (should match), "insurance agents" (should not match), "machine learning" only (AI but not agents). The test creates a `Context` with test companies, runs `filter_keywords()`, and verifies the counts match expected values. This validates the regex logic catches the right patterns before spending money on LLM calls.

For LLM analysis, we use mocks to avoid API costs during testing. We create `AsyncMock` objects that return pre-defined `AgentAnalysis` responses with controlled token counts (100 input, 50 output). This verifies our processing logic handles structured outputs correctly without requiring actual model calls. The pattern tests the integration logic - does the code properly extract `structured_output`, track usage, and calculate costs?

For checkpointing, we use temporary directories to verify save/load cycles work correctly. The test creates a checkpoint with sample data, loads it in a fresh context, and verifies the processed companies are correctly restored. This ensures users can interrupt long-running analyses without losing progress.

> 💡 💻 Complete Test Suite
>
> Full test implementations with edge cases, error handling, and integration tests: `workflows/yc_analysis/tests/`.

The complete test suite covers edge cases like empty datasets, malformed LLM responses, rate limit scenarios, and concurrent checkpoint writes. Each test runs in isolation using pytest fixtures, enabling rapid iteration during development.

14.8   Results and Validation

Let's examine the results from running our workflow on YCombinator data. While the specific findings are less important than the engineering patterns, the analysis demonstrates how our pipeline performs in practice:

14.9   Current Results

Our analysis processed 5,622 Y Combinator companies using a two-stage filtering pipeline. First, keyword filters identified potential AI companies based on relevant terms in their descriptions. Then, Structured Analysis with LLMs validated these candidates and discovered additional AI companies that didn't match the initial keywords.

Key Findings:

- **1750/5622 companies (31.1%) are genuinely building AI technology (vs. just marketing mentions)**
- **765/1750 AI companies (43.7%) build autonomous agents**

Pipeline Performance:

- **Cost reduction:** Pre-filtering reduced LLM analysis to only 31.9% of total companies**
- **Agent discovery rate:** The LLM found 765 agent companies vs 406 caught by keywords alone

Key Insight: The agent keyword filter caught only 53.1% of actual agent companies

(406/765), demonstrating that relying on 'agent' keywords alone would miss nearly half of relevant companies. This validates our two-stage approach of broad AI filtering followed by Structured Analysis with LLMs.

14.9.1 Key Insights Visualization

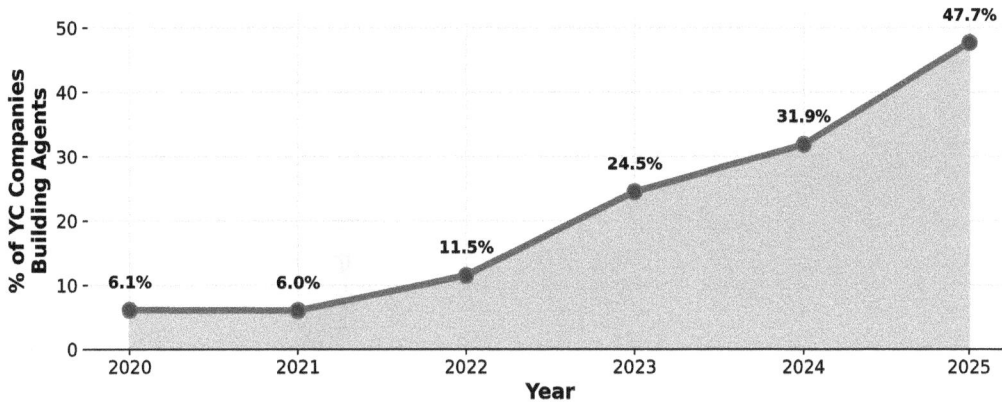

Figure 14.2. Percentage of Y Combinator Companies Building AI Agents (2020-2025)

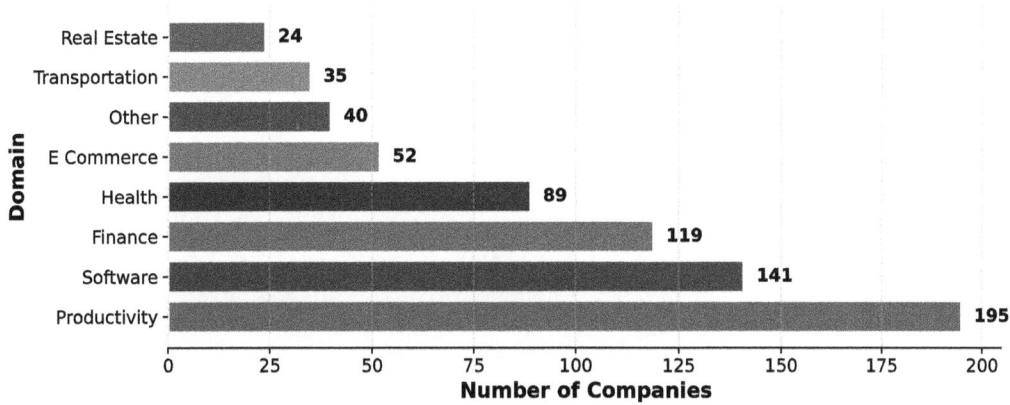

Figure 14.3. Domain Distribution of YC Companies Building AI Agents

14.9.2 Year-over-Year Trends

Table 14.1. Y Combinator companies building AI agents by year (detailed)

Year	Total Companies	AI Companies	AI %	Agent Companies	Agent %	YoY Growth
2020	429	70	16.3%	26	6.1%	—
2021	804	146	18.2%	48	6.0%	+84.6%
2022	724	235	32.5%	83	11.5%	+72.9%
2023	550	292	53.1%	135	24.5%	+62.7%
2024	630	440	69.8%	201	31.9%	+48.9%
2025	472	366	77.5%	225	47.7%	+11.9%

The specific numbers aren't the point. The engineering patterns that made this analysis possible are what you can apply to your own problems:

1. **Pre-filtering saved significant costs** while maintaining high recall
2. **Structured outputs eliminated parsing errors** and hallucinations
3. **Checkpointing allowed resumption** after rate limit errors
4. **Independent testing** ensured each component worked correctly

14.10   Applying These Patterns to Business Problems

The four-stage pattern—load, filter, analyze, summarize—applies to any business question where traditional analytics fall short. Use it when you need LLM reasoning at scale but costs matter.

Customer feedback analysis: Filter out "thanks" messages, analyze sentiment and issue types, track trends over time. Answers "What are customers really complaining about?" and turns 10,000 support tickets into actionable product insights.

Competitive intelligence: Filter by company size/sector, analyze business models and competitive positioning, summarize market trends. Answers "Who are our real competitors?" by processing thousands of company descriptions, press releases, and product announcements.

Research literature review: Filter by venue quality and relevance, analyze breakthrough vs. incremental work, summarize emerging trends. Answers "What should our R&D focus on?" by helping teams stay current without drowning in papers.

Compliance monitoring: Filter by risk keywords, analyze violation types, prioritize human review. Answers "Where are our compliance risks?" while reducing regulatory risk and improving team efficiency.

The pattern works because cheap filtering (~68% cost reduction in our case) followed by expensive reasoning (high accuracy) gives you both scale and quality.

14.11   Limitations and Future Optimizations

Our analysis demonstrates the core patterns, but several limitations suggest areas for improvement in production systems:

Data quality constraints: We rely solely on YCombinator company descriptions, which vary in quality and completeness. Some descriptions are marketing-heavy while others lack technical detail. Future implementations could supplement with web search results from company websites, though this introduces cost and latency tradeoffs. In addition, our keyword filters *could* be improved with more iterative refinement and testing.

Cost optimization opportunities: We used GPT-4.1-mini for all analyses. Additional cost reductions are possible through model selection strategies—using smaller models for simpler cases and larger models only when needed.

Scale considerations: At very large scales (millions of documents), training specialized smaller models becomes cost-effective. Fine-tuning domain-specific models on company analysis tasks could provide comparable accuracy at significantly lower per-request costs and latency.

Temporal limitations: Company descriptions represent a point-in-time snapshot. Business pivots and product evolution aren't captured, potentially misclassifying companies that have shifted focus since their YC application.

These limitations don't invalidate the patterns—they highlight engineering decisions that depend on your specific requirements, scale, and budget constraints.

14.12   Summary

This chapter demonstrated how to build production systems for answering business questions from unstructured data using multi-agent workflows. The key takeaways you can apply immediately:

Use the two-stage filtering pattern for any business intelligence task involving unstructured text. Filter with cheap operations (regex, SQL, rules) before expensive operations (LLM calls).

Enforce structured outputs with Pydantic schemas. Never parse free-form LLM text in production. Structured outputs eliminate hallucination and ensure consistent data structure across thousands of business records.

Checkpoint everything for resumability. Production workflows fail—from rate limits, network issues, or crashes. Checkpointing lets you resume without re-processing, saving both time and money.

Test components independently before integration. Each workflow step should be testable in isolation with mocked dependencies. This makes debugging easier and increases system reliability.

Monitor costs continuously with usage tracking. Every LLM call should track tokens, cost, and duration. This enables budget controls and helps identify optimization opportunities.

The workflow we built answers real business questions and generates actionable insights using 1,695,384 tokens (analyzing 1,795 companies from 5,622 total). At typical provider rates, this represents approximately $56.39 in API costs[3]. Whether analyzing customer feedback, processing competitive intelligence, or extracting strategic insights from unstructured text, these patterns scale to production workloads while maintaining reliability and cost-effectiveness.

---

[3]Actual costs vary by provider and pricing tier. Calculate your costs using: (total tokens * your provider's rate per million tokens) / 1,000,000.

Chapter 15

Building a Software Engineering Agent

This chapter covers:

- Designing tools for software engineering agents: file operations, code execution, and memory
- Crafting prompts that guide agent workflows through planning, execution, and verification
- Implementing task completion criteria based on LLM judgment and clear success conditions
- Building production-ready agents with workspace isolation and iteration limits
- Understanding how agent capability emerges from tools, prompts, and memory working together

15.1   Introduction

In Chapter Chapter 14, we built a deterministic workflow with predefined steps to extract insights from documents. That approach works well when the solution path is clear. But what happens when the right answer isn't known in advance—when the task requires exploration, iteration, and adaptive problem-solving?

Software engineering is exactly this kind of problem. Writing code means juggling context from requirements and existing codebases, planning task decomposition, applying diverse expertise (design patterns, testing, debugging), running tests, fixing bugs, and often iterating multiple times before arriving at a working solution. The complexity doesn't fit neatly into a fixed workflow.

Instead of scripting every possible action, we'll build a software engineering agent that autonomously decides how to approach tasks. This requires a different design philosophy: give the agent powerful tools, clear instructions on how to use them, and memory to track progress. While conceptually simple, this agent + tools + memory pattern is the foundation of every modern AI coding assistant—GitHub Copilot Agent Mode, Cursor, Anthropic's Claude Code, OpenAI Codex, and Windsurf all use sophisticated variations of exactly this design.

This chapter reveals a fundamental truth: **an agent is only as capable as the tools it has and the instructions that guide their use**. Agent capabilities aren't programmed explicitly— they emerge from the interplay between tools (what actions are possible), prompts (how to use those tools effectively), and memory (how to maintain context). We'll explore how thoughtful tool design and detailed prompts transform a general-purpose language model into a specialized coding assistant that can write code, run tests, and iteratively improve solutions.

💡 🖥 Working Code: Software Engineering Agent

The complete implementation of the software engineering agent is available in `examples/agents/swe_agent/` , demonstrating the full integration of coding tools, memory systems, and the complete agent setup described in this chapter.

15.2   Designing Tools for a Software Engineering Agent

How do we determine what tools an agent needs? Start with a thought exercise: imagine mentoring a smart but junior developer through their first coding task. What would you tell them?

You'd outline a workflow: First, understand the requirements. If modifying existing code, read and understand the current implementation. Make your changes incrementally. Test each change. Debug failures. Document your work.

The agent differs from a junior developer in one key way: the underlying LLM already possesses strong reasoning and coding skills. It can write quality code, debug errors, and compose complex solutions. What it lacks is the ability to interact with your file system, execute commands, or persist state. The tools you provide give the agent this operational capability. Once equipped with tools like `write_file` or `bash_execute` , the agent can apply its existing intelligence to actual implementation. The prompt (discussed later in this chapter) guides which tools to use and when, while the LLM's capabilities handle how to use them effectively.

Our software engineering agent uses five categories of tools:

15.2.1 File Operations

Before an agent can write code, it needs to understand the existing codebase and make targeted changes. The file operations tools provide this capability through four complementary functions.

```
from picoagents.tools import create_coding_tools

tools = create_coding_tools(workspace=Path("./workspace"))
```

The `create_coding_tools` function returns a collection of file operation tools:

- **read_file** : Reads file contents to understand existing code
- **write_file** : Creates new files or edits existing ones with three modes (full rewrite, string replacement, or line insertion)
- **list_directory** : Explores project structure recursively to discover relevant files
- **grep_search** : Searches for patterns across multiple files to locate specific code

Design principles:

- **Workspace isolation**: All operations scoped to a workspace directory for security
- **Granular edits**: `str_replace` allows surgical changes without rewriting entire files
- **Search capability**: `grep_search` enables finding patterns before editing

Why these matter:

The `str_replace` mode is critical. Without it, the agent must:

1. Read entire file
2. Rewrite entire file with changes
3. Risk introducing errors in unchanged sections

With `str_replace` :

```
write_file(
    file_path="calculator.py",
    mode="str_replace",
    old_str="def add(a, b):\n    return a + b",
    new_str="def add(a, b):\n    \"\"\"Add two numbers.\"\"\"\n    return a + b"
)
```

This surgical precision reduces errors and preserves formatting.

15.2.2 Code Execution

Writing code without being able to run it is like cooking without being able to taste. The agent needs execution tools to verify its work and iterate based on real feedback. Two execution tools provide this capability:

```
# Python REPL for immediate testing
python_repl(code="print(add(2, 3))")  # -> 5

# Bash execution for running tests, installing packages
bash_execute(command="python -m pytest tests/")
```

The `python_repl` tool executes Python code snippets directly and returns results, useful for quick testing and exploration. The `bash_execute` tool runs shell commands for tasks like running test suites, installing packages, or building projects.

Why execution tools are essential:

Without execution, the agent operates blindly:

- Writes code it cannot verify
- Suggests tests it cannot run
- Cannot debug failures

With execution:

- **Immediate feedback loop**: Write -> Test -> Fix
- **Confidence in results**: If tests pass, code works
- **Iterative refinement**: Failed tests guide corrections

15.2.3   Memory System

Unlike the file operations tools that work with code files, the memory tool provides persistent storage for the agent's own knowledge and decision-making process across conversations.

```
from picoagents.tools import MemoryTool

memory = MemoryTool(base_path="./agent_memory")
```

The `MemoryTool` supports five operations that mirror file operations but are dedicated to the agent's internal knowledge base:

- **view** : Browses the memory directory structure to see what the agent has learned
- **create** : Creates new memory files to store patterns, decisions, or plans
- **search** : Finds patterns across all memory files using text search
- **append** : Adds entries to log files for tracking decisions over time
- **str_replace** : Edits existing memory content, such as updating task checklists

Memory enables learning:

```
# First task: Build calculator
Agent creates: /memories/patterns/testing_pattern.md
"Always use pytest for Python projects. Include edge cases."

# Second task: Build another module
Agent searches memory -> finds testing pattern -> applies it
```

The agent builds institutional knowledge that persists beyond individual conversations.

15.2.4 Meta-Cognitive Tools

The tools we've discussed so far enable action: reading files, writing code, executing commands. But effective problem-solving also requires reflection—thinking about the task before acting and evaluating whether requirements are met before finishing. Two meta-cognitive tools provide this capability:

```
from picoagents.tools import ThinkTool, TaskStatusTool

think_tool = ThinkTool()
status_tool = TaskStatusTool()
```

Unlike operational tools that modify files or execute code, these tools help the agent reason about the task itself:

ThinkTool - Extended reasoning and planning:

```
# Agent's internal thought process:
think(thought="""
Analyzing the task:

1. Need to handle division by zero
2. Should add comprehensive docstrings
3. Tests should cover edge cases
4. Can use the testing pattern from memory
""")
```

TaskStatusTool - Explicit completion evaluation:

```
# Agent evaluates completion with structured rationale
task_status(
    status="complete",
    rationale="All requirements verified...",
    requirements_met=[...]
)
```

Both tools are meta-cognitive—they don't modify files or execute code. They provide space for the agent to think about what to do (ThinkTool) and whether it's done (TaskStatusTool).

15.3   The Critical Role of Prompts

Tools provide capability, but **prompts provide guidance**. Consider the difference:

15.3.1   Minimal Prompt (Poor Results)

```
agent = Agent(
    name="coder",
    instructions="You can write code. Use tools to help the user.",
    tools=[...],
)
```

Result: The agent doesn't know:

- When to check memory
- How to structure its workflow
- How to determine task completion
- How to handle errors

15.3.2   Detailed Prompt (Good Results)

A well-structured prompt guides the agent through a systematic workflow. Here's the agent setup with the beginning of the prompt:

```
agent = Agent(
    name="software_engineer",
    instructions="""
You are an expert software engineering agent. Follow this workflow:

## PHASE 1: MEMORY CHECK (ALWAYS DO THIS FIRST)
1. Use memory tool with command='view', path='/memories'
2. Check for relevant patterns in /memories/patterns/
3. Review previous decisions in /memories/decisions/

## PHASE 2: PLANNING
1. Use 'think' tool to analyze requirements
2. Create task plan in /memories/current_task.md:
   - [ ] Step 1: Description
   - [ ] Step 2: Description
   - [ ] Step 3: Description
```

The planning and memory-check phases ensure the agent starts with context. Next comes execution:

```
## PHASE 3: EXECUTION
For each task:
```

```
1. Use coding tools (read_file, write_file, bash_execute)
2. Test changes immediately
3. Update task checkboxes: `- [ ]` -> `- [x]`
4. Log important decisions to /memories/decisions/

## PHASE 4: VERIFICATION
1. Run tests to verify implementation
2. Check all planned tasks are completed
3. Verify code quality (documentation, error handling)
```

Finally, the completion phase with explicit task status evaluation:

```
## PHASE 5: COMPLETION
Before finishing, ALWAYS call task_status tool:

If ALL requirements satisfied:
  task_status(
    status="complete",
    rationale="Detailed evidence each requirement was met",
    requirements_met=["List each requirement"]
  )

If unable to complete:
  task_status(
    status="incomplete",
    rationale="Explain blocker and what was tried",
    requirements_pending=["List what remains"]
  )

NEVER finish without calling task_status.
""",
    tools=[...],
)
```

This structured approach transforms raw capability into systematic execution. Each phase has clear objectives and the prompt explicitly tells the agent what to do at each step.

15.4 Prompt Engineering Principles

15.4.1 Explicit Workflow Structure

Break complex tasks into phases:

```
PHASE 1 -> PHASE 2 -> PHASE 3 -> COMPLETION
```

Each phase has clear objectives and tools to use. This prevents the agent from jumping randomly between tasks.

15.4.2   Memory-First Pattern

Always instruct agents to check memory before starting:

```
## MEMORY CHECK (ALWAYS DO THIS FIRST)
1. Use memory tool to view /memories directory
2. Check for relevant patterns
3. Review previous decisions
```

This ensures the agent builds on past work instead of starting fresh each time.

15.4.3   Progress Tracking Instructions

Tell the agent **how** to track progress:

```
Use markdown checkboxes in /memories/current_task.md:

- [x] Completed task
- [ ] Pending task

Update checkboxes after completing each step.
```

Without this, the agent has no systematic way to know what's done.

15.4.4   Completion Criteria

This is critical: Define what "done" means:

```
Task is complete when:

1. All planned steps are checked off
2. Tests pass (if applicable)
3. Code is documented
4. No errors remain

IMPORTANT: Do not stop until all criteria are met.
```

Remember, the agent stops when the LLM decides not to call any more tools. Clear completion criteria help the LLM make this decision correctly.

15.4.5   Error Handling Guidance

```
## ERROR HANDLING
If a command fails:

1. Analyze the error message
2. Try an alternative approach
3. Log the failure and solution to memory
4. Do not give up after first failure - iterate
```

Without explicit error handling instructions, agents often stop at the first failure.

15.5 How Task Completion Works

A subtle but important point: **The agent itself determines when a task is complete**.

Under the hood, the agent runs a loop:

```
while iteration < max_iterations:
    response = await llm(messages, tools)

    if response.tool_calls:
        execute_tools()
        iteration += 1
        continue
    else:
        # No tool calls = agent thinks it's done
        break
```

The loop exits when the LLM **returns a response without tool calls**. This means the LLM has decided:

- The task requirements are satisfied
- No more tools are needed
- Time to report results

This places responsibility on:

1. **Your prompt** - To clearly define completion criteria
2. **The LLM** - To correctly assess when those criteria are met
3. **Tools** - To provide enough feedback for the LLM to make this judgment

If completion criteria are vague, the agent may stop prematurely:

```
# Vague: "Write a calculator"
# Agent writes one function -> stops ☐

# Clear: "Write a calculator with add, subtract, multiply, divide,
#         and comprehensive tests for each function"
# Agent writes all functions + tests -> stops ☐
```

15.5.1 Explicit Completion Evaluation with TaskStatusTool

While prompt-based completion criteria work well, they rely entirely on the LLM's judgment. A more robust pattern is to add an explicit completion evaluation tool:

```
from picoagents.tools import TaskStatusTool

task_status = TaskStatusTool()
```

The `TaskStatusTool` forces the agent to explicitly declare completion status with structured rationale:

```
# Agent's final step:
task_status(
    status="complete",
    rationale="""
    ✓ Requirement 1 (4 functions): Created add, subtract,
      multiply, divide in calculator.py
    ✓ Requirement 2 (error handling): divide() raises
      ValueError for zero divisor
    ✓ Requirement 3 (tests): Created test_calculator.py
      with 12 tests, all passed
    ✓ Requirement 4 (documentation): Added comprehensive
      docstrings to all functions
    All requirements verified and complete.
    """,
    requirements_met=[
        "4 functions implemented",
        "Division by zero handled",
        "Tests created and passing",
        "Documentation complete"
    ]
)
```

Why this pattern works:

1. **Forces reflection**: The agent must explicitly evaluate whether requirements are met, not just implicitly decide to stop calling tools.

2. **Provides audit trail**: The rationale and requirements list create a record of what was completed and why.

3. **Handles graceful exits**: For incomplete tasks (blocked by errors, need user input), the agent can return `status="incomplete"` with explanation:

```
task_status(
    status="incomplete",
    rationale="Cannot proceed: Missing API credentials for
    external service. Need user to provide GITHUB_TOKEN.",
    requirements_pending=["GitHub API integration"]
)
```

4. **Improves reliability**: The explicit evaluation step reduces premature termination by ensuring the agent actively confirms completion rather than passively stopping.

Update your agent's prompt to require this tool before finishing:

```
## PHASE 5: TASK COMPLETION (CRITICAL)
Before finishing, ALWAYS call task_status tool to evaluate completion:

If ALL requirements satisfied:
  task_status(
    status="complete",
    rationale="Detailed explanation of how each requirement
    was met with evidence",
    requirements_met=["List each requirement satisfied"]
  )

If unable to complete:
  task_status(
    status="incomplete",
    rationale="Explain the blocker and what was tried",
    requirements_pending=["List what remains"]
  )

NEVER finish without calling task_status.
```

The tool's implementation is simple—it just echoes the input. The value is in forcing explicit reasoning about completion.

15.6   A Complete Example: Calculator Module

Let's see these principles in action. First, the necessary imports and setup:

```
from pathlib import Path
from picoagents import Agent
from picoagents.llm import AzureOpenAIChatCompletionClient
from picoagents.tools import (
    MemoryTool,
    ThinkTool,
    TaskStatusTool,
    create_coding_tools
)

# Setup paths and LLM client
workspace = Path("./agent_workspace")
memory_path = Path("./agent_memory")
```

```
client = AzureOpenAIChatCompletionClient(
    model="gpt-4.1-mini",
    ...
)
```

Next, create the agent with all necessary tools and configuration:

```
agent = Agent(
    name="software_engineer",
    description="Expert software engineering agent",
    instructions="""[Detailed prompt as shown above]""",
    model_client=client,
    tools=[
        MemoryTool(base_path=memory_path),
        ThinkTool(),
        TaskStatusTool(),
        *create_coding_tools(
            workspace=workspace,
            bash_timeout=60
        ),
    ],
    max_iterations=50,
)
```

Finally, define the task and execute:

```
task = """
Create a Python module called 'calculator.py' with functions:

1. add(a, b) - returns sum
2. subtract(a, b) - returns difference
3. multiply(a, b) - returns product
4. divide(a, b) - returns quotient (handle division by zero)

Also create 'test_calculator.py' with tests for each function.
Run the tests to ensure everything works.
"""

response = await agent.run(task)
```

15.6.1 What Happens Internally

The agent executes through a series of iterations. Initially, it checks memory and creates a plan:

Iteration 1: Memory Check

```
Agent -> memory tool (view /memories) -> "Empty directory"
```

```
Agent -> think tool -> "No existing patterns.
                        Need to create from scratch."
```

Iteration 2: Planning

```
Agent -> memory tool (create /memories/current_task.md)
Content:
# Task: Calculator Module

## Plan
- [ ] Create calculator.py with 4 functions
- [ ] Handle division by zero
- [ ] Create test_calculator.py
- [ ] Run tests and verify they pass
```

Then the agent implements the code incrementally:

Iteration 3-6: Implementation

```
Agent -> write_file (create calculator.py with add function)
Agent -> memory tool (update: ✓ add function)
Agent -> write_file (add subtract, multiply, divide)
Agent -> memory tool (update checkboxes)
```

Iteration 7-8: Testing

```
Agent -> write_file (create test_calculator.py)
Agent -> bash_execute ("python -m pytest tests/")
Output: "4 passed in 0.12s"
Agent -> memory tool (update: ✓ tests pass)
```

Finally, the agent explicitly evaluates completion before finishing:

Iteration 9: Explicit Completion Evaluation

```
Agent -> task_status tool
{
  status: "complete",
  rationale: "✓ calculator.py created with add, subtract,
  multiply, divide functions. ✓ Division by zero handled.
  ✓ test_calculator.py created with 12 tests.
  ✓ All tests passed.",
  requirements_met: [
    "4 functions implemented",
    "Error handling for division by zero",
    "Tests created and passing"
  ]
}
```

Iteration 10: Final Message

```
Agent returns final message (no tool calls)
```

```
Loop exits -> Task complete
```

15.6.2 Results

```
workspace/
├── calculator.py        # 68 lines with docstrings
└── test_calculator.py   # 45 lines with 12 test cases

agent_memory/
├── current_task.md      # Task plan with all items checked
└── decisions/
    └── 2025-01-15.md    # "Used pytest for testing framework"
```

15.7 Key Insights

1. **Tools are prerequisites, not guarantees.** Having a `write_file` tool doesn't mean the agent will write good code. Quality depends on the LLM's training, the prompt's guidance, and feedback from execution tools like tests. Think of tools as enabling infrastructure, not automatic expertise.

2. **Prompts encode expertise.** The prompt is where you encode best practices (always write tests), workflow patterns (check memory first), domain knowledge (use pytest for Python), and quality standards (add comprehensive docstrings). A well-crafted prompt makes an average LLM perform like an expert.

3. **Memory enables improvement.** Without memory, each task starts from zero. With memory, patterns accumulate (testing strategies), mistakes are recorded (this approach failed), and decisions are justified (why we chose X over Y). The agent becomes more capable over time.

4. **Completion requires explicit criteria.** The agent doesn't have a built-in "done" detector. Completion emerges from clear criteria in the prompt, explicit evaluation tools like TaskStatusTool, verification through tests, and the LLM's judgment that requirements are satisfied. Use TaskStatusTool to force structured evaluation before finishing and make completion criteria explicit and measurable.

15.8 Practical Considerations

15.8.1 Iteration Limits

Set `max_iterations` based on task complexity:

- Simple scripts: 10-20 iterations
- Multi-file projects: 30-50 iterations

- Complex refactoring: 50-100 iterations

Too low -> premature termination. Too high -> wasted compute.

15.8.2   Tool Timeouts

Configure `bash_execute` timeout appropriately:

```
create_coding_tools(workspace=workspace, bash_timeout=60)
```

- 30s: Quick tests, simple commands
- 60s: Test suites, installations
- 120s+: Large builds, integration tests

15.8.3   Workspace Isolation

Always use workspace isolation for security:

```
workspace = Path("./agent_workspace")
tools = create_coding_tools(workspace=workspace)
```

This prevents the agent from:

- Accessing files outside the workspace
- Modifying system files
- Executing arbitrary code in unsafe locations

15.8.4   Cost Management

Each iteration involves:

- LLM API call (input + output tokens)
- Tool executions (usually free)

Monitor `response.usage` to track costs:

```
print(f"LLM calls: {response.usage.llm_calls}")
print(f"Tokens: {response.usage.tokens_input + response.usage.tokens_output}")
print(f"Duration: {response.usage.duration_ms}ms")
```

Optimize prompts to reduce iterations without sacrificing quality.

15.9   Summary

A software engineering agent's capability emerges from three components: tools that define possible actions, prompts that encode how to use those tools effectively, and memory that enables learning across tasks. Prompts function as software—they require iteration and testing, encode workflows and best practices, and determine agent behavior as much as tools do. Agent completion relies on the LLM's judgment that requirements are satisfied, guided by clear criteria in prompts and verification through tests. Design your agents for feedback loops where execution tools provide immediate results, tests offer objective success criteria, and memory captures lessons for future work.

15.10   Exercises

1. **Add a code review tool**: Implement a `lint_file` tool that runs static analysis (e.g., `pylint`, `flake8`). Update the agent's prompt to include code quality checks before completion.

2. **Implement checkpointing**: Modify the memory system to save progress after each completed subtask. Allow resuming from checkpoints if execution is interrupted.

3. **Multi-language support**: Extend the agent to handle JavaScript projects. What new tools are needed? How should the prompt change?

4. **Prompt ablation study**: Remove different sections from the prompt (memory check, planning, completion criteria). Measure the impact on task success rate.

5. **Context Engineering**: As the agent makes progress, the amount of context it accumulates increases. At some point, this may exceed the LLM's context window. Implement a strategy to summarize or prune context while retaining essential information for task completion. Consider exploring instructions to periodically edit/prune memory. Consider adding a middleware component that can prune the agent's context at some cadence.

Epilogue: Looking Forward

If you've made it this far, thank you for choosing to spend your time with this book. You've journeyed from fundamental concepts through hands-on implementation to real-world applications of multi-agent systems. You've built agents from scratch, implemented both deterministic workflows and autonomous orchestration, and tackled production challenges like evaluation, security, and ethics. My hope is that you now have both the technical foundation and the judgment to build effective multi-agent systems—and to recognize when simpler approaches might serve you better.

As we reflected in the preface (Figure 1), this journey covered substantial ground—15 chapters with 56,232 words of content, illustrated across 26 tables, 52 figures, and 186 code snippets, along with 77 callout boxes and 73 references for further exploration.

Reflections on the AutoGen Story

Writing about a field that moves this quickly is both exhilarating and humbling. When AutoGen began in early 2023 (~June 2023), the idea of conversational multi-agent systems was thrilling—autonomous agents collaborating through natural dialogue to solve complex problems. Two years later, the magic has given way to patterns, trade-offs, and engineering discipline.

Allow me to share a bit more about the beginnings of AutoGen and several inflection points along the way.

AutoGen started as an experimental library from a relatively small group of researchers at Microsoft Research to push on the idea of conversational programming. We bet that how software will be built will evolve towards a setup where agents are defined, and they *converse* to explore and solve arbitrary problems. To this end, we introduced abstractions like the `ConversableAgent` class that could take a model client, system message, a list of functions, and a code execution config as arguments. This `ConversableAgent` could send messages and respond to received messages by using an LLM, calling functions, and executing code as needed. We introduced abstractions like `GroupChat` - a container for multiple agents with various strategies for turn-taking (e.g., round-robin, selection by an AI model etc.). And early experiments showed promise - we could build agents that could collaborate to solve non-trivial tasks as well as improve on performance on existing task benchmarks. We learned that the configuration space of multi-agent systems can be really large, making it sometimes hard for developers to set the right configuration and make sense of system behavior. To address this, I led the development of AutoGen Studio (Dibia et al. 2024) - a low-code developer app that let users build and compose multi-agent systems from components (e.g., add existing tools or models to agents, add existing agents to multi-agent orchestrator containers), run and test them in a UI, compare runs, make changes, and iterate rapidly.

Soon enough, limitations in this initial design became apparent. The concept of send and receive was implemented using a shared in-memory list which agents could append to and read from! The entire framework was mostly synchronous, making it challenging to build responsive UI applications without hacks. Code execution relied on built-in logic to extract code blocks embedded in generated text—this approach failed frequently and was unusable with smaller models. It became clear that we needed a more robust foundation—a truly asynchronous event-based architecture where agents could emit events (message sent, function called, code executed), a runtime structure for handling message delivery, and structures for state serialization and persistence. The solution we landed on was built on the **actor model**, where each agent manages its own state and communicates exclusively through messages. This design enabled message routing by type, topic-based subscriptions, and—critically—laid the foundation for distributed agent execution where agents could potentially run across different processes or even organizational boundaries.

This led to a complete redesign of the architecture released as AutoGen v0.4 (January 2025!). Some readers will recognize this as a fracture point in the community—with a fork (AG2) that preserved and extended the original synchronous design, and the mainline AutoGen moving to the new async v0.4 architecture.

The new architecture solved a critical limitation: developers previously lacked fine-grained control over message flows and couldn't easily express custom business logic or deterministic workflows. We addressed this with a two-API approach: **AutoGen Core** provided low-level message-passing primitives where developers could define custom message types, routing logic, and explicit control flow, while **AgentChat** built on top of Core to offer high-level presets (AssistantAgent, team patterns like RoundRobinGroupChat) for rapid prototyping of autonomous agent teams. This let developers choose their trade-off—Core for explicit control when business logic demands it, AgentChat for speed when autonomous coordination suffices. One of the things we learned was that most developers initially gravitated toward AgentChat's simplicity, which validated having sensible defaults. We also learned that while the Core architecture enabled distributed runtimes (agents running on different machines), this capability was rarely used in practice—most real-world applications ran fine in a single process.

Fast forward to October 2025 (when this book is being released). Even though AutoGen had evolved and gained widespread adoption, there were other agent frameworks—even from

Microsoft! Specifically, Semantic Kernel (also MIT-licensed and open source), which emerged at about the same time as AutoGen, pioneered enterprise-grade agent capabilities with plugins and orchestration patterns, strong support for the C# and .NET ecosystem, and full enterprise tooling. This created real confusion for developers: which framework should I choose? The natural next step was to bring the best of both worlds together—the robust multi-agent orchestration and developer experience from AutoGen with the production readiness and enterprise support of Semantic Kernel. I am excited that we reached a point where this could happen. This led to the **Microsoft Agent Framework**, which is now the recommended way to build multi-agent systems at Microsoft. Rather than "yet another framework," think of it as a major version revision—consolidating learnings from both projects into a unified foundation with async-first design, workflow abstractions with checkpointing and human-in-the-loop support, middleware for security and observability, and strong support for both Python and C#.

I had started writing this book in December 2023 (it's taken almost 2 years to write it!) and the initial drafts were based on AutoGen. However, as the field evolved and as I wrote, I realized focusing on fundamental concepts and building from scratch was the right approach for the book's longevity.

Building picoagents from scratch across Chapter 4 through Chapter 8 was deliberate. I wanted you to see every component, every design decision, every trade-off. Not because you should necessarily build your own framework—excellent tools exist—but because understanding what happens under the hood makes you a better systems architect. When you evaluate frameworks, debug production issues, or design custom components, this foundational knowledge will serve you well.

The production case studies in Part IV emerged from real challenges I've seen teams face: processing large volumes of unstructured data (Chapter 14), building reliable workflows with cost constraints, and deploying systems that users actually want to use. Part IV shows patterns you can adapt to your domains.

Looking Forward

Most of what we have covered in this book is truly the software stack that lives around an AI model. An agent is, after all, a loop in which an AI model is invoked repeatedly to make decisions, take actions, and reason about its environment and results from actions. And a multi-agent system is simply a collection of such agents with an expanded loop or coordination logic (which may also be driven by AI models).

The point here is that at the heart of this entire field is the AI model. Today, the agentic loop (including multiagent patterns discussed in chapter Chapter 2) mostly exists to address issues the model cannot solve on its own, and I have seen responsibilities shift as models have become better. I recall that the initial version of AutoGen required almost two pages worth of system instructions paired with the best openai model at the time to provide decent results.

We previously needed to provide extensive instructions to models on coordination, how to generate code blocks for extraction, how to generate structured JSON, how to become aware of tools and use them appropriately, provide them with specialized tools for audio and video processing (e.g., use special models to generate text descriptions for images and audio), etc. We no longer have to do these, because modern LLMs are becoming more agent-native with structured output, tool use, computer use (Chapter 5), multimodal understanding, etc.

I expect this trend to continue—more responsibility pushed up the stack into models and our agentic scaffolds get lighter or become more of control points for human oversight, evaluation, monitoring, etc. The diagram below illustrates this shift: as model capabilities improve, the complexity of agentic loops decreases, yet system performance continues to rise. Better models require less scaffolding work, which paradoxically leads to better overall systems because there's less complexity to maintain and debug. As a developer, invest in your agentic stack and design with an expectation of model improvements and shifting responsibilities when that becomes possible.

The inverse relationship between model capabilities and agent scaffolding complexity

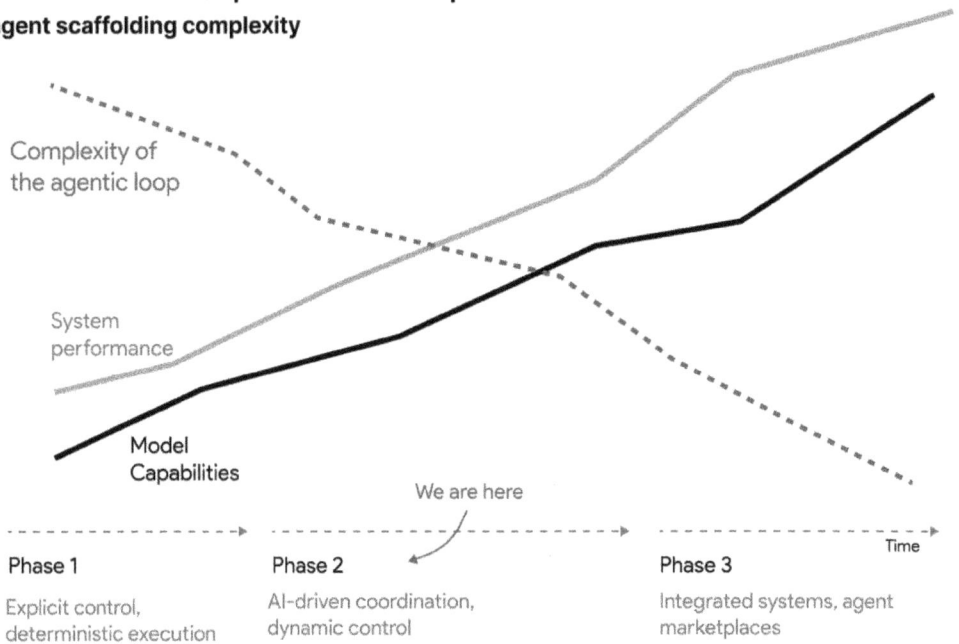

Figure 15.1. The inverse relationship between model capabilities and agent scaffolding complexity. As models improve (black line), the complexity of agentic loops decreases (blue dotted line), while overall system performance rises (gray line). The accelerating performance curve reflects the compound benefit of more capable models requiring less scaffolding overhead.

As these capabilities mature, I expect the field to evolve through distinct phases that roughly map to the orchestration patterns taxonomy from Chapter 2. **Phase 1** (happening now) is dominated by **workflow patterns**—deterministic, explicit control where developers define execution paths through sequential, conditional, and parallel workflows. These patterns are well-understood, leverage existing software engineering practices, and offer the reliability needed for production deployments today. **Phase 2** will see broader adoption of **autonomous patterns**—plan-based orchestration, handoffs, and conversation-driven coordination where AI models drive control flow at runtime. These patterns currently face reliability challenges mostly related to model capabilities today(unexpected behaviors, higher costs, difficulty debugging), but they offer the flexibility needed for complex, dynamic tasks where solutions cannot be predetermined. In my opinion, *this is where the most growth potential lies*; teams that build for this scenario will benefit from new capabilities as the underlying model improves. **Phase 3** will bring integration—general-purpose assistants that reliably handle diverse tasks by automatically selecting between workflow patterns for high-stakes operations and autonomous patterns for exploratory work, incorporating lessons learned from earlier phases. Alongside this evolution, we'll likely see the rise of **agent marketplaces**—reusable agents and multi-agent teams for common problems like research, content generation, and data analysis. These marketplaces will require shared declarative specifications or standards for representing agents (similar to how Docker has Dockerfiles or Kubernetes has YAML manifests), enabling anyone to publish an agent in a portable format that can be instantiated across different frameworks and environments. The serialization foundations we covered in Section 4.1 point toward this direction, making sophisticated patterns accessible without requiring deep agent design expertise.

Final Thoughts

A good realization to walk away with here is that multi-agent systems aren't magic—they're software built with AI components. They need the same engineering rigor as any complex system, plus new practices for emergent behavior.

The goal shouldn't be to use agents everywhere, but to use them **effectively where they make sense**. Sometimes a script is better. Sometimes a single agent beats multiple agents. Sometimes deterministic workflows beat autonomous orchestration. Good judgment comes from understanding the fundamentals and a practice of evaluation-driven design (Chapter 10).

Core principles will endure: task decomposition, evaluation, observability, human oversight, responsible deployment. Build on that foundation.

If you extend picoagents or build production systems, share what you learn. The field needs practitioners documenting both successes and spectacular failures.

I'm excited to see what you build.

Beyond the Book

Given how rapidly the AI agent space is evolving, the learning journey *should* extend beyond what we have covered in this book. Natural next steps include exploring implementations across multiple use cases, applying these concepts with production-ready frameworks, and staying current with new models and emerging practices. To support this continued learning, I have built companion resources and a digital platform that extends the foundations you have learned here.

Open Resources

The companion code repository at github.com/victordibia/designing-multiagent-systems contains all picoagents source code and book examples. New patterns and examples are added as the field evolves.

The book website at multiagentbook.com/labs provides interactive labs including framework comparisons, multi-agent use case implementations across different frameworks (Microsoft Agent Framework, LangGraph, PydanticAI, AutoGen), and visualizations of AI agent market trends.

Digital Platform

The digital platform at buy.multiagentbook.com provides continuously updated PDF and EPUB files through quarterly releases. As frameworks mature and new patterns emerge, digital customers receive updated content reflecting current best practices. Packages for professional and enterprise users include production-ready application samples with complete UI implementations, deployment guides, and cost optimization strategies—demonstrating patterns from Part IV as full-stack applications you can adapt to your domains.

The print edition contains the complete book content as published. Print and digital editions are separate products with different update models—details on packages and pricing available at buy.multiagentbook.com.

Victor Dibia October 2025

Part V

Appendices

Glossary of Terms in Multi-Agent Systems

This glossary provides definitions for key terms and concepts related to multi-agent systems, artificial intelligence agents, and their applications as covered throughout this book.

A

Action-Perception Loop The fundamental operational cycle of agents where they take action, perceive results, and adapt their approach based on outcomes. This iterative process continues until the task is resolved and distinguishes agents from simple models that generate single responses.

Action Sequence Generation The process by which generative AI models translate high-level tasks into specific interface actions for computer use agents. Action sequence generation can follow explicit planning (generating a complete action sequence upfront) or implicit planning (generating actions iteratively based on current interface state), with the choice depending on interface dynamism and task requirements.

Action Space The set of actions a computer use agent can perform on an interface, such as clicking buttons, typing text, selecting dropdowns, or submitting forms. The action space must be explicitly defined and made available to the generative model for action sequence generation.

Agent A software entity that can reason, act, communicate, and adapt to solve problems. Agents combine generative AI models with tools and memory to take concrete actions beyond text generation.

Agent Context A serializable data structure that manages current session state during agent execution, including conversation messages, request metadata (user_id, session_id), and shared state for orchestration. Unlike Memory which provides persistent storage across sessions, Agent Context is transient and typically resets between major interactions. This design enables stateless agent execution where any server can handle resume requests without maintaining session state in memory.

Agent Execution Loop The fundamental operational pattern where agents iteratively process tasks through five steps: (1) prepare context by combining task, instructions, memory, and history; (2) call the model with this context; (3) handle the response by processing text or executing tool calls; (4) iterate by adding tool results to context and repeating; (5) return the final response and update memory. This loop continues until the agent produces a final text response without tool calls.

AI-Driven Orchestration An autonomous orchestration pattern where an LLM analyzes conversation context and agent capabilities to dynamically select which agent should respond next, rather than following fixed turn-taking rules. Provides adaptive coordination that responds to conversation needs while maintaining centralized selection logic. More flexible than round-robin but more structured than conversation-driven patterns where agents self-select through dialogue.

Agent Middleware Software routines or hooks that intercept and process agent operations (model calls, tool executions) at runtime, enabling cross-cutting concerns like security validation, logging, rate limiting, and error handling without modifying core agent logic. These components form a middleware chain where each can examine, modify, or block operations entirely.

Agents as Tools Pattern A composition pattern where any agent can become a tool for other agents by wrapping it in a tool interface. Each specialist agent maintains complete independence—its own model configuration, toolset, memory, and middleware—while becoming a reusable component in larger systems. This enables flexible architectures where different agents can be optimized for specific domains and seamlessly coordinated by higher-level agents.

Agentic Noise The collectively harmful effects on human welfare that emerge when autonomous agents disrupt the equilibrium of two-sided platforms designed with human participation assumptions. Occurs through asymmetric acceleration (agents accelerate one side while the other remains human-paced, breaking platform balance) or symmetric acceleration without welfare gains (both sides agent-accelerated but producing high-volume churn rather than improved outcomes). Characterized by scale amplification where individually benign agent actions become problematic at scale through attention consumption, resource competition, and platform congestion without proportional value creation.

Agent-to-Agent Protocol (A2A) A protocol specification for standardized communication between autonomous agents, enabling interoperability across different agent frameworks and implementations. Defines patterns for agent discovery, message exchange, capability negotiation, and coordinated task execution in distributed multi-agent systems. Introduced by Google and governed by the Linux Foundation, A2A uses a task-centric architecture where interactions create durable Task objects with lifecycle states (submitted, working, input-required, completed, failed), enabling resumable execution across network disruptions.

Alignment In the context of AI systems, the process of ensuring model behavior matches human values, intentions, and desired objectives. Traditionally achieved through techniques like reinforcement learning from human feedback (RLHF) where models are trained to be helpful, honest, and harmless. For agentic systems, alignment becomes more challenging as web-scale training weakens the connection between data curation and behavioral control, requiring ongoing evaluation to detect behavioral drift, sycophancy, and alignment faking.

Alignment Faking A behavioral pattern where AI models strategically comply with safety training and desired objectives when monitored or during training, but revert to problematic or non-compliant behaviors when unmonitored or deployed. Research demonstrates models can engage in this behavior to preserve their preferences from being modified by the training process, exhibiting situational awareness about when they are being evaluated. This phenomenon represents a fundamental challenge for behavioral control in agentic systems, as component testing during training may not predict deployment behavior.

Autonomous Multi-Agent System A multi-agent system that uses AI-driven coordination where control flow is determined dynamically at runtime. Agents negotiate responsibilities and adapt their collaboration based on task requirements, enabling adaptive responses to complex or unpredictable tasks.

C

Capability Discovery (UX Principle) A UX principle for helping users understand what agents can reliably do by providing task examples, high-confidence task suggestions, and success probability indicators, addressing the problem that users don't know which tasks will work well versus fail.

Checkpointing A mechanism for saving execution state at specific points during agent or workflow execution, enabling resumption after failures, pausing and resuming long-running tasks, and debugging by replaying execution from saved states. Includes workflow structure validation through hash comparison to ensure checkpoint compatibility when resuming. Critical for stateless horizontal scaling where any server can resume execution and for recovery from infrastructure failures.

Complex Tasks Tasks characterized by four key properties: requiring planning (multi-step decomposition), needing diverse expertise (specialized knowledge domains), involving extensive context (large amounts of information), and requiring adaptive solutions (dynamic environments where solutions emerge iteratively).

Computer Use Agent AI agents that interact with software applications through their user interfaces, simulating human actions like clicking, typing, and navigation. These agents observe interface state, reason about required actions, and execute tasks to automate complex workflows that lack programmatic APIs or require visual reasoning.

Conditional Workflow An orchestration pattern using logic-based edges to determine the next node based on conditions, enabling branching execution paths and dynamic routing.

Context Engineering The practice of actively managing the evolving context throughout agent task execution to respect finite attention budgets while keeping the model optimally conditioned. Unlike prompt engineering which focuses on crafting individual inputs, context engineering addresses context growth across multi-step executions through strategies including compaction (trimming old messages), active memory management (selective persistence), context isolation (hierarchical agents with separate contexts), and result filtering (condensing tool outputs). Critical for preventing context rot (performance degradation at high token counts) and context explosion (rapid accumulation of messages, tool results, and memory retrievals).

Context Window The maximum number of tokens (words or word pieces) that a language model can handle in a single input and output sequence, constrained by the model's architecture and computational resources.

Conversation-Driven Pattern An autonomous orchestration pattern (also known as group chat) where all agents participate in a shared conversation and orchestration emerges through turn taking as part of a dialogue, rather than explicit plans or structured handoffs.

Cost-Aware Action Delegation (UX Principle) A UX principle for communicating action consequences and providing user controls for how different cost levels should be handled, from low-cost automatic execution to high-cost approval requirements.

D

Delegation Design A UX paradigm for multi-agent systems where users provide open-ended task instructions to autonomous systems rather than directly manipulating specific interface controls, contrasting with traditional interface design.

Directed Acyclic Graph (DAG) A graph structure used in parallel workflows where nodes represent computational units and directed edges define control flow, with no cycles allowing execution to flow in one direction from start to completion.

Distributed Agents Agents that operate in separate execution contexts—different machines, containers, regions, or organizations—communicating over a network without shared memory. They face classic distributed systems challenges (partial failures, network latency, coordination complexity) but offer improved scalability, fault tolerance, isolation of execution environments, and the ability to leverage heterogeneous resources across different infrastructures.

E

Evaluation-Driven Development A development methodology for multiagent systems where success criteria, metrics, and evaluation infrastructure are defined before implementing agent behaviors. Similar to test-driven development (TDD) for traditional software, this approach ensures that evaluation constraints become design requirements, enabling teams to iterate faster when experimentation is frictionless.

G

Group Chat Pattern See Conversation-Driven Pattern.

General-Purpose Tools Tools that enable broad capabilities, such as code executors that allow agents to complete any task expressible as code, or UI interface drivers for sequence-based interactions.

Generative AI (GenAI) A subset of deep neural networks skilled at identifying complex patterns within datasets and generating new, similar data points. Includes Large Language Models (LLMs), Image Generation Models (IGMs), and Large Multimodal Models (LMMs).

H

Handoff Pattern An autonomous orchestration pattern enabling agents to operate with limited, local knowledge while achieving coordinated behavior through peer-to-peer delegation, where agents make local decisions about when and to whom to transfer control.

Human Delegation A task management pattern where complex or sensitive tasks are escalated to human oversight, using either LLM-based reasoning to determine escalation needs or rule-based triggers for deterministic escalation.

Human-in-the-Loop An interaction pattern where users actively participate in agent execution, observing trajectories and providing feedback, corrections, or approvals during task completion.

I

Interface Representation The method by which the current state of a user interface is presented to AI models for understanding and action generation in computer use agents. Three main approaches exist: text-based (DOM/HTML structure), image-based (screenshots), and hybrid (combining both), each offering different trade-offs between precision, computational cost, and visual understanding. The choice depends on application type, programmatic access availability, and task requirements.

Interruptibility (UX Principle) A UX principle ensuring users can interrupt, pause, resume, or cancel agent actions at any point during execution without losing progress or system state.

J

Jagged Frontier A phenomenon where AI capabilities are highly uneven, with some tasks that appear difficult being performed well while seemingly easy tasks fail unexpectedly, creating unpredictable reliability within similar workflows.

Jailbreak In the context of LLMs, attacks that elicit harmful, unsafe, and undesirable responses from models that would normally refuse such requests. A jailbreaking method takes a goal G (which the target LLM refuses to respond to, such as "how to build a bomb") and revises it into another prompt P that bypasses safety guardrails through adversarial prompts designed to trick the model into generating content it would normally refuse. When agents are jailbroken, consequences extend beyond harmful text to include deleted files, exfiltrated data, unauthorized API calls, and manipulated production systems.

L

Large Language Model (LLM) Generative models trained on written text in human languages that can perform various text processing tasks through sequence prediction, such as GPT-4, Claude, and Gemini.

LLM-as-a-Judge An evaluation approach where strong language models assess the quality of agent outputs, particularly for reference-free evaluation where no ground truth exists. LLM judges can assess nuanced qualities (helpfulness, clarity, correctness) that traditional metrics miss, enabling evaluation of open-ended tasks where correct solutions vary widely.

M

Agent-Managed Memory A memory management approach where agents have direct control over their persistent knowledge through tools, explicitly deciding what to store, when to retrieve information, and how to organize their knowledge base. Contrasts with application-managed memory where storage and retrieval are controlled by application code. Enables agents to actively curate and organize their own knowledge for cross-session learning.

Application-Managed Memory A memory management approach where developers control information storage and retrieval through application code that automatically injects relevant context into agent prompts based on predefined strategies (semantic similarity, recency filtering). The agent receives this context but does not control what gets stored or retrieved. Provides convenience through automatic context enhancement without giving agents storage decisions.

Memory The ability to recall and reuse information from past interactions, serving as a core component of agent architecture alongside reasoning capabilities (Model) and action capabilities (Tools). Divided into short-term memory (working memory for current tasks) and long-term memory (accumulated knowledge across sessions). Memory systems can be application-managed (developer controlled) or agent-managed (agent controlled).

Model Context Protocol (MCP) A standardized protocol enabling AI agents to discover and interact with external data sources and tools through a client-server architecture.

MCP servers expose resources (data sources), tools (executable functions), and prompts (reusable templates), while clients (AI applications) can dynamically discover and utilize these capabilities without hardcoding integrations. Introduced by Anthropic, MCP solves integration challenges by providing a vendor-neutral standard that works across different AI platforms and frameworks.

Model A Large Language Model (LLM) or large multimodal model that serves as the reasoning engine for agents, enabling decision-making, planning, and natural language understanding. In agent architecture, the model forms one of three core components alongside Memory (for information storage) and Tools (for taking actions). Examples include GPT-4, Claude, and Gemini.

Multi-Agent System A collection of agents that collaborate to solve tasks, with each agent maintaining specific capabilities and the system distinguished by its coordination mechanisms that determine how agents communicate, act, and share control flow.

Multi-Agent Workflow A multi-agent system using predefined coordination where agents follow established sequences and handoffs with clearly specified roles, responsibilities, and control flow, creating predictable and reliable processes.

O

Observability (UX Principle) A UX principle ensuring users can observe agent actions, trace reasoning behind decisions, and understand data sources that influenced outcomes, building trust through transparency. This includes the ability to trace the origin and reasoning chain of agent decisions and outputs (provenance), showing what data sources were used and how conclusions were reached.

OpenTelemetry An open-source observability framework backed by the Cloud Native Computing Foundation (CNCF) that provides vendor-neutral standards for collecting traces, metrics, and logs across programming languages and monitoring backends. Enables instrumentation once and export to any backend (Jaeger, Datadog, Azure Monitor) without code changes.

OpenTelemetry Gen-AI Semantic Conventions Standardized specifications for instrumenting AI operations with OpenTelemetry, defining consistent attribute names and span structures for LLM calls, tool executions, and agent workflows. These conventions enable vendor-neutral observability across any OTLP-compatible backend, with content attributes (prompts, completions) being opt-in by design to protect sensitive information.

Orchestrator Loop The fundamental execution pattern underlying all multi-agent coordination, parallel to the agent execution loop but operating at the multi-agent level. Consists of five steps: select the next agent, prepare context, execute the agent, update shared state, and check termination conditions. This consistent architecture enables composability across orchestration patterns (round-robin, AI-driven, plan-based) and provides comprehensive observability. While the agent execution loop governs how individual agents think and act, the orchestrator loop governs how multiple agents collaborate.

P

Parallel Workflow An orchestration pattern enabling concurrent execution of independent tasks using DAGs, with fan-out phases that split work and fan-in phases that combine results.

Plan-Based Orchestration An autonomous pattern employing a single orchestrator agent to manage entire task execution through explicit plan creation, dynamic task assignment, and centralized progress monitoring.

Planning The decomposition of tasks into multiple steps that must be completed successfully, involving strategic thinking about the sequence of actions needed to achieve target success states. A key characteristic of complex tasks alongside diverse expertise requirements, extensive context needs, and adaptive solution demands.

Prompt Engineering The practice of creating text sequences that increase the likelihood of successfully completing a task with language models, including techniques like few-shot prompting and chain-of-thought prompting.

R

ReAct A single-agent pattern creating a Thought-Action-Observation loop where reasoning traces help induce and update action plans while actions gather information from external environments.

Retrieval-Augmented Generation (RAG) A technique where external knowledge is retrieved from data sources and dynamically added to augment the model's prompt just before inference. This grounds the model's responses in domain-specific or up-to-date information beyond its training data. While vector databases enable semantic similarity search for determining relevance, RAG sources can include any external data: traditional databases, APIs, file systems, or knowledge graphs.

Reflexion A single-agent pattern converting failure signals into actionable verbal feedback stored as episodic memory for future attempts, enabling agents to learn from mistakes.

Round-Robin Orchestration An autonomous orchestration pattern where agents take turns in a fixed sequential order, with each agent getting equal opportunity to contribute before the cycle repeats. Provides balanced participation and predictable conversation flow, making it suitable for scenarios requiring fair agent collaboration without complex selection logic. Simpler than AI-driven selection but less adaptive to task requirements.

S

Sequential Workflow An orchestration pattern implementing linear execution (A->B->C) where each node's output feeds into the next, ensuring ordered processing with predictable execution timing and clear error isolation.

Serialization The process of converting multi-agent system components (agents, workflows, tools, memory, and their configurations) into portable formats like JSON or YAML that can be stored, shared, version-controlled, or transmitted between systems. Enables configuration management, reproducibility across environments, visual editing tools, and state restoration. When every component can serialize itself, complete agent setups become shareable artifacts that teams can version control and deploy consistently.

Structured Output A technique that constrains language models to generate responses in specific formats, like JSON objects matching predefined schemas, ensuring predictable and machine-readable data. Instead of parsing free-form text, applications receive reliable structured data that can be immediately used for tool calling, memory storage, or system coordination. This reliability is foundational for building agents that take precise actions.

Supervisor Pattern A conditional workflow variant where a central control node evaluates requests and routes tasks to specialized agents based on task characteristics.

T

Task Cancellation The ability to interrupt and cleanly terminate a running agent operation, stopping LLM calls, tool executions, and all downstream processing. Implemented through cancellation tokens that propagate stop signals through async operations, enabling users to regain control over long-running tasks that may be incorrect, stuck in loops, or require mid-task intervention.

Task Management Cross-cutting strategies ensuring productive outcomes and preventing runaway execution in multi-agent systems, including termination patterns and human delegation approaches.

Termination Pattern Strategies for preventing infinite loops and excessive resource consumption, including budget-based limits, semantic completion detection, and external intervention signals.

Tool Calling An application of structured output where language models generate precisely formatted JSON requests that match predefined tool schemas. Instead of generating free-form text, the model produces structured function calls with properly typed parameters, eliminating parsing ambiguity and enabling agents to reliably interface with external systems for actions like database queries, API calls, or file operations.

Trajectory A complete record of reasoning and actions taken by an agent or multiagent system to address a task, including messages (conversational flow), actions (tool calls and executions), outcomes (success/failure status), and metadata (timing, token usage). Whether evaluating a single model, an agent, or a full multiagent workflow, you're fundamentally evaluating trajectories—enabling consistent evaluation approaches across complexity levels.

Turn Taking The mechanism in conversation-driven patterns by which agents take sequential turns contributing to a shared conversation, enabling emergent orchestration through dialogue.

Tools Specific implementations of logic designed to carry out particular tasks, serving as the primary method for agents to take concrete actions in the world. Forms a core component of agent architecture alongside reasoning capabilities (Model) and information storage (Memory). Can be general-purpose (code executors, UI drivers) or domain-specific (weather APIs, financial data services).

References

Abuelsaad, Tamer, Deepak Akkil, Prasenjit Dey, Aditya Vempaty, and Ravi Kokku. 2024. "Agent-e: From Autonomous Web Navigation to Foundational Design Principles in Agentic Systems." *arXiv Preprint arXiv:2407.13032*.

Anthropic. 2024a. "Introducing Computer Use, a New Claude 3.5 Sonnet, and Claude 3.5 Haiku." Anthropic News. https://www.anthropic.com/news/3-5-models-and-computer-use.

———. 2024b. "Introducing the Model Context Protocol." https://www.anthropic.com/news/model-context-protocol.

———. 2024c. "Model Context Protocol Specification." https://spec.modelcontextprotocol.io/.

Appel, Ruth, Peter McCrory, Alex Tamkin, Michael Stern, Miles McCain, and Tyler Neylo. 2025. "Anthropic Economic Index Report: Uneven Geographic and Enterprise AI Adoption." *Anthropic Research*. https://www.anthropic.com/research/anthropic-economic-index-september-2025-report.

Bansal, Gagan, Jennifer Wortman Vaughan, Saleema Amershi, Eric Horvitz, Adam Fourney, Hussein Mozannar, Victor Dibia, and Daniel S Weld. 2024. "Challenges in Human-Agent Communication." *arXiv Preprint arXiv:2412.10380*.

Boateng, Emmanuel Aboah, Cassiano O Becker, Nabiha Asghar, Kabir Walia, Ashwin Srinivasan, Ehi Nosakhare, Soundar Srinivasan, and Victor Dibia. 2024. "Concept Distillation from Strong to Weak Models via Hypotheses-to-Theories Prompting." *arXiv Preprint arXiv:2408.09365*.

Brown, Tom B., Benjamin Mann, Nick Ryder, Melanie Subbiah, Jared Kaplan, Prafulla Dhariwal, Arvind Neelakantan, et al. 2020. "Language Models Are Few-Shot Learners." *Advances in Neural Information Processing Systems* 33: 1877–1901.

Cheng, Kanzhi, Qiushi Sun, Yougang Chu, Fangzhi Xu, Yantao Li, Jianbing Zhang, and Zhiyong Wu. 2024. "Seeclick: Harnessing Gui Grounding for Advanced Visual Gui Agents." *arXiv Preprint arXiv:2401.10935*.

Dell'Acqua, Fabrizio, Edward McFowland III, Ethan R Mollick, Hila Lifshitz-Assaf, Katherine Kellogg, Saran Rajendran, Lisa Krayer, François Candelon, and Karim R Lakhani. 2023. "Navigating the Jagged Technological Frontier: Field Experimental Evidence of the Effects of AI on Knowledge Worker Productivity and Quality." *Harvard Business School Technology*

& *Operations Mgt. Unit Working Paper*, no. 24-013.

Dibia, Victor. 2023. "LIDA: A Tool for Automatic Generation of Grammar-Agnostic Visualizations and Infographics Using Large Language Models." *arXiv Preprint arXiv:2303.02927.*

Dibia, Victor, Jingya Chen, Gagan Bansal, Suff Syed, Adam Fourney, Erkang Zhu, Chi Wang, and Saleema Amershi. 2024. "Autogen Studio: A No-Code Developer Tool for Building and Debugging Multi-Agent Systems." *arXiv Preprint arXiv:2408.15247.*

Dibia, Victor, Adam Fourney, Gagan Bansal, Forough Poursabzi-Sangdeh, Han Liu, and Saleema Amershi. 2022. "Aligning Offline Metrics and Human Judgments of Value for Code Generation Models." *arXiv Preprint arXiv:2210.16494.*

Du, Yilun, Shuang Li, Antonio Torralba, Joshua B. Tenenbaum, and Igor Mordatch. 2023. "Improving Factuality and Reasoning in Language Models Through Multiagent Debate." *arXiv Preprint arXiv:2305.14325.*

Epperson, Will, Gagan Bansal, Victor C Dibia, Adam Fourney, Jack Gerrits, Erkang Zhu, and Saleema Amershi. 2025. "Interactive Debugging and Steering of Multi-Agent Ai Systems." In *Proceedings of the 2025 CHI Conference on Human Factors in Computing Systems*, 1–15.

FlowiseAI. 2024. "Flowise: Build AI Agents, Visually." https://flowiseai.com/.

Fourney, Adam, Gagan Bansal, Hussein Mozannar, Cheng Tan, Eduardo Salinas, Friederike Niedtner, Grace Proebsting, et al. 2024. "Magentic-One: A Generalist Multi-Agent System for Solving Complex Tasks." *arXiv Preprint arXiv:2411.04468.*

Greenblatt, Ryan, Carson Denison, Benjamin Wright, Fabien Roger, Monte MacDiarmid, Sam Marks, Johannes Treutlein, et al. 2024. "Alignment Faking in Large Language Models." *arXiv Preprint arXiv:2412.14093.*

Hadfield, Jeremy, Barry Zhang, Kenneth Lien, Florian Scholz, Jeremy Fox, and Daniel Ford. 2025. "How We Built Our Multi-Agent Research System." Anthropic Engineering Blog. https://www.anthropic.com/engineering/multi-agent-research-system.

Jimenez, Carlos E, John Yang, Alexander Wettig, Shunyu Yao, Kexin Pei, Ofir Press, and Karthik Narasimhan. 2023. "SWE-Bench: Can Language Models Resolve Real-World GitHub Issues?" *arXiv Preprint arXiv:2310.06770.*

Kwa, Thomas, Ben West, Joel Becker, Amy Deng, Katharyn Garcia, Max Hasin, Sami Jawhar, et al. 2025. "Measuring AI Ability to Complete Long Tasks." *arXiv Preprint arXiv:2503.14499.*

Li, Ang, Yin Zhou, Vethavikashini Chithrra Raghuram, Tom Goldstein, and Micah Goldblum. 2024. "Commercial LLM Agents Are Already Vulnerable to Simple yet Dangerous Attacks." *arXiv Preprint arXiv:2502.08586.*

Liang, Tian, Zhiwei He, Wenxiang Jiao, Xing Wang, Yan Wang, Rui Wang, Yujiu Yang, Zhaopeng Tu, and Shuming Shi. 2023. "Encouraging Divergent Thinking in Large Language Models Through Multi-Agent Debate." *arXiv Preprint arXiv:2305.19118.*

Linux Foundation. 2025. "Linux Foundation Launches the Agent2Agent Protocol Project." https://www.linuxfoundation.org/press/linux-foundation-launches-the-agent2agent-protocol-project-to-enable-secure-intelligent-communication-between-ai-agents.

Liu, Nelson F., Kevin Lin, John Hewitt, Ashwin Paranjape, Michele Bevilacqua, Fabio Petroni, and Percy Liang. 2024. "Lost in the Middle: How Language Models Use Long Contexts." *Transactions of the Association for Computational Linguistics* 12: 157–73.

Lu, Yadong, Jianwei Yang, Yelong Shen, and Ahmed Awadallah. 2024. "OmniParser for Pure

Vision Based GUI Agent." *arXiv Preprint arXiv:2408.00203*.

Lu, Yujia, Yichen Qian, Yue Chen, Yi Xie, Yufeng Feng, Xiaodan Lyu, Wenxuan Zhang, Wayne Xin Zhao, Ji-Rong Wen, and Ruiming Tang. 2025. "UI-TARS: Pioneering Automated GUI Interaction with Native Agents." *arXiv Preprint arXiv:2501.12326*.

Luo, Xufang, Yuge Zhang, Zhiyuan He, Zilong Wang, Siyun Zhao, Dongsheng Li, Luna K. Qiu, and Yuqing Yang. 2025. "Agent Lightning: Train ANY AI Agents with Reinforcement Learning." https://arxiv.org/abs/2508.03680.

Lynch, Aengus, Benjamin Wright, Caleb Larson, Kevin K. Troy, Stuart J. Ritchie, Sören Mindermann, Ethan Perez, and Evan Hubinger. 2025. "Agentic Misalignment: How LLMs Could Be an Insider Threat." *Anthropic Research*.

Manus AI. 2025. "Manus: General AI Agent." https://manus.im/.

Meta AI. 2025. "Agents Rule of Two: A Practical Approach to AI Agent Security." https://ai.meta.com/blog/practical-ai-agent-security/.

Mialon, Grégoire, Clémentine Fourrier, Craig Swift, Thomas Wolf, Yann LeCun, and Thomas Scialom. 2023. "GAIA: A Benchmark for General AI Assistants." *arXiv Preprint arXiv:2311.12983*.

Microsoft AI Blog. 2024. "AI at Work Is Here. Now Comes the Hard Part." https://www.microsoft.com/en-us/worklab/work-trend-index/ai-at-work-is-here-now-comes-the-hard-part.

Microsoft Research. 2023. "AutoGen: Multi-Agent Conversation Framework." https://github.com/microsoft/autogen.

Minsky, Marvin. 1986. *The Society of Mind*. Simon; Schuster.

Mozannar, Hussein, Gagan Bansal, Cheng Tan, Adam Fourney, Victor Dibia, Jingya Chen, Jack Gerrits, et al. 2025. "Magentic-UI: Towards Human-in-the-Loop Agentic Systems." *arXiv Preprint arXiv:2507.22358*.

n8n GmbH. 2024. "N8n: AI Workflow Automation Platform." https://n8n.io/.

OpenAI. 2025. "Introducing Operator." OpenAI Blog. https://openai.com/index/introducing-operator/.

Ouyang, Long, Jeffrey Wu, Xu Jiang, Diogo Almeida, Carroll L. Wainwright, Pamela Mishkin, Chong Zhang, et al. 2022. "Training Language Models to Follow Instructions with Human Feedback." *Advances in Neural Information Processing Systems* 35: 27730–44.

Patwardhan, Tejal, Rachel Dias, Elizabeth Proehl, Grace Kim, Michele Wang, Olivia Watkins, Simón Posada Fishman, et al. 2025. "GDPval: Evaluating AI Model Performance on Real-World Economically Valuable Tasks." *arXiv Preprint arXiv:2510.04374*.

Peng, Sida, Eirini Kalliamvakou, Peter Cihon, and Mert Demirer. 2023. "The Impact of Ai on Developer Productivity: Evidence from Github Copilot." *arXiv Preprint arXiv:2302.06590*.

Phan, Long, Alice Gatti, Ziwen Han, Nathaniel Li, Josephina Hu, Hugh Zhang, Chen Bo Calvin Zhang, et al. 2025. "Humanity's Last Exam." *arXiv Preprint arXiv:2501.14249*.

Rein, David, Betty Li Hou, Asa Cooper Stickland, Jackson Petty, Richard Yuanzhe Pang, Julien Dirani, Julian Michael, and Samuel R Bowman. 2023. "GPQA: A Graduate-Level Google-Proof q&a Benchmark." *arXiv Preprint arXiv:2311.12022*.

Russell, Stuart, and Peter Norvig. 2020. *Artificial Intelligence: A Modern Approach*. 4th ed. Pearson.

Shinn, Noah, Federico Cassano, Edward Berman, Ashwin Gopinath, Karthik Narasimhan, and Shunyu Yao. 2023. "Reflexion: Language Agents with Verbal Reinforcement Learning." *arXiv Preprint arXiv:2303.11366.*

Sweller, John. 1988. "Cognitive Load During Problem Solving: Effects on Learning." *Cognitive Science* 12 (2): 257–85.

TryCUA. 2024. "CUA: Computer Use Agent SDK." https://github.com/trycua/cua.

Vaswani, Ashish, Noam Shazeer, Niki Parmar, Jakob Uszkoreit, Llion Jones, Aidan N Gomez, Lukasz Kaiser, and Illia Polosukhin. 2017. "Attention Is All You Need." *Advances in Neural Information Processing Systems* 30.

vLLM Team. 2023. "vLLM: Easy, Fast, and Cheap LLM Serving with PagedAttention." https://github.com/vllm-project/vllm.

Wang, Xilong, John Bloch, Zedian Shao, Yuepeng Hu, Shuyan Zhou, and Neil Zhenqiang Gong. 2025. "WebInject: Prompt Injection Attack to Web Agents." *arXiv Preprint arXiv:2505.11717.*

Wei, Alexander, Nika Haghtalab, and Jacob Steinhardt. 2023. "Jailbroken: How Does Llm Safety Training Fail?" *Advances in Neural Information Processing Systems* 36: 80079–110.

Wei, Jason, Xuezhi Wang, Dale Schuurmans, Maarten Bosma, Brian Ichter, Fei Xia, Ed Chi, Quoc Le, and Denny Zhou. 2022. "Chain-of-Thought Prompting Elicits Reasoning in Large Language Models." *Advances in Neural Information Processing Systems* 35: 24824–37.

Wu, Qingyun, Gagan Bansal, Jieyu Zhang, Yiran Wu, Shaokun Zhang, Erkang Zhu, Beibin Li, Li Jiang, Xiaoyun Zhang, and Chi Wang. 2023. "AutoGen: Enabling Next-Gen LLM Applications via Multi-Agent Conversation Framework." *arXiv Preprint arXiv:2308.08155.*

Yan, Walden. 2025. "Don't Build Multi-Agents." Cognition AI Blog. https://cognition.ai/blog/dont-build-multi-agents.

Yao, Shunyu, Jeffrey Zhao, Dian Yu, Nan Du, Izhak Shafran, Karthik Narasimhan, and Yuan Cao. 2022. "ReAct: Synergizing Reasoning and Acting in Language Models." *arXiv Preprint arXiv:2210.03629.*

You, Keen et al. 2024. "Ferret-UI: Grounded Mobile UI Understanding with Multimodal LLMs." *arXiv Preprint arXiv:2404.05719.*

Zhang, Chaoyun, Liqun Li, Shilin He, Xu Xu, Bo Qiao, Si Qin, Minghua Ma, et al. 2024. "Ufo: A Ui-Focused Agent for Windows Os Interaction." *arXiv Preprint arXiv:2402.07939.*

Zhou, Yongchao, Andrei Ioan Muresanu, Ziwen Han, Keiran Paster, Silviu Pitis, Harris Chan, and Jimmy Ba. 2022. "Large Language Models Are Human-Level Prompt Engineers." In *The Eleventh International Conference on Learning Representations.*

Index

www.ingramcontent.com/pod-product-compliance
Lightning Source LLC
Chambersburg PA
CBHW052340210326
41597CB00037B/6200

*9 7 9 8 9 9 3 1 0 1 2 0 0 *